Food Choice and the Consumer

Food Choice and the Consumer

David MARSHALL

*Lecturer in Marketing and
Consumer Behaviour,
University of Edinburgh*

BLACKIE ACADEMIC & PROFESSIONAL
An Imprint of Chapman & Hall

London · Glasgow · Weinheim · New York · Tokyo · Melbourne · Madras

Published by
Blackie Academic & Professional, an imprint of Chapman & Hall,
Wester Cleddens Road, Bishopbriggs, Glasgow G64 2NZ

Chapman & Hall, 2–6 Boundary Row, London SE1 8HN, UK

Blackie Academic & Professional, Wester Cleddens Road, Bishopbriggs, Glasgow G64 2NZ, UK

Chapman & Hall GmbH, Pappelallee 3, 69469 Weinheim, Germany

Chapman & Hall USA, 115 Fifth Avenue, Fourth Floor, New York, NY 10003, USA

Chapman & Hall Japan, ITP-Japan, Kyowa Building, 3F, 2-2-1 Hirakawacho, Chiyoda-ku, Tokyo 102, Japan

DA Book (Aust.) Pty Ltd, 648 Whitehorse Road, Mitcham 3132, Victoria, Australia

Chapman & Hall India, R. Seshadri, 32 Second Main Road, CIT East, Madras 600 035, India

First edition 1995

© 1995 Chapman & Hall

Typeset in 10/12pt Times by Acorn Bookwork, Salisbury, Wilts
Printed in Great Britain by The University Press, Cambridge

ISBN   0 7514 0234 6

A catalogue record for this book is available from the British Library

Library of Congress Catalog Card Number: 95-77459

♾Printed on acid-free text paper, manufactured in accordance with ANSI/ NISO Z39.48-1992 (Permanence of Paper)

Preface

Food Choice and the Consumer fulfils two needs. First, it captures the inter-disciplinary aspects of food choice and advocates an appreciation for other perspectives on the subject in an attempt to discourage some of the disciplinary parochialism which surrounds this area. Second, it accommodates a range of different approaches to domestic food choice in a coherent way by encouraging the reader to see food choice as comprising a set of key tasks, such as shopping, preparing, cooking, etc. Furthermore, it illustrates the way in which the antecedents of choice vary according to which stage in the 'decision process' the 'enigmatic' consumer finds him or herself.

Food Choice and the Consumer is written for a wide audience including: academics and students interested in food related topics; policy makers, nutritionists and health educators striving to improve the nation's diet; food manufacturers and retailers keen to gain an insight into some of the underlying motivations, concerns and constraints on consumers' food choice. This is not about specific brands, but about consumers and the many factors that influence their choice. Rather than an ABC of food choice, this book aims to stimulate interest while offering the commercial sector, suffering from increasing competition and brand myopia, a fresh perspective on consumer food choice. I hope that this book will contribute to the ongoing debate on food choice and bring us a little closer to understanding how and why consumers choose food.

My thanks are due to Professor Anne Murcott for her comments on the common misuse of the term 'ethnic' food. There is a growing interest in what has been termed 'ethnic' food, although in classifying food and dishes which originate elsewhere, as such, we display our ethnocentrism and tend to forget that 'ethnic' applies to everyone, including ourselves.

I would also like to thank all the contributors for their timely submissions and putting up with constant haranguing from myself. Special thanks are due to the publishers, Blackie Academic and Professional, for getting this project up and running and keeping it on schedule. Alison Gault provided the front cover illustration and inspiring fireside conversations. Thanks to Dr. Les Gofton, University of Newcastle upon Tyne and colleagues in the Department of Business Studies, University of Edinburgh for helpful comments and encouragement. I would like to thank

Lisbeth Lindsey for helping with the index, and Linda Hogg for checking the references. Finally to all those who warned against the perils of attempting to put together an edited collection I am glad on this occasion that I did not take your advice. That, however, has all to do with the contributors.

David Marshall

Contributors

Annie Anderson Research Fellow, Department of Human Nutrition, University of Glasgow, Royal Infirmary, Queen Elizabeth Building, Glasgow G31 2ER, UK

Rick Bell Consumer Research Scientist, Behavioral Sciences Division, R D & E Center, US Army Soldier Systems Command, Natick, MA 01760-5020, USA

David Buisson Professor of Marketing, Department of Marketing, University of Otago, PO Box 56, Dunedin 9001, New Zealand

Beatrice Daillant-Spinnler Sensory Co-ordinator, Europe, Griffith Laboratories, 10 Boulevard du Parc, 92521 Neuilly-sur-Seine, Cedex, France

John Dawson Professor of Marketing, Department of Business Studies, University of Edinburgh, William Robertson Building, 50 George Square, Edinburgh EH8 9JY, UK

Rosires Deliza EMBRAPA-Brazilian Enterprise of Agricultural Research, Av das Americas 29501, 23020-470 Rio de Janeiro-RJ, Brazil

Nick Fiddes Postdoctoral Research Fellow, Department of Social Anthropology, University of Edinburgh, William Robertson Building, 50 George Square, Edinburgh EH8 9JY, UK

Les Gofton Lecturer in Behavioural Science, Department of Agricultural Economics and Food Marketing, The University, Newcastle upon Tyne NE1 7RU, UK

Richard Hutchins Lecturer in Marketing, Department of Agricultural Economics and Food Marketing, The University, Newcastle upon Tyne NE1 7RU, UK

Michael Lean Rank Professor of Human Nutrition, Department of Human Nutrition, Royal Infirmary, Queen Elizabeth Building, Glasgow G31 2ER, UK

Hal MacFie Deputy Director, Department of Consumer Sciences, AFRC, Insitute of Food Research, Earley Gate, Whiteknights Road, Reading RG6 6BZ, UK

David Marshall Lecturer in Marketing and Consumer Behaviour, Department of Business Studies, University of Edinburgh, William Robertson Building, 50 George Square, Edinburgh EH8 9JT, UK

Herbert L. Meiselman Senior Research Scientist, Science and Technology Directorate, R D & E Center, US Army Soldier Systems Command, Natick, MA 01760-5020, USA

Kathryn Milburn Specialist Development and Evaluation Officer (Sexual, Reproductive and Family Health), Health Education Board for Scotland, Woodburn House, Canaan Lane, Edinburgh EH10 4SG, UK

Rolland Munro Reader in Accountability, Department of Management, Keele University, Newcastle-under-Lyme, Staffordshire ST5 5BG, UK

Anne Murcott Professor of the Sociology of Health, School of Education and Health Studies, South Bank University, 103 Borough Road, London SE1 0AA, UK

Monique Raats Research Scientist, Department of Consumer Sciences, AFRC, Institute of Food Research, Earley Gate, Whiteknights Road, Reading RG6 6BZ, UK

Christopher Ritson Professor of Agricultural Marketing, Department of Agricultural Economics and Marketing, The University, Newcastle upon Tyne NE1 7RU, UK

Contents

Cooking   217

Eating   237

Introduction

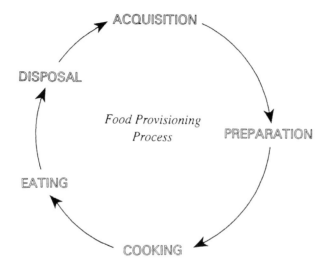

ACQUISITION

DISPOSAL

Food Provisioning Process

PREPARATION

EATING

COOKING

Introduction: food choice, the food consumer and food provisioning

David Marshall

Consumers exercise choice every day, but when it comes to asking consumers why they eat what they do the answer is disarmingly simple – they like the taste, end of story. Or is it? This book looks at how food choice is shaped at each of the different stages in the food provisioning process[1] which involves buying, preparing, cooking, eating and disposing of food. It does not pretend to provide a prescriptive explanation of consumer choice but offers an insight into some of the fundamental concerns that drive consumer choice at each of these stages. In doing so it focuses on those constraints on food choice which are often implicit, seldom questioned, taken for granted and simply part of everyday activity.

1.1
Food choice

Food choice is a luxury not afforded to all, and we should neither take it for granted, nor reduce it to some simple biological reaction of our sense organs to a physical substance (nor indeed ignore its material properties and regard it as pure symbolic gesture). For too long we have simply regarded 'food as feed' and focused on the nutritional aspects of eating. Food is both substance and symbol, material and aesthetic and certainly worthy of further investigation. Furthermore, consumers' food choices affect not only their health, but the state of the economy, agricultural production, the balance of trade and employment in the food sector, as well as the fortunes of many companies. Yet despite the economic, political, social and nutritional importance of these choices relatively little is known about food choice, other than the fact that individuals are conservative in their selection given the abundance of edible foods (Turner, 1979; Fischler, 1979).

Food has been the 'object-matter' in a wide range of disciplines includ-

---

[1]This draws on the notion of the 'human food cycle' introduced by Goody in *Cooking Cuisine and Class* (1982).

ing, for example, biology, nutrition, economics, psychology, sociology, anthropology, history, agriculture, home economics, food science and catering, but there has been little in the way of 'transdisciplinary' debate (Murcott, 1988; Symons, 1991; see also Ritson *et al.*, 1986; National Consumer Council, 1992). The dominant discourse has been in the area of (sensory) food science, psychology and nutrition (Barker, 1982; McFie and Thomson, 1994; Shepherd, 1989; Thomson, 1988; Turner, 1979), although food has been used as a means to study other phenomena in sociology and anthropology (Beardsworth and Keil, 1990; Douglas, 1984; Goody, 1982, 1986; Harris, 1986; Mennell, 1985; Mintz, 1985; Murcott, 1983). Interdisciplinary communication is virtually non-existent and differences, even intradisciplinary, exist as much in terms of vocabulary as perspective and method.[2]

The implicit rationale which seems to drive much of the (commercial) research into why consumers choose what they do (see Thomson, 1988) is that food choice can be explained by looking at some inherent physical property or product characteristic. Moreover, manufacturers and food scientists continue to measure consumers' reaction to the taste, textures, flavours and even smells of their products in an attempt to explain why consumers choose what they do. Food acceptability and consumer preference offer a more 'scientific' explanation and serve as surrogate measures of food choice. Even some sensory scientists are questioning the validity of using purely sensory measures as accurate predictors of consumer choice. It is relatively easy to measure consumer reaction to chemical stimuli but there is little work on taste and smell interactions, and, indeed, it has been suggested that real foods may even be too complex for this type of analysis, not to mention meals (Rolls, 1987, quoted in Meiselman, 1992).

Furthermore, the transition from laboratory to the domestic table is something of a quantum leap by any stretch of the imagination. Indeed, taste perception may have a relatively weak influence on food habits and be comparatively unimportant in controlling daily food habits (Lawless, 1991; Meiselman, 1992). The use of 'consumer' panels is on the increase, partly in an attempt to bring the consumer into the product development process, but the problem of how to 'relate sensory properties to non-sensory properties' remains problematic (Booth, 1992; Meiselman, 1992, 1994). The psychology and physiology of taste are well understood, but in trying to understand consumer choice there appears to be little attempt to take the explanation beyond the olfactory. Outside the sensory laboratory, taste, narrowly defined as a sensory–psychological reaction, plays a relatively limited role when choosing in the store, unless of course free samples are provided.

[2]One solution may be to problematise food choice and in doing so remove some of the antagonism aroused by the prospects of an integrated interdisciplinary approach.

It has been argued for a long time that 'people like what they eat' rather than 'eat what they like' (Lewin, 1943) and food choice is moulded by 'cultural representation' which dictates what is eaten long before food reaches the mouth (Falk, 1994). As Falk reminds us:

> a cultural order in which an alimentary code (food taboos, ritual rules) defines that which may be eaten, by whom, how and when, does not leave much room for individual matters of taste. The sense of taste is surely there, but the 'judgement' is located primarily at the boundaries of culture, in the 'mouth' of the community, as it were. Only when these boundaries grow weaker, is the judgement of taste transposed to the level of the individual self, body and mouth – still, however, related and conditioned by cultural representations. (Falk, 1994: p. 13)

This notion of 'taste' needs to be redefined to include social and cultural perspectives as well as biological. McCorkindale reminds us that 'individual differences in taste are acknowledged but accepted only if they do not exceed the parameters of social acceptability. The rules governing taste may operate at the level of an unstated, shared understanding of what is appropriate' (McCorkindale, 1992: 9). On the broader topic of consumption *per se*, 'understanding some of society's characteristics could provide the consumer researcher with the context necessary to help the study of individual consumer choices' (Nicosia and Mayer, 1976: 65). This reiterates the need to broaden the concept of 'taste'. Quoting again from Falk:

> taste preference is a multi-relational concept which cannot be reduced to a relationship between objective properties of foodstuffs and the sensory–physiological reaction of the human ingestive and digestive organism. At the sensory level taste preferences are necessarily also related to and even determined by the symbolic principles which translate the material universe into representations of the edible vs inedible, which are then further specified into different sub-categories according to taboos and ritual rules. (Falk, 1994: p. 68)

But even these taboos vary according to temporal, spatial and social dimensions. Further evidence of the limitations of relying solely on sensory characteristics to explain consumer choice is found in individuals' acceptance of foods to which they have an innate aversion, for example coffee, or chilli (Rozin, 1982). Little is known about this transformation (Falk, 1994). A justification for food choice on the basis that something simply 'tastes nice' if taken at face value contributes little to an understanding of why people choose the foods that they do – the very idiosyncrasies of taste render such evaluations meaningless.

While there is a recognition of external influences such as availability and economic factors, most food choice models focus on the interaction between the individual and the food product (McEwan, 1990; McEwan and Thomson, 1988; Shepherd and Sparks, 1994). In consumer behaviour many of the schematic multi-attribute models depict choice as the penultimate stage in a decision process which requires the consumer to move through a hierarchical sequence of stages from problem recognition through search, alternative evaluation, choice and finally post-purchase evaluation. This decision process is facilitated by information-processing mechanisms and conditioned by psychological social, cultural and situational influences (Andreasan, 1965; Engel *et al.*, 1978, 1993; Howard and Sheth model, 1969; Nicosia model, 1966). There is, however, an increasing body of evidence which suggests that consumer choice may be much less involved than previously supposed and product evaluation may take place post-purchase (Foxall, 1990; Nord and Peter, 1980; Olshavsky and Granbois, 1979). The focus, however, is still on a rational consumer making a hierarchical decision in which social, cultural and situational influences are afforded a peripheral role.

Food choice is apparent as both an outcome (an end) of a decision process and as a mechanism or process (a means to that end). As an 'end' it represents the selection of food from whatever is available and 'disappearance estimates' offer a convenient, if not entirely accurate, measure of 'choice' (Lesser *et al.*, 1986). But this tells us little about the consumer. Is it enough to simply know *what* we eat? There has been a call to examine the mechanisms of choice as well as the outcome motivated by a desire 'to uncover native understandings, where they exist, of the very expression "choice"; to identify culturally derived obstacles and opportunities for its exercise; to examine its politico-economic context' (Mennell *et al.*, 1992).

Choosing, as a mechanism, has been defined as 'making up one's mind with regard to a particular object, action or state of affairs, in a context of alternatives (where) the particular choice is' (Daveney, 1964, quoted in Straughan, 1992). As Straughan recognises, however, this implies a highly rational, deliberate, decision maker (directly descended from *homo economicus* and first seen somewhere in the USA around the 1950s!). Nevertheless, the decision process is likely to be a complex one. In marketing food products manufacturers attempt to make the task less perplexing by creating brands which promise to deliver consistent quality (Doyle, 1990; Merrett and Whithell, 1994), although some writers have questioned their real contribution to consumer welfare (Casson, 1994). But marketeers and consumers have different perspectives on the same decision. While consumers are interested in their individual needs, a range of products or brands and controlling marketing influences, manufacturers are interested in their target market, segments of buyers, specific products and influencing choice. To understand how consumers choose, marketeers have to think like consumers (Wilkie, 1986).

Consumers often find it difficult to explain why they make particular food choices that are dependent on a multitude of factors, besides taste. Food choices in the developed world, at least for the better-off sections of the community, are driven by more than hunger, and can be triggered by a range of conditions from physiological needs through to observance of others eating (Blundell, 1979). As many nutritionists and policy makers are only too aware, choice is not necessarily driven by nutritional concerns. The fear of new foods (neophobia) has to be balanced with a desire for variety (neophilia) which is increasingly manifest in a systematic search for new and unexperienced pleasurable tastes (Falk, 1994; Rozin, 1982). Choice is also dependent on who one is talking to: the food caterer, responsible for buying, preparing and cooking the food, or the eater (King, 1979). Taken together it makes the task of finding out why consumers choose what they do a complex and challenging task.

Food choice in this book is concerned with daily routine of domestic food provisioning, and what Murcott calls 'the ordinary, the familiar, and the mundane and ... not ... the newsworthy, unusual, or exotic' (Murcott, 1986: 32). It examines food availability, the range of choice and the restrictions imposed on that choice. Food choice decisions are not confined to the retail store but extend through the distribution, preparation, consumption and disposal stages in this provisioning process. It includes related decisions on where to buy the food, how to cook and prepare it, how to eat it and who to eat it with as well as what to do with the leftovers. It is above all an interdisciplinary matter and requires an interdisciplinary approach.

1.2
The food consumer

Faced with low rates of population growth in developed economies, 0.3% in Europe and 1.0% in the USA in the period 1990–1995, the number of food consumers is unlikely to grow and may even start to decline in Great Britain by 2028 (CSO, 1994). Holding onto existing customers and attracting new ones is likely to become increasingly difficult. To win market share, international food manufacturers and retailers are wooing customers with an abundance of new products, therefore increasing consumer choice and perhaps adding to consumer confusion. While some feel that industry is meeting consumer 'wants', others argue that much of this new product development is wasteful and really about stealing market share from the competitors (Hughes, 1994). Faced with more 'choice', today's consumers are more discerning, better informed and discriminatory, as they select from the multitude of brands on the supermarket shelf. What they choose to eat has repercussions throughout the whole of the food system, affecting retail, manufacturing, processing and farm production. Food sales in the USA were worth $628 billion in 1988 (Senauer *et al.*, 1991), while UK consumers spent £45bn on food in 1992 (CSO, 1994).

The population is ageing. Consumers over pensionable age, the 'grey' or 'third age' market, could account for around 25% of the UK population by the year 2000 and represent a market of 68 million US consumers by 2040. These consumers are more 'conservative' in their food consumption and can remember the restrictions on their choice imposed by rationing and post-war food shortages. They have different wants to the younger generation of consumers who eat more, eat out more often and are more likely to follow food fads (CSO, 1994; Senauer *et al.*, 1991; *Wall Street Journal*, 1989).

Marriage is on the decline and divorce rates are up on both sides of the Atlantic. One consequence has been a change in the size and composition of households. The average size of US households was 2.6 in 1988 and married couples with children represented 27% of households (US Department of Commerce, 1987, cited in Senauer *et al.*, 1991). There has been a significant rise in the number of single-person households in the UK; for example, the number of households containing one person is up from 14% in 1961 to 27% in 1992 (CSO, 1994). A high proportion of these contain single women over 65 years old, reflecting the ageing population. The food industry has responded with single food portions, smaller packs, ready meals, even cookery books for one person. Eating out among the younger singles is on the increase. Companies are increasingly recognising the diversity of consumer wants and segmenting the market into distinct groups in terms of their shopping habits (*Marketing*, 1994), food habits (Pillsbury Co. in Senauer *et al.*, 1991), and even attitudes towards health (DMBB, 1987).

While the number of food consumers is unlikely to grow, per capita real disposable income[3] is rising and the proportion of consumer expenditure on food has fallen from around 20 to 12 per cent in the UK and the USA over the past 20 years (CSO, 1994; Hughes, 1994; Putnam, 1990). There is a trend towards buying better quality produce and added value products. In the USA consumers are substituting poultry for red meat, vegetable oils for animal fats, processed fruit juices for fresh fruit, artificial sweeteners for sugar and semi-skimmed milk for full fat milk (Senauer *et al.*, 1991). In the UK, breakfast cereals, poultry and to a lesser degree fresh fruit consumption is up while eggs, sugar, beef and veal consumption is down (CSO, 1994). More food is processed, packaged, labelled, branded and 'positioned' in this 'industrial' cuisine (Symons, 1993). While the average disposable income is increasing, the gap between the 'haves' and the 'have nots' is widening and the lower fifth of the UK population has a lower real net income than they did over 15 years ago (see Leather, 1992).

Perhaps the most important single influence on food provisioning has been the changing role of women in the work force. Across Europe an increasing number of women (under 55 years) are employed outside the home, figures ranging from 90% of women in Sweden to 50% in Italy. In the UK women now account for 56% of the civilian workforce, although more women tend to be employed part time.[4] One consequence has been

[3]United Kingdom per capita real disposable income averaged 3.6 per annum growth between 1971 and 1991 (Hughes, 1994).
[4]This is not such as recent a phenomenon as it might seem. Historically a high proportion of women have worked either in the cottage industries or outside the home, traditionally as domestic servants or factory labour. Victorian values which emphasised the importance of the family and the women's role in the home seem to have created the impression that it is only recently that women have been going out to work. A major difference of the recent trend is that more women are now pursuing careers.

greater economic independence for women and a corresponding rise in the number of 'dual income' households. However, the amount of time available for domestic 'chores' has fallen. Despite these changes women are still responsible for the majority of domestic tasks, spending double the time cooking, cleaning, shopping and looking after children compared to males in full-time employment (Burros, 1988; CSO, 1990; Henley Centre for Forecasting, 1992, cited in Hughes, 1994). The number of meals eaten at home has been falling and lighter meals appear to be taking the place of more formal meals during the week in line with these changes (Taylor Nelson, 1990, 1993a,b; Morris, 1988; Ver Meulen *et al.*, 1987).

The, proclaimed, effects on food choice have been greater use of convenience foods/ready prepared meals for everyday use, increasingly 'casual' meals, a growing distinction between foods for nourishment (more recently manifest in the growth of functional foods) and foods for fun, and an increase in the number of meals eaten outside the home (McKenzie, 1980). A parallel trend has been the increase in ownership of 'time-saving' kitchen 'equipment': the microwave oven can be found in 80% of US and 60% of UK households, and most households (85%) have a freezer. As food manufacturers and retailers endeavour to develop more 'convenience' products the trend towards convenience seems likely to continue.

Consumers have long divested responsibility for processing to the food manufacturing sector but are becoming increasingly concerned with what happens to their food before it reaches the supermarket shelf. While this does not appear to have radically altered food choice, excluding the short-term effects of recent food scares, it represents a 'latent' concern. Over-, rather than under-nutrition is a major health issue in developed countries as policy makers and health educators warn the public to reduce fat, sugar and salt intake. Despite this the evidence suggests that the majority of consumers are reluctant to change. In the USA 'fat' intake is currently top of the hit list and while there are an abundance of low fat products consumers continue to over-consume as the 'workout–pigout' mentality takes over (Senauer *et al.*, 1991). The health issue is especially prevalent in the USA where one-third of all new food products launched in 1991 claimed some health benefit (Market Intelligence Inc., USA, cited in Hughes, 1994: p. 19). In the UK there is some evidence of a trend towards a healthier diet but the twin trends of health and convenience seem somewhat at odds (*Independent*, 1995; Gofton and Ness, 1991).

The ethnic gourmet has been stimulated through increased travel, and an ever-increasing exposure to ethnic cuisine via the restaurant, take-away and finally retail trade. Further impetus to the demand for more exotic foods comes from the increasingly ethnic diversity in many countries, as immigrants bring their own flavourings and foods, none more so than in Australia with its 'Pacific Rim' cuisine mixing East and West in a process of culinary acculturation (Ripe, 1993).

Animal welfare and environmental issues surrounding food are increasingly becoming consumer issues (Hughes, 1994). These trends, undoubtedly fuelled by the media interest and highlighted by government policy, are likely to influence choice for some consumers.

Ten major trends in consumer attitudes and their likely impact on food choice were identified for the US Food Marketing Institute by Faith Popcorn (Iggers, 1987, cited in Senauer *et al.*, 1991). The report suggested a trend towards better quality produce with minimal processing (1. Neo-traditionalism) and a renewed desire for increased variety, particularly among smaller households and adult women (2. Adventure). The occasional treat (3. Indulgence) is likely to continue as individuals reward themselves for all that hard work and take on more responsibility for what they eat (4. Individuality) with single-serve portions and ready meals to meet the demands of single-person households. The growth in 'Take Home ready To Eat' (TOTE) foods reflected a new focus on the home (5. Cocooning). Eating on the move (6. Grazing) was a further response to working away from home and pursuing leisure interests which restricted the amount of time available for cooking and eating at home. Consumers were reported to be adopting a holistic approach to food which included a renewed interest in diet and exercise (7. Wellness). With less time available consumers would find convenience goods and convenience shopping more attractive (8. Controlling time) and faced with an abundance of choice exercise their rights as consumers to 'exit, voice, or loyalty' in the marketplace (9. Selectivity). And, finally, Popcorn predicted a declining narcissism and a search for moral stability (10. Ethics) in food choice decisions.

1.3 The food provisioning process

Food consumption can be seen as part of a wider process linked to production and preparation, recognising its social aspects along the way (Richards, 1932; see also Beardsworth and Keil, 1990). Goody (1982),[5] looking at the development of high and low cuisine as a marker of social stratification, recognised the need to look beyond the functional aspects of food, beyond the biological to the total process. For Goody, 'the study of the process of providing and transforming food covers four main areas, that of growing, allocating, cooking and eating, which represent the phases of production, distribution, preparation and consumption ... to which can be added a fifth phase, often forgotten, disposal' (Goody, 1982: p. 37; see also Powers and Powers, 1984). From this we have the basis of food provisioning.

[5]He was essentially interested in why traditional African cultures are largely lacking in a differentiated cuisine, even across great states with differentiated political structures, but recognised the need to understand the structure of the society within which the consumption takes place, and then compare that with other societies. In the process he rejected what he called the outdated 'mind–body' dichotomy.

Process	Phases	Locus
Growing	Production	Farm
Allocating/storing	Distribution	Granary/market
Cooking	Preparation	Kitchen
Eating	Consumption	Table
Clearing up	Disposal	Scullery

(*Source*: adapted from Goody, 1982, p. 37)

Goody outlined in further detail different phases and aspects of production, distribution, preparation and consumption. The first stage is outside the scope of this book. Distribution, for example, might include reference to the nature of transactions, within and without the group, the quality of distribution, the technology of transport and storage and the periodicity of distribution. Preparation might include all aspects of preliminary work,[6] cooking and dishing up. Different aspects of preparation included membership of the cooking group and the consuming group, the levels of technology available, etc. Cooking, usually allocated to women even in our 'modern' society, and preparation, once a domestic chore, are increasingly done outside the home. The consumption phase included assembling the 'consumers', serving the food, eating it and clearing away, the structure of the meal, different ways of eating, eating implements and accoutrements, the eating group, and differentiation in the cuisine.

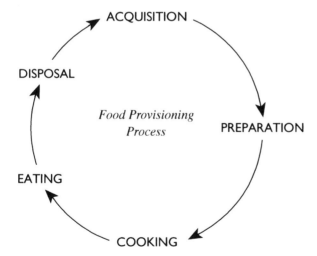

Figure 1.1 The food provisioning process.

[6]For Goody's African tribe this would include butchering the meat, shelling nuts, roasting and pickling food. In our modern cuisine it is more likely to involve washing and cleaning food, slicing and cutting food or opening a tin or packet.

Most modern analysis of food choice examines only one, or possibly two, of these phases. No wonder we find it somewhat difficult to explain. This book looks at the sequence of stages which the consumer goes through in the process of 'domestic food provisioning' (Gofton and Marshall, 1992; Marshall, 1988). Rather than simply focusing on 'eating', in the narrowest sense of consumption, it follows each of the stages from shelf to stomach, and beyond, and looks at various factors that affect consumer choice.

1.4
Food choice and the food consumer

Few books have examined all the stages involved from the initial decision to buy the food through to the final stage of disposing of the food, preferring to concentrate on specific aspects such as the psychology, physiology, psychobiology, sociology and anthropology of food choice. This book adopts a transdisciplinary approach, drawing material from a variety of disciplines to encourage further discourse on the topic of food provisioning. It also avoids some of the inherent problems of confining the investigation of food choice to a single discipline. Through this 'integrated' process approach it runs the risk of pleasing no one in an attempt to cater for all needs. However, consumers make food choices every day under the constraints imposed by social, cultural, historical, nutritional, market and economic factors, at all of the stages. By highlighting the changing nature of food choice at each of the different stages in the food provisioning process it aims to bridge some of the interdisciplinary gaps in the discussion on food choice. It is divided into the five stages of acquisition, preparation, cooking, eating and disposal (Figure 1.1).

At the early stages of food acquisition, consumers have relatively little direct control over what appears on the supermarket shelves. It is up to the state to ensure that our food supplies are safe and to regulate the industry through bodies such as MAFF and FDA. However, the institutional structures that decide on the controls at a European or international level have other objectives, which include pressures to improve trade between countries, and one question is whether more choice comes at the expense of less protection (Jukes, 1992). Stage 1 focuses on the factors that determine what food is made available to consumers, the changing nature of demand, the role that food retailers play in shaping consumer choice and finally the problems associated with promoting a healthy diet.

In chapter 2 Ritson and Hutchins examine how the biological basis of production, the structure of production and agricultural policies affect food prices, availability, quality and consumer attitudes. They draw on a variety of examples to illustrate the extent to which the supply chain not only responds to, but directly influences, consumer choice. In chapter 3 they consider the role of price and incomes on patterns of food consumption and the need to investigate what would have happened had these remained stable. This underlying trend in demand reveals changes in consumption that arise because of shifts in consumer attitudes and behaviour that standard economic theory cannot explain.

In chapter 4 Dawson considers the recent changes in the retailing sector and questions the extent to which we are free to choose as consumers given the increasing concentration in this sector. Particular attention is devoted towards the retailing methods employed to 'help us' choose.

Healthy eating is fuelling much of the debate on food choice and the commercial sector has been keen to exploit any advantages in this area. This question on freedom of choice has emerged in the recent debate on healthy eating in the UK as a consequence of the Department of Health report (1994) which, in a radical change of policy, has made recommendations for healthy eating based on actual foods as opposed to nutrients. In chapter 5 Anderson, Lean and Milburn look at nutritional issues in food choice and some of the inherent problems associated with changing consumer behaviour towards a healthier diet. Their discussion on 'lay perceptions' serves as a reminder of the need for health educators to take account of how consumers conceptualise health.

Stage 2 is concerned with food preparation. In our 'industrial' cuisine much of the preparation is undertaken before food reaches the domestic arena and Fiddes, in chapter 6, examines the cultural meanings of food and the conflicting dilemma of continuity and change as preparation becomes the responsibility of the food industry. He highlights the increasing gulf between production and consumption which is creating a crisis of confidence for the consumer faced with conflicting and confusing messages from the food industry. Gofton, in chapter 7, looks at the rise of 'convenience' and asks what it means to the consumer. In taking the discussion beyond some simple notion of convenience as simply 'time-saving', he reveals the way in which our attitudes towards convenience have changed in line with broader economic, social and cultural changes. Lastly in this section, Buisson looks at the complex process of new product development in chapter 8. He sees an increasingly globalised food market, a more educated consumer and new technology impacting on product development. In many ways he is more accepting of our industrial cuisine than Fiddes but stresses the need for producers to involve the consumer at all stages in the process and accommodate local tastes. Finally he examines some of the emerging opportunities in the area and points to likely future developments.

Stage 3 concerns cooking. In chapter 9 Murcott looks at the ubiquity of cooking and details the diverse range of cooking methods employed across social groups. She examines the cultural significance of cooking in transforming non-food into food and making 'proper' meals before discussing the division of labour in the kitchen and the impact of new cooking technology and practices on domestic cooking and food choice.

Stage 4 looks at the sensory aspects, eating occasions and situational influences on choice. Raats, Daillant-Spinnler, Deliza and MacFie, in chapter 10, examine the role of sensory preference in predicting food acceptance and the mediating effects of consumer expectations and appropriateness. In chapter 11 Marshall looks at the structure and pattern of domestic meals and proposes that, despite the changes taking place, meals continue to play an important role in shaping food choice. Bell and Meiselman continue with this 'occasions' theme in their discussion of eating environments in chapter 12. They move outside the laboratory and beyond the home to look at what individuals bring to the eating environment and how consumer choice can be manipulated situationally. As their chapter reveals, context and appropriateness play an important, and often neglected, role in food choice.

Finally stage 5 looks at disposal as Munro brings the debate full circle in chapter 13 with his discussion of conduits of disposal. He shows how the final stage, disposal, has implications for all of the preceding stages in the food provisioning process. Beyond the more contemporary issues relating to recycling he highlights the way in which disposal offers a theoretical perspective which supersedes the polarised views of production or consumption paradigms of choice. As marketing becomes increasingly aware of the need to consider consumer behaviour post-purchase, Munro challenges us to look at food choice post-consumption.

The book draws together a number of authors from across a range of disciplines, including the natural and social sciences, to address one particular aspect from these stages drawing, where possible, on material from outside their own field in order to encourage a transdisciplinary perspective. Consumers do not think as economists, sociologists, nutritionists or scientists. They are consumers, and each field has something to offer those who wish to understand what consumers do when it comes to choosing food.

References

Andreasen, A. R. (1965) Attitudes and Consumer Behaviour: a decision model. In L. E. Preston (ed.), *New Research in Marketing*, University of California, Berkeley. pp. 1–16.

Barker, L. M. (1982) *The Psychobiology of Human Food Selection*, AVI Publishing, Westport.

Beardsworth, A. and Keil, T. (1990) Putting Menu on the Agenda. *Sociology*, **24** (1), 139–151.

Blundell, J. (1979) Hunger, Appetite and Satiety – Constructs in Search of Identities. In M. Turner, *Nutrition and Lifestyles: Proceedings from the First Annual Conference, The Royal Society London, May 15–16*, Applied Science Publishers, London. pp. 121–142.

Booth, D. (1992) Towards Scientific Realism in Eating Research. *Appetite*, **19**, 56–60.

Burros, M. (1988) Women: Out of the House But Not out of the Kitchen. *New York Times*, February 1994.

Casson, M. (1994) Brands: Economic Ideology and Consumer Society. In G. Jones and N. J. Morgan (eds), *Adding Value: Brands and Marketing in Food and Drink*, Routledge, London. pp. 162–190.

Central Statistical Office (1990) *Social Trends*, HMSO, London.

Central Statistical Office (1994) *Social Trends*, HMSO, London.

Daveney, T. F. (1964) Choosing. *Mind*, LXXIII, 524. Cited in R. Straughan (1994) Freedom of Choice. In *Your Food: Whose Choice*, HMSO, London. pp. 135–156.

Department of Health (1994) *Nutritional Aspects of Cardiovascular Disease*, HMSO, London.

DMBB (1987) *The Branwagon's Further Progress: Healthy Eating Attitudes*, 2 St. James Square, London, SW1Y 4JN.

Douglas, M. (ed) (1994) *Food in the Social Order: Studies in Food and Festivities in Three American Communities*, Russell Sage Foundation, New York.

Doyle, P. (1990) Building Successful Brands: The Strategic Options. *Journal of Marketing Management*, **5**, 1, 77–95.

Engel, J. F., Blackwell, R. D. and Miniard (1993) *Consumer Behaviour*, 7th edition, Dryden Press, Chicago.

Engel, J. F., Kollat, D. T. and Blackwell, R. D. (1978) *Consumer Behaviour*, 1st edition, Holt, Rinehart and Wilson, New York.

Falk, P. (1994) *The Consuming Body*, Sage, London.

Fischler, C. (1979) Gastro-nomie and Gastro-anomie: Sagesse du Corps et Crise Bioculturelle de L'alimentation Moderne. *Communications*, **31** (Automme), 189–210.

Foxall, G. (1990) *Consumer Psychology in a Behavioural Perspective*, Routledge, London.

Gofton, L. R. and Marshall, D. W. (1992) *Fish: Consumer Attitudes and Preferences – A Marketing Opportunity*, Horton Publishing, Bradford.

Gofton, L. R. and Ness, M. (1991) Twin Trends; Health and Convenience in Food Change or Who Killed the Kazy Housewife. *British Food Journal*, **93**, 7, 17–23.

Goody, J. (1982) *Cooking, Cuisine and Class: A Study in Comparative Sociology*, Cambridge University Press, Cambridge.

Harris, M. B. (1986) *Good to Eat: Riddles of Food and Culture*, Simon and Schuster, New York.

Howard, J. A. and Sheth, J. N. (1969) *The Theory of Buyer Behaviour*, John Wiley and Sons, New York.

Hughes, D. (ed.) (1994) *Breaking With Tradition: Building Partnerships and Alliances in the European Food Industry*, Wye College Press, Wye.

Iggers, J. (1987) Food Marketeers are told of Emerging Consumer Trends. *Minneapolis Star Tribune*, May 9. Cited in B. Senauer, E. Asp and J. Kinsey (1991) *Food Trends and the Changing Consumer*, Eagan Press, Minnesota.

The *Independent* (1995) A Nation Both Richer and Poorer, 26 January.

Jukes, D. (1992) Food Law and Regulation: Is the Consumer Voice Heard? In *Your Food: Whose Choice*, edited by the National Consumer Council, HMSO, London. pp. 157–178.

King, S. (1979) Presentation and the Choice of Food. In M. Turner, *Nutrition and Lifestyles: Proceedings from the First Annual Conference, The Royal Society London, May 15–16*, Applied Science Publishers, London.

Lawless, H. T. (1991) Bridging the Gap Between Sensory Science and Product Evaluation. In H. T. Lawless and B. P. Klein (eds), *Sensory Science Theory and Applications*, Institute of Food Technologies, Chicago. pp. 1–36.

Leather, S. (1992) Less Money, Less Choice: Poverty and Diet in the UK Today. In *Your Food: Whose Choice*, edited by the National Consumer Council, HMSO, London. pp. 72–94.

Lesser, D., Hughes, D. and Marshall, D. (1986) Researching The Food Consumer – Techniques and Practice in the UK and North America. In C. Ritson, L. Gofton and J. McKenzie (eds), *The Food Consumer*, John Wiley & Sons, London. pp. 171–198.

Lewin, K. (1943) Forces behind food habits and methods of change. In *The Problem of Changing Food Habits*, Bulletin No. 108, National Academy of Science, National Research Council, Washington, D.C. pp. 35–65.

McCorkindale, L. (1992) What is Taste? *Nutrition and Food Science*, **6**, November/December, 8–12.

McEwan, J. A. (1990) *Food Choice Models: A Literature Review*, Campden Food and Drink Research Association, Technical Bulletin No. 74, August.

McEwan, J. A. and Thomson, D. M. H. (1988) A Behavioural Interpretation of Food Acceptability, *Food Quality and Preference*, **1**, 4–11.

MacFie, H. J. H. and Thomson, D. M. H. (eds) (1994) *Measurement of Food Preferences*, Blackie Academic & Professional, Glasgow.

McKenzie, J. (1980) The Eating Environment. In G. Glew (ed.) *Advances in Catering Technology*, Applied Science Publishers, London. pp. 474–481.

Marketing (1994) Top of the Shops, August 25, 14–17.

Marshall, D. W. (1988) Behavioural Variables Influencing the Consumption of Fish and Fish Products. In D. M. H. Thomson (ed.) *Food Acceptability*, Elsevier, London. pp. 219–232.

Meiselman, H. L. (1992) Critical Evaluation of Sensory Techniques. *Paper presented at Advances in Sensory Food Science: Rose Marie Pangborn Memorial Symposium, 2–6 August, Jarvenpaa, Finland.*

Meiselman, H. L. (1994) A Measurement Scheme for Developing Institutional Products. In H. J. H. MacFie and D. M. H. Thomson, *Measuring Food Preferences*, Blackie Academie and Professional, Glasgow. pp. 1–24.

Mennell, S. (1985) *All Manners of Food: Eating and Taste in England and France from the Middle Ages to the Present*, Basil Blackwell, Oxford.

Mennell, S., Murcott, A. and van Otterloo, A. H. (1992) *The Sociology of Food: Eating, Diet and Culture*, Sage, London.

Merrett, D. and Whithell, G. (1994) The Empire Strikes Back: Marketing Australian Beer and Wine in the United Kingdom. In G. Jones and N. J. Morgan (eds), *Adding Value: Brands and Marketing in Food and Drink*, Routledge, London. pp. 162–190.

Mintz, S. (1985) *Sweetness and Power: The Place of Sugar in Modern History*, Penguin, New York.

Morris, B. (1988) Are Square Meals Headed for Extinction? *Wall Street Journal*, April 29, 16.

Murcott, A. (1983) It's a Pleasure to Cook for Him: Food Mealtimes and Gender in some South Wales Households. In E. Garmarinkow *et al.* (eds), *The Public and the Private*, Heinemann, Oxford.

Murcott, A. (1986) You are What You Eat – Anthropological factors Influencing Food Choice. In C. Ritson, L. Gofton and J. McKenzie (eds), *The Food Consumer*, John Wiley and Sons, London. pp. 107–127.

Murcott, A. (1988) A Finger in Every Pie: The Variety of Approaches to the Study of Food Habits. In A. S. Truswell and M. L. Wahlqvist (eds), *Food Habits in Australia*, Proceedings of the First Deakin/Sydney Universities Symposium on Australian nutrition, René Gordon, Victoria.

National Consumer Council (1992) *Your Food: Whose Choice*, HMSO, London.

Nicosia, F. M. and Mayer, R. N. (1976) Toward a Sociology of Consumption. *Journal of Consumer Research*, **3**, September, 65–75.

Nicosia, F. M. (1966) *Consumer Decision Processes*, Prentice Hall, Englewood Cliffs, NJ.

Nord, W. R. and Peter, J. P. (1980) A Behavioral Modification Perspective on Marketing. *Journal of Marketing*, **44**, Spring, 36–47.

Olshavsky, R. W. and Granbois, D. H. (1979) Consumer Decision Making – Fact or Fiction? *Journal of Consumer Research*, **6**, September, 93–100.

Pillsbury Company (1988) What's Cookin'. A Pillsbury Study of Trends in American Eating Behaviour. *Consumer Communications*. The Pillsbury Company, Minneapolis, Minnesota. (Cited in Senauer, 1991.)

Powers, W. K. and Powers, M. N. (1984) Metaphysical Aspects of an Oglala Food System. In M. Douglas (ed) *Food in the Social Order: Studies in Food and Festivities in Three American Communities*, Russell Sage Foundation, New York.

Putnam, J. J. (1990) Food Consumption, Prices and Expenditure 1967–88. *Stat. Bull. 804*, US Dep. Agric., Econ. Res. Serv., Washington, DC.

Richards, A. (1932) *Hunger and Work in a Savage Tribe: A Functional Study of Nutrition Among the Southern Bantu*, Routledge, London.

Ripe, C. (1993) *Goodbye Culinary Cringe*, Allen and Unwin, Sydney.

Ritson, C., Gofton, L. and McKenzie, J. (eds) (1986) *The Food Consumer*, Wiley, London.

Rozin, E. (1982) The Structure of Cuisine. In L. M. Barker (ed.), *The Psychobiology of Human Food Selection*, AVI, Westport. pp. 189–203.

Senauer, B., Asp, E. and Kinsey, J. (1991) *Food Trends and the Changing Consumer*, Eagan Press, Minnesota.

Shepherd, R. (ed.) (1989) *Handbook of the Psychobiology of Human Eating*, Wiley, London.

Shepherd, R. and Sparks, P. (1994) Modelling Food Choice. In H. J. H. MacFie and D. M. H. Thomson, *Measuring Food Preferences*, Blackie Academic and Professional, Glasgow. pp. 202–226.

Straughan, R. (1992) Freedom of Choice: Principles and Practice. In *Your Food: Whose Choice*, edited by the National Consumer Council, HMSO, London, 135–156.

Symons, M. (1991) Eating into Thinking: Explorations in the Sociology of Cuisine. *PhD Thesis*, Flinders University of South Australia.

Symons, M. (1993) *The Shared Table: Ideas for Australian Cuisine*, Australian Government Press Publication, Canberra.

Taylor Nelson (1990) *What's for Breakfast, Lunch, Tea-time, Evening Meal*, Family Food Panel Special Report, February, Taylor Nelson House, 4–46 Upper High Street, Epsom, Surrey KT17 4QS.

Taylor Nelson (1993a) *Family Food Panel Management Summary*, Winter/Spring, August, Taylor Nelson House, 44–46 Upper High Street, Epsom, Surrey KT17 4QS.

Taylor Nelson (1993b) *Light Meals: The Growth of Informal Eating Occasions*, Family Food Panel Special Report, November, Taylor Nelson House, 44–46 Upper High Street, Epsom, Surrey KT17 4QS.

Thomson, D. M. H. (ed.) (1988) *Food Acceptability*, Elsevier Applied Science, London.

Turner, M. (1979) Nutrition and Lifestyles. *Proceedings from the First Annual Conference, The Royal Society, London, May 15–16*, Applied Science Publishers, London.

US Department of Commerce (1987) Households, Families, Marital Status and Living Arrangements: March 1987 (adv. rep.), *Curr. Pop. Rep.*, Ser. P-20, No. 417, Washington, DC. (Cited in Senauer, 1991.)

Ver Meulen, M., Ryan, M., Clements, M. and Ubell, E. (1987) What America Eats. *Parade*, October 25, 4–16.

Wall Street Journal (1989) People Patterns, Feb. 7, p. B1.

Wilkie, W. (1986) *Consumer Behaviour*, Wiley International, New York.

Acquisition

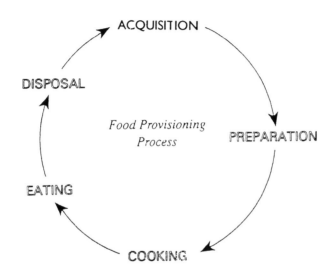

ACQUISITION

DISPOSAL

*Food Provisioning
Process*

PREPARATION

EATING

COOKING

Supply and food availability 2
Christopher Ritson and Richard Hutchins

2.1
Introduction

It is obvious that people can only choose foods which are available, and that their choice is constrained by what they can afford (or believe they can afford) to pay. It is superficially attractive to assume that the range of foods available, and the prices at which these foods are offered for sale, reflect merely the reaction of the food provisioning system to freely determined consumer demand. This is clearly not the case in a rigidly planned economy, where diets may be more to do with planning decisions than consumer preferences. But even in a market economy it is possible to identify a series of ways in which food choice may be influenced by independent events within the food supply chain which cannot be described as the food provisioning system 'responding to consumer demand'. This chapter therefore considers how food choice may be influenced by these supply-orientated factors.

In his famous 'General Theory', Keynes set out (in part), to refute 'Say's Law' – that 'Supply Creates its own Demand'. According to the classical economist Say, an economy would naturally reach full employment, because whatever was produced would find a market. Keynes demonstrated the possibility of an economy stagnating in a position of simultaneously experiencing substantial underemployment of productive resources *and* consumer shortage, and the 'Keynesian revolution' introduced the idea, at the macro level, of demand management of an economy.

At the micro level, much the same thing subsequently happened to the subject of marketing, with the traditional view that marketing was about 'selling' what had been produced being replaced by the 'modern marketing concept'.

For example:

> The marketing concept is a philosophy, not a system of marketing or an organisational structure. It is founded on the belief that sales and satisfactory returns on investment can only be achieved by identifying, anticipating and satisfying customer needs and desires – in that order. It is a philosophy which rejects the proposition that production is an end in itself, and that the products manufactured to the satisfaction of the manufacturer merely remain to be sold. (Barwell, 1965)

This modern marketing view is consistent with the economic theory of competitive agricultural and food markets. It is the price mechanism which is responsible for conveying messages from consumers to producers. The consumer will of course be influenced by a range of forces – convenience, health, and so on – as discussed in other chapters of this book; and the level and pattern of consumption will be constrained by technical knowledge and the fact that productive resources are limited. But, essentially, it is the consumer not the producer who determines what is eaten.

Suppose, for example, that a particular vegetable becomes – for whatever reason – more popular. This is expressed in the market as an increase in demand which will push up prices. This in turn will induce producers to allocate more productive resources (land, buildings, machines, labour, etc.) to the production of this particular vegetable, moderating the rise in market prices, and more is consumed.

Of course, because resources are limited (and as far as food production is concerned, particularly land) there may be reductions in the production (and thus consumption) of other food products (particularly other vegetables). But it will be consumers who have dictated this.

The high prices (and, at least for a time, the high profitability) of producing the vegetable may even lead to increased research into more productive varieties, production systems, etc., and this will eventually increase supplies further. But again, this is not 'supply creating demand', it is supply responding to changes in the pattern of consumer requirements.

Why, then, is it necessary to have a chapter on 'Supply and food availability' in this book? We should perhaps draw attention to the rather obvious point that limits to the availability of productive resources and technical knowledge will, via the price mechanism, influence a nation's diet. The more demanding the production of a particular food product on the use of resources, and the more difficult it is to produce, the higher its price and the less of it will be consumed.[1] But why an entire chapter, when the focus of the book is on consumption and the food consumer?

The answer is, of course, that the supply side of the food provisioning system is not always as straightforward and consumer-oriented as implied by conventional economics and marketing theory. It is not just limits to the availability of productive resources and technical knowledge which influence food choice. A variety of other supply-orientated factors can also be important. These influence food choice because of their effects on, respectively, food prices, food availability, food quality and consumer attitudes. When considering those effects which work through into patterns of food choice via the price mechanism, it is helpful to group the

[1]In low-income countries this may be expressed as overall malnutrition and even, in extreme cases, by starvation.

supply factors into three categories, namely: the biological basis of food production; the structure of production; and agricultural policies. (Agricultural policies can also have important effects on food quality and food availability.)

2.2
**Food prices and the
biological basis of
production**

Most fresh food products display seasonal variations in consumption. These are partly consumer oriented, and partly the consequences of seasonal variations in production costs expressing themselves as changes in consumption via the price mechanism. The price of salad vegetables, for example, would be *higher* in summer than in winter if it were not for the fact that the higher cost of production in winter dominates the higher demand in summer. The growth in the provision of 'out of season' vegetables in Northern Europe from Mediterranean and tropical countries provides an excellent example of the price mechanism inducing a productive response to consumer demand, as discussed above.[2]

Seasonal consumption patterns (even when supply determined) are therefore best thought of as a time-related aspect of the consumption effects of production costs. There are, however, two further implications of the biological nature of food production which impinge upon food provisioning. First, all agricultural production, to a greater or lesser extent, is influenced by natural factors – pests, diseases, rainfall, etc. – which means that production plans are rarely fulfilled. Often many of these factors apply to a large number of producers coincidentally, such that the amount of produce becoming available for consumption during any period may fall short of, or exceed, the aggregate of production plans. The fact that many agricultural products are either impossible or very costly to store means that these unplanned fluctuations in production are reflected in corresponding changes in consumption – usually, these days, via the price mechanism but in principle also via availability.

These changes may well pass with little comment. For example, the consumption of apples in the European Union was much higher in 1992/93 than in the previous marketing year. The explanation was nothing to do with changes in consumer preferences, production costs, retailing systems, market promotion, or anything like that.[3] It was merely the consequence of the fact that poor climatic conditions for the European apple crop (late frosts, and so on) was followed by spectacularly favourable conditions in 1992 (see Figure 2.1) with an abundance of low-priced apples available. As far as 'food provisioning' was concerned, the change

[2]Though changes in the structure of retailing (see Dawson, chapter 4) have also been important.
[3]Though there is doubtless a market research company report somewhere which draws attention to 'an *x*% growth' in the market for apples.

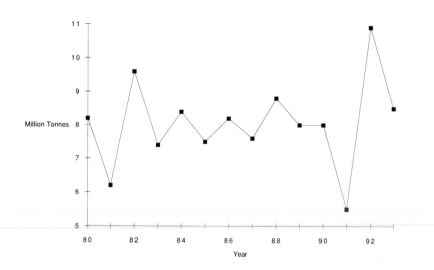

Figure 2.1   Total production of apples in the European Union (1980–1993). (*Source:* Calculated from European Union statistics)

was little noticed as consumers appear to move fairly freely between different fruits in response to price changes.[4]

Where consumers are less flexible, the result can be severe price instability, as consumers, reluctant to reduce purchases of a particular food, force up prices in times of reduced supplies; and a corresponding reluctance to eat more of it, simply because certain factors have dictated an increase in supply, forcing prices down. For example, the so-called 'potato shortage' in the UK during the mid-1970s involved only a modest (5–10%) reduction in the amount of potatoes eaten, but a colossal increase in prices.

An example of a food product somewhere between the two is provided by Figure 2.2, which shows the supply-oriented monthly variations in the amount of Scottish-farmed salmon marketed. With much of this salmon destined for the fresh market, consumption also 'has' to fluctuate. Prices are also very unstable, twice during the period covered falling by 50% in little more than one year. The price is not just influenced by Scottish production – in fact Norwegian supplies dominate the European market. Fluctuating supplies from different sources sometimes cancel each other out, leading to a degree of smoothing in the pattern of price movements. Thus the 'food choice' effect is partly to cause a month-by-month variation in total salmon consumption, and partly a switch between supplies from different sources.

The second way in which the biological nature of agricultural produc-

[4]What *was* noticed was the destruction of 1.5 million tonnes of the crop, a point discussed later in the chapter.

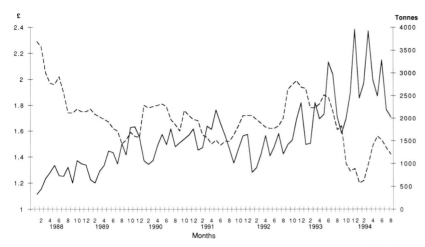

Figure 2.2 Scotish-farmed salmon: (——) total monthly volumes (tonnes); and (– –) average monthly delivery price (£/lb). (*Source:* Calculated from Salmon Growers' Association
statistics)

tion can influence the character of food provisioning is a consequence of
the fixed time period between the date of production decisions and produce becoming available for sale onto the market. The effect is that large
numbers of small producers may decide to expand production when
market prices are high, only to find that their produce is selling for much
less when all of these production plans are fulfilled. They all (or at least a
significant number of them) then decide to reduce production because of
low prices, only of course to find that prices are high when they come to
sell.

An example of this is provided by Figure 2.3, which shows annual total
production and consumption of beef in the European Union (excluding
recent new members) over the past 20 years. The quite marked cycle in
consumption is nothing to do with consumers or production costs; the
quantity of beef passing through the food provisioning system varies
because of the 18 months or so it takes beef animals to reach maturity,
and the coincidence of a large number of small producers independently
reacting to contemporary market prices.

During the 1970s, the EC was a significant beef importer, and the
rather gentle cycle in consumption followed international price patterns.
The sharp increase in production in 1977–79 therefore mostly had the
effect of replacing imports rather than increasing consumption. However,
since 1980, the EU has become more than self-sufficient in beef and veal;
European prices have become more sensitive to European production variations; and consumption has therefore fluctuated more. (Figure 2.2
shows that salmon prices also display something of a cyclical pattern.)

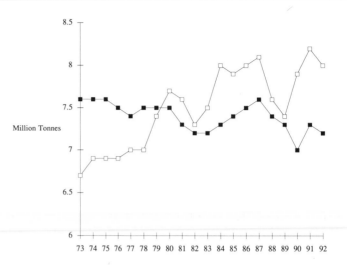

Figure 2.3   Beef and veal production (□) and consumption (■) in the EC. (*Source:* Calculated from European Union statistics)

2.3
Food prices and the
structure of production

The cycle in beef production and consumption referred to above is dependent on a competitive structure of production, with large numbers of producers taking independent decisions. It is possible to speculate about the way a less competitive production structure might have implications for food provisioning via the price mechanism. In principle, if a part of the nation's food supply should become concentrated, and thus dominated by a production interest which has the power to control supplies (and thus market prices), then less of that product will be available for consumption (and higher prices will be paid). In truth, in Western Europe anyway, it is very difficult to come up with concrete examples of where the exploitation of monopoly power by a concentrated source of supply has done this. Much of the concentration has occurred in food manufacturing and retailing. But intense competitive pressure has restricted the possibilities for pushing up prices (though there have been accusations that the concentrated food marketing sector has exploited less concentrated sources of agricultural product supply).

The major example of where this effect might be detected for agricultural products is the control over supplies that may be exercised by a monopoly marketing board, and since such a board needs government support (or at least acquiescence) there is an element of policy here, as well as structure. One good example of this was the activities of the British Milk Marketing Boards. The story is complicated, but for the purposes of this section, the essential point is that the Boards used their monopoly control over the supply of milk from farms to charge a higher price for milk destined for liquid consumption (where consumers were inflexible and the market would tolerate a higher price) than milk destined for manufacture of butter and cheese, where competition from imports required lower price. (The structure of milk supply in the UK also had various other effects on the 'dairy product provisioning system'. For full discussion, see Ritson and Swinbank (1991).)

The British government eventually decided to end the price-controlled milk marketing system in the UK, but the Milk Marketing Board in England and Wales was replaced in 1994 by a voluntary cooperative, known as Milk Marque. This led to claims by the dairy trade that a new monopoly had been created from which it was no longer protected:

> The [dairy] companies, along with consumer groups and high street stores, are protesting that Milk Marque, which begins operation on Nov. 1, has created a monopoly by signing up 65 per cent of milk suppliers in England and Wales at a time of national milk shortages. They say they face increases of between 5 and 20 per cent in the cost of their basic raw product, which will be passed on to the consumer. (Spencer, 1994)[5]

2.4 Food prices and agricultural policies

The distinction between an agricultural and food policy is a subtle one, and is explored in detail in Ritson (1983). For the purposes of this chapter, however, we are concerned with ways in which government policies which have been developed in order to meet objectives connected with farmers and the rural sector impinge upon food provisioning. Food policies which aim to influence the quality and composition of the diet are covered elsewhere in the book. We shall use the European Union's Common Agricultural Policy (CAP) as an example.

There is a very commonly held view that the CAP does have a major impact on the food we eat, and that this is something to do with 'food mountains'.[6] This view is not restricted to the general public, but is also held within sections of the media and by specialists in food and nutrition, where it is also often contended that the policy discourages healthy eating. Here is an example of this view.

The front page headline story in a popular Scottish Sunday newspaper in April 1994 went as follows:

WHAT A WASTE
The scandalous destruction of vast quantities of fruit and vegetables in the European Common Market has shocked MPs. In a single year, European growers destroy enough fruit and veg to keep a city the size of Glasgow going for years.

An MP is quoted as pointing out:

the health loss involved, since fruit and vegetables are essential to healthy eating.

---

[5]At the time of writing, it seems possible that the new Milk Marque will be considered by the British Monopolies and Mergers Commission.
[6]Pictures of some of the 1.5 million tonnes of apples referred to in a previous note were shown in the press being bulldozed into 'mountains'.

Well, of course, if it really was the inhabitants of Glasgow who had to do without fruit and vegetables for years, this example would be appearing in the section below on 'availability'! What happens is that produce withdrawn from the market sustains market prices, and the impact on consumption is spread throughout Europe via the price mechanism. In the main, as a proportion of total consumption, the amounts withdrawn are relatively modest. But fruit and vegetables attract attention because, unlike most other intervention products, they cannot be stored and subsequently sold or exported; they have to be destroyed, and the destruction of food is a very emotive issue.[7]

There are, however, no supermarkets complaining about lack of supplies. The problem is that perishable products like fruit and vegetables when in abundance can push market prices down to ruinously low levels, driving some growers out of business. Without the withdrawal system, the variation in apple consumption referred to above would have been greater – and that is the main impact on the food provisioning system – evenning out supplies, not denying consumers the opportunity to consume healthy products.

It is far from clear, anyway, that all of the destroyed produce would have passed through the food provisioning system in the absence of the CAP withdrawal and compensation mechanism. A common experience with many fruits and vegetables is that, because of the high cost of harvesting, when market prices collapse, produce is left 'to rot on the trees or in the fields'.

However, patterns of food consumption *are* influenced by supply-orientated manipulation of market prices. By a mixture of minimum import prices, intervention and export subsidies, the Common Agricultural Policy sustains the prices of most food products at higher levels than would apply in a free market. It is, of course, extremely difficult to estimate to what extent the prices of individual food products are pushed up by these CAP policies, and to what extent and in what way patterns of food consumption are distorted. Elsewhere, Ritson (1991) has published a 'league table' relating to this, reproduced here as Table 2.1. The products are grouped into four categories which might reasonably be described as: high price raising effect; moderate price raising effect; neutral effect and price depressing effect. Recent changes in the Policy are having the effect of increasing the gap between the 'top three' and bread and the cereal-based products; and have removed what were in effect consumer subsidies for lamb and processed tomatoes. But the 'damping-down' of consumption of (in particular) dairy products, because of the price raising policy, continues.

Thus the manipulation of market prices under the Common Agri-

[7]As evidenced by the fact that, on the same Sunday, other newspapers led either on the conflict in the Balkans, or the death of former US President, Richard Nixon.

Table 2.1 Impact of CAP on food prices ('League table')

Butter (and other milk products)
Sugar
Beef

Bread
Pork
Poultry and eggs

Vegetable oils
Most fruits and vegetables

Lamb
Canned tomatoes

Source: Ritson (1991)

cultural Policy provides a clear example of a way in which the balance of consumption of individual foods in Europe is influenced by a supply-driven factor – that is, the price levels fixed by the Policy which are motivated by objectives associated with farmers and farming, not with consumers or consumption. But, as hinted at earlier when introducing the fruit and vegetables example, although this point in itself is widely accepted, among certain sections of the media and interest groups, there is often a complete failure to distinguish between a policy which *encourages* the production of a product and a policy which *encourages* its consumption. In fact, by virtue of its reliance on price support, the CAP tends to *discourage* the *consumption* of those products for which it gives the greatest stimulus to production. Coincidentally, it has succeeded in pushing the diet of European consumers in the direction which the medical profession would now regard as more healthy. For example, a recent study of the impact of the CAP on the consumption of food products in Greece shows a decline in consumption of sugar, meat and dairy products, and increases for citrus fruit, vegetable oils and vegetables (with no change for bread and cereals) – broadly consistent with the implications of Table 2.1 (Georgakopoulos, 1990).

Figure 2.4 illustrates the power of supply-induced changes in relative prices influencing diets. Until the early 1980s, the *combined* consumption of butter and margarine was virtually constant, but the relative position showed dramatic changes which can be ascribed almost entirely to relative prices.

A coincidence of dry Northern and Southern Hemisphere summers around 1970 caused a worldwide fall in milk yields and a dramatic rise in international butter prices. The UK then imported most of its butter (particularly from New Zealand) and UK butter prices went up, and consumption down. Butter prices then came down again, just as world vege-

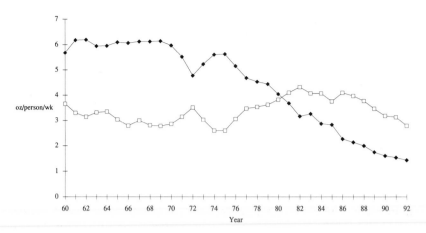

Figure 2.4 Butter (◆) and margarine (□) purchases (1960–1992). (*Source:* Calculated from National Food Survey data)

table oil prices were caught up in the 1974 international commodity boom (or 'World Food Crisis', as it was sometimes called). When vegetable oil prices came down, the full effect of British adoption of the CAP on butter prices was beginning to take effect. (During the 1980s, it was, however, other factors which sustained the fall in butter consumption, as discussed by Ritson and Hutchins in chapter 3).

2.5 Availability

Philosophically, it is quite difficult to distinguish between the idea of a food product *not* being consumed because of, on the one hand, high price, and on the other, non-availability. When prices rise, the product drops out of the diet completely for some consumers and so, in one sense, the non-availability of a product is merely an extreme version of consumption falling because of rising prices. It is nevertheless in practice helpful to make a distinction between the kind of effects discussed in the previous sections – where supply-orientated factors modify the relative prices of different foods and this alters the balance of the diet – and those cases where supply-orientated factors influence the availability of a food product in a particular country or region.

In the developed world, the past few decades have witnessed an enormous increase in the range of food products available to the typical consumer (see Buisson, chapter 8). The new products which have become available can be separated into four categories:

1. Products previously available in the locality of production become available elsewhere (e.g. Kiwi fruit, or South American table wine, in Western Europe).
2. New products manufactured from existing raw materials (e.g. convenience ready meals).

3. Animal products from (usually intensive) farmed sources available previously in 'wild' form (e.g. ducks, venison, salmon).
4. Genuine 'new' foods (see below).

The point about all these newly available foods is that they cannot originate as a consequence of supply responding to an increase in demand via the price mechanism. Rather, supply is responding to a perceived demand and someone, somewhere in the food chain, will take the initiative to market the new product. In that sense supply *is* creating its own demand. However, we need to draw a line between cases where the new product is nevertheless demand-led in a broader sense, and where it originates more as an accident of a supply-led change.

Before the Kiwi fruit appeared in Western Europe, it would not have been possible to draw a demand curve for it, since consumers were largely unaware of its existence. (Kiwi fruit was anyway an export-orientated renaming of the 'Chinese gooseberry'.) But in a broader sense, its appearance in Europe was a market response to a measurable increase in the demand for novelty and variety in fruit and vegetable purchases. Much the same can be said of the appearance of new wines. In the UK, anyway, much of the initiative for making products available in that sense has been taken by the retail sector. Similarly, new manufactured food products can be seen as a market response to the experience of increasing demand for convenience in the overall pattern of food purchases (as discussed by Gofton, chapter 7). In the case of (3) and (4) above, however, forces originating the development of the new product may sometimes lie much further back in the food chain.

Take again the case of farmed salmon. Figure 2.5 shows the phenomenal growth in the consumption of farmed Atlantic salmon since the pro-

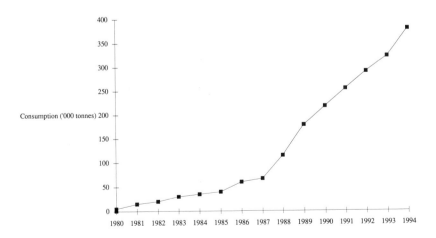

Figure 2.5 Consumption of farmed (Atlantic) salmon (1980–1994). (*Source:* Ritson, 1993)

duct first appeared in the 1970s. The new product was of course the consequence of technological innovation which made it possible to parody in captivity the complicated life of wild salmon. The governments of the relevant countries (particularly Norway, the major producer, still supplying more than two-thirds of the European market) were keen to support this new development, because of its importance in sustaining employment in remote regions. For most of the 1980s the product seemed 'to sell itself', with market prices being maintained in the face of increasing supplies. Prices did eventually collapse in 1989, and subsequently the product began to display some of the other supply-originating consumption effects discussed above – but nevertheless the market continues to absorb rapidly increasing supplies.

Analysis of the market (Ritson, 1993) suggests that most of the growth in consumption has occurred because of a curious, coincidental, surge in demand, and not messages passed through the price mechanism. In part this has probably been because farmed salmon happened to fall into a broad product category subject to a positive underlying trend in demand (a mixture of convenience, at least relative to other fish, a healthy image and suitability for the catering trade). But in the main, the growth in demand seems primarily to have been simply a consequence of availability. This is the word which is mentioned most often by people in the trade when discussing salmon consumption. 'Wild' salmon was a luxury product available in specialist shops and on high-class restaurant menus. As supplies of farmed salmon increased, consumers became aware of its existence and it entered the diet. As awareness of salmon availability spreads, so demand grows and prices are sustained – whereas a surge in supplies for a product already widely available would have involved an increase in consumption via lower prices.

The growth in consumption in France is perhaps the most remarkable. The French are not exactly renowned for a positive attitude towards imported food products; yet, by the early 1990s 'Saumon' was becoming almost as common in the catering trade as 'Steak Haché' and 'Poulet Rôti'. Supply does seem to have created its own demand.

With respect to category (4), it is quite difficult to identify genuinely new foods. One candidate for inclusion, however, is the mycoprotein food developed during the 1980s in the UK, jointly by ICI and Rank Hovis McDougal, and marketed under the brand name Quorn. The original mycelium source was found (so it is claimed, anyway) accidentally in a field in England, and the fermentation technology for its growth was a spin-off from a production plant developed for producing animal feed. The outcome is a high-fibre, high-protein, fat-free, vegetable raw material, which is tasteless, and can be 'spun' to parody the mouth feel of different meats.

The question then confronting the originators of the 'new food' was what exactly had been created when considered from the perspective of the food consumer – was it a cheap substitute for meat; a healthy alternative to meat; a vegetarian product; or what? Is it possible to market, simultaneously, to consumers all the various options which are technologically feasible? This issue is returned to subsequently when we consider influencing consumer tastes and preferences.

Looking to the future, the main possibilities for new foods (as opposed to new food products) seems to be associated with plant biotechnology – that is, genetic modification. For example, an American genetics firm has:

> successfully devised new tomato strains which soften less rapidly and are more resistant to deterioration when frozen. (Pearsall, 1993)

By 1995, the first product from this innovation had appeared on the US market – the 'Flavr Savr' tomato.

Whether this kind of development should be described as a newly available food, or merely an improvement in quality, is open to debate, but the point is that the fundamental research which led to the discovery of gene technology was not driven by the requirements of food consumers. Someone in the food chain is applying new techniques developed as a consequence of the general advance in scientific knowledge, which is leading to new foods becoming available. Up to a point, therefore, the consumption of new foods can be supply-led. At some stage they have to be matched to a latent demand.

It is, of course, also possible that consumers might be denied access to food products which are potentially available. All the examples which come to mind are the consequence of government intervention. Such intervention may be directed towards the interests of consumers – for example prohibiting the marketing of products in the interests of food safety.

For example, in the UK, a Food Advisory Committee is charged with the task of assessing 'the risk to humans of chemicals which occur in or on food and to advise Ministers on the exercise of powers in the Food Safety Act 1990 relating to the labelling, composition and chemical safety of food' (MAFF, 1994). As well as safety aspects, the Committee considers 'need' in the sense of justification on the basis of economic or quality improvements. In the USA, the Food and Drug Administration (FDA) has prime responsibility for a complex system of food safety regulations. In particular, since 1958 the burden of proving that an additive is safe is placed on the manufacturer before introduction. (For a discussion, see Senauer et al., 1991.)

It is also possible for a government to translate a supply-orientated shortage of a product into lack of availability by imposing price controls. The market mechanism rations such products by price increases. Price controls mean ration books, or queues, and sometimes the product simply disappears from the shelves to be sold 'under the counter'.

Government action may, however, also restrict the availability of food products as a consequence of a policy adopted in order to support domestic farmers. This usually involves pushing up the price of imported products, as discussed above, but may sometimes involve a complete ban (usually justified on grounds of plant health or animal diseases). Where the prohibited imports are of an agricultural raw material – as is often the case – then all that will be experienced by consumers will be an increase in the average price paid, as the source of supply does not differentiate at the retail level. But in some cases – particularly with fruits and vegetables – consumers *may* well differentiate between the imported and domestic product and in these cases the action is translated into non-availability. For example, there is a small banana industry in Crete and Cyprus, and the governments concerned have prohibited imports. But the European bananas are so different that the effect is not just to push up the price of bananas in these countries, but to make unavailable what is, in effect, a different product.

The CAP for fruit and vegetables can also have this effect. Once import taxes are triggered, they can rise rapidly to prohibitive levels against supplies from a particular country. (The way the mechanism works is described in Ritson and Williams (1987).) For example, in May 1994 the import tax on Moroccan tomatoes *exceeded* the wholesale market price of tomatoes within the EU. In other words, Moroccan producers would have had to *pay* for the privilege of European consumers being able to eat their produce! Obviously they chose not to do so, and Moroccan tomatoes were driven from the European market. (Of course this is only a real issue of availability to European consumers if they distinguish Moroccan produce from tomatoes grown in European Union member states.)

2.6
Food quality

Just as it is difficult to make a clear distinction between prices and availability as influences on food choice, so it is a matter of debate at what point a high quality version of a particular product becomes regarded as, in effect, a different product. Quality is, anyway, a rather illusive and subjective feature of food consumption. A product may pass with flying colours on every imaginable technological criterion, and achieve excellent scores in blind taste panel tests, but nevertheless fail to find consumer acceptance, as has frequently been discovered at the Aberdeen Ministry of Agriculture Fish Research Station, which has been attempting to develop new fish products from unfamiliar fish species. And when quality and price are combined as two inputs into food choice, then there is no reason to believe that a consumer might not prefer to purchase a lower quality, lower priced, food product. But is there any evidence to suggest that the balance of the range of quality and price available to consumers is restricted by factors within the food chain? Again, it seems most likely that any examples will occur as a consequence of government policies. As we have seen, the CAP has a major influence on diets, by influencing patterns of food consumption *via the price mechanism*. In principle, some of the policy mechanisms might also influence the 'quality' of individual foods, in terms of product composition.

Within the meat sector, two examples have been quoted as evidence of the CAP affecting product quality. First, the fact that fat class 4 qualified for the sheepmeat variable premium in the UK (whereas 'the market requires a less fat product') implies that the lamb consumed in the UK was of a higher fat content than would otherwise apply. (The scheme ended in 1991.)

The other example concerns the intervention standards for beef, which ensure that it is high-quality beef which is removed from the market. Thus, it is argued, the overall average quality of beef consumed within the European Union is reduced.

Withdrawal of fruit and vegetables from the market might be regarded as analogous to the beef example – as an inspector has to certify that produce is of a *sufficiently high quality* before it can be destroyed! In addition, some traders argue that withdrawal induces EC growers to produce to minimum quality standards, rather than meeting the requirements of the market.

When it comes to imported produce, in contrast, there are strong grounds for believing that the Reference (minimum import) prices may have improved quality. Ritson and Williams (1987) quote one Convent Garden importer who was quite explicit on the subject, claiming that British membership of the EEC had provided a clear signal to third countries to improve their marketing in the UK. Their reaction was:

> to improve their quality standards – thus securing the highest prices by having a quality premium, and so avoiding countervailing charges. The event of reference prices has had many bad effects but one good thing is this quality upgrading.

Another way in which the CAP is supposed to affect quality of fruit and vegetables is by quality standards – imposed rigorously on produce imported from non-EEC countries, but much less so for produce sold locally in Southern Europe. Ideally, such a system should be consumer driven. Kohls and Uhl (1990) comment:

> The food grading system sets up a channel of communication between food producers and consumers. Ideally, grades will result in a perfect match between the diverse wants of consumers (according to their means and preferences) and the heterogeneous quality of commodities produced and marketed. This sorting and matching process has the potential of increasing both consumer satisfaction and farm profits.

However:

> [A] criticism of present [USA] food grades is that they are convenient for traders, but are not consumer-orientated.

The problem is, of course, that many of the properties of such products which consumers value are difficult (or at least expensive) to measure objectively and reliance is placed on colour, size, shape and so on. This only has an *adverse* effect on food quality if it leads to a suppression of other quality characteristics – in particular taste. But this can happen if the officials responsible for food grades are insensitive to genuine consumer attitudes. There is a view that the European Commission is obsessed with the need for straight cucumbers and uniform, blemish-free (but tasteless) apples, and that the grading system has a negative (and not as intended, a positive) impact on food quality.

Barker (1990) refers to a study which showed that only 3 per cent of strawberries chosen by consumers at a self-pick farm would have met EEC standards:

> There appeared to be very little relationship between grading standards and consumer preferences.

The implication is that the EEC quality standards might have eliminated the 'preferred' strawberries from the market.

2.7
Consumer attitudes

On the face of it, consumer attitudes, tastes and preferences would seem the least likely of our four categories to be influenced by supply-orientated factors. The consumer is supposed to be king, and good marketing is meant to be about identifying and meeting consumer requirements, not creating them. However, a little thought reveals that certain aspects of consumer food preferences originate in supply factors. The fact that Greek cuisine includes a lot of olive oil, and Norwegian does not, is not unconnected with the geographical distribution of olive trees. The point is that many international differences in dietary patterns are the consequence of availability (and price) in the locality. Rozin *et al.* (1986) comment that 'Specific systems of food production have been implemented in different parts of the world to exploit specific resources or ecosystems, and this has led to the establishment of cuisines based on food resources with particular limits.' But consumers (and cultures) progressively acquire a taste for these products such that, even following the geographical spread of availability in the modern world, international differences in tastes and preferences remain. Olive oil is now widely available in Northern Europe, at prices which no longer exceed by very much those in Mediterranean countries, and consumption levels now differ more because of tastes and preferences than they do because of price and availability.

The same kind of forces are manifested in what is known as the 'vintage effect'. This can be illustrated by reference to data from the British National Food Survey (NFS) – the longest running continuous survey of household food consumption in the world. The stage in the family life cycle, characterised by the recorded age of the housewife in the NFS, provides one of the best explanations of differences in patterns of household

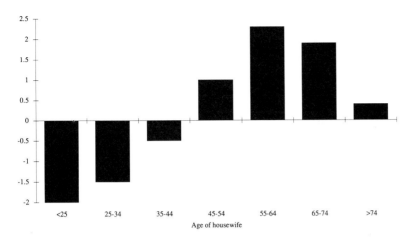

Figure 2.6 Total food expenditure by age of housewife (£ per person per week compared with national average). (*Source*: Ritson and Hutchins, 1991)

consumption for some foods. The typical pattern, illustrated in Figure 2.6 by total food expenditure, is one of less than average consumption at early stages (young single people and young marrieds), rising to a peak when the housewife is in her fifties (the children have left the household and the disposable income is high) and then declining towards the average for pensioner households.

The consumption variations according to age can be attributed to two effects. First, the structure of the household (size, number of children, etc.), proportion of meals eaten outside the home and income of the household will be closely connected with the age of the housewife, and this will influence patterns of consumption between different products – say milk or biscuits – as well as the typical pattern shown in Figure 2.6. Second, consumption habits are formed by children and young adults and they carry these habits through with them as they grow older. This is the 'vintage effect'. The food habits acquired in this way will be partly supply-orientated – dependent on what is available (and cheap) at the time of habit formation.

Figure 2.7 shows the consumption of mutton and lamb, and poultry, in the UK from 1940 to 1990. Historically, mutton had been the predominant meat for many British households and local production, together with a well developed trade in frozen lamb from New Zealand, meant that supplies were plentiful, even during the War, and more so in the early post-war period. The technological revolution which affected the poultry industry began to have an effect on the market during the 1960s, influencing habit formation for those now in their 40s and 50s (but not 60s and 70s) by the mid-1980s. Now compare Figures 2.8 and 2.9. Consumption of poultry roughly follows the typical pattern up to middle age,

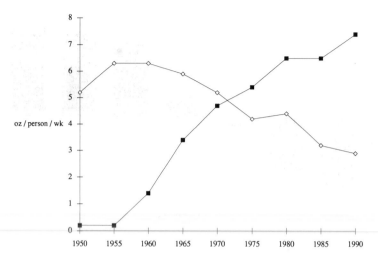

Figure 2.7 Consumption of lamb (◇) and poultry (■) in the UK. (*Source*: Calculated from Natural Food Survey data)

but indicates lower than expected consumption for the age groups over 55. In contrast, consumption of mutton and lamb increases progressively with age.

Even seasonal shifts in food preferences may originally be supply orientated. The (sensible) cultural tradition develops to demand certain foods when they are available. Modern production and marketing techniques reduce supply variation, but the seasonal pattern of demand lives on. It has for example been argued that certain feast days developed because of

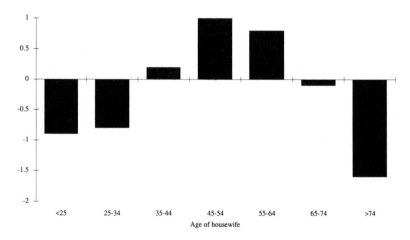

Figure 2.8 Consumption of poultry by age of housewife (oz per person per week compared with national average). (*Source*: Ritson, 1988)

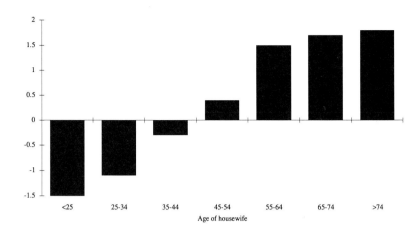

Figure 2.9 Consumption of mutton and lamb by age of housewife (oz per person per week compared with national average). (*Source*: Ritson, 1988)

the need to consume surplus food – to eat animals which could not be kept through the winter, for example; and to eat in spring the ones that had been kept through the winter, once offspring begin to mature. Modern supply then has to meet the established, consumer preference-based surges in demand; for example turkeys at Christmas, or Thanksgiving.

Another interesting example of how consumer tastes might be shifted by a supply event, is if a (temporary) shortage leads them to acquire new tastes. Part of the explanation for the growth in consumption of pasta and rice in the UK is thought to be that consumers were forced (or induced) to try them as alternatives to potatoes during the period of high prices mentioned earlier, and then continued to purchase a new product which a supply factor had induced into the diet.

Thus there are many examples of how consumer attitudes and preferences for foods have been moulded by supply and availability. The other way in which supply orientated forces might be argued as influencing consumer preferences is via advertising and promotion. This leads to the debate over the extent to which advertising is informative or persuasive', and takes us beyond the confines of this chapter. However, the issue is well summarised by Kohls and Uhl (1990):

[A] criticism of advertising is that it distorts consumer choice and interferes with consumer sovereignty. Producers and marketing firms are charged with not only creating products but with creating the demand for their products. Proponents of advertising, on the other hand, argue that advertising simply helps consumers to identify their needs and to match them with available products.

Whichever interpretation is taken, there are cases of attempts to alter consumer food preferences where the motivation originates well back in the supply chain. For example, olive oil is regarded as an important source of income for small farmers in Southern Europe. Domestic sales are increasingly being threatened by other (cheaper, imported) vegetable oils. The European Community is financing an Olive Oil Council which is undertaking generic promotion with the intention of stimulating demand in Northern Europe.

However, even if producers are 'not only creating products but creating the demand for their products', the resulting shift in consumer attitudes and preference need not be regarded as a distortion. Let us return to the example of Quorn.[8] Developing the market for this product would have been impossible without the supplier influencing consumer attitudes and preferences. Decisions over marketing variables automatically influence consumer attitudes for something of which they had no previous knowledge. Even the brand name itself would influence consumer attitudes to the new food.

Quorn, a bit like 'corn', provided something of a vegetable/rural image, and the hard consonants were thought to aid memory. But research indicated that consumers were broadly neutral to the name – there were no major preconceptions. Similarly, one of the reasons for creating the new company – Marlow Foods – was to dissociate the new food from the 'chemical image' of ICI.

Next comes price. Economic theory would like to see 'price' and 'tastes and preferences' as independent influences on food choice; but in practice the sequence of events tends to be that the forces of supply and demand dictate that some food products (e.g. caviar) are expensive and some (e.g. potatoes) are cheap. But then the scarce and expensive product acquires the attribute of being a luxury and becomes valued more than the cheap product over and above its sheer scarcity value. For most products competition and the availability of the raw material constrain the eventual price of the food, but in this case there was considerable flexibility over the choice of price and obvious concern that the price would influence consumer perception of the product. As one of the marketing executives commented to one of the authors at the time:

> It is easy to reduce a price which is too high; much more difficult to raise one which is too low.

The issue over how to establish consumer attitudes to the new product relative to other major protein foods can be illustrated by the use of product perception maps. What follows is purely hypothetical to illustrate – not based on consumer research. Figure 2.10 does this by identifying four dimensions on which consumer preference for the new product might be based, and suggests where some major protein foods might be located in the minds of consumers. (The pairings are purely arbitrary.)

---

[8]This section draws on unpublished work by one of the authors for the company involved.

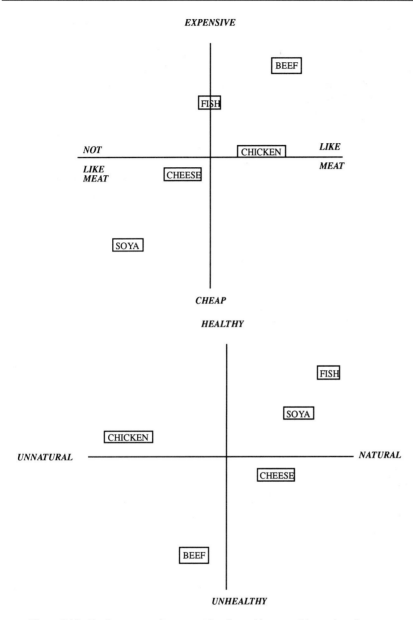

Figure 2.10   Product perception maps showing arbitrary pairings of preferences.

Expensive/cheap refers to consumer perceptions, but of course will be heavily influenced by prevailing prices. Like/not like meat was important. In this particular case it was possible to give the product a cooking structure and image which was meat-like – as opposed to beans, for example, which are known to be vegetable protein products. Natural/unnatural would be a mixture of the way the product is farmed (i.e. the reputation of chicken, as opposed to beef, as an intensive product produced in an unnatural environment) and degree of manufacture. Healthy is again a matter of perception, but there is a view of red meat and dairy products as less healthy than white meat, vegetable products, and fresh fish.

In the event, it seems that Quorn was marketed as a healthy, natural, alternative to meat – but not an exclusively vegetarian product, and not a cheap substitute. But, arguably, its market position is still a little unclear.

2.8 Conclusion

This chapter has shown how various aspects of the food supply chain can influence patterns of food consumption in ways which cannot be regarded as simply the food provisioning system responding to freely determined consumer demand. In particular it has highlighted the way consumption will follow biologically determined movements in production, and the role of agricultural policies in influencing the balance of a nation's diet.

References

Barker, J. W. (1990) *Agricultural Marketing*, OUP. Oxford.

Barwell, C. (1965) The Marketing Concept. In A. Wilson (ed.), *The Marketing of Industrial Products*, Hutchinson, London.

Georgakopoulos, T. A. (1990) The Impact of Accession on Food Prices, Inflation and Food Consumption in Greece. *European Review of Agricultural Economics*, **17**, 4.

Kohls, R. and Uhl, J. (1990) *Marketing of Agricultural Products*, 7th edition, Macmillan, New York and London.

MAFF (1994) Press Release FAC 12/94.

Pearsall, D. (1993) *Public Perception of Plant Biotechnology: A Detailed Study*, The Chamberlain Partnership, Huntingdon.

Ritson, C. (1983) A Coherent Food and Nutrition Policy: the Ultimate Goal. In Burns, McInerney and Swinbank (eds), *The Food Industry: Economics and Politics*, Heinemann, London.

Ritson, C. (1988) Special Study on Meat and Meat Products. *Section 3, Household Food Consumption and Expenditure, Annual Report of the National Food Survey*, HMSO, London.

Ritson, C. (1991) The CAP and the Consumer. In C. Ritson and Harvey (eds), *The Common Agricultural Policy and the World Economy*, CAB International, Wallingford.

Ritson, C. (1993) The Behaviour of the Farmed Salmon Market in Europe: A Review. *Centre for Rural Economy Research Report*, University of Newcastle upon Tyne.

Ritson, C. and Hutchins, R. (1991) The Consumption Revolution. In J. M. Slater (ed.), *Fifty Years of the National Food Survey*, HMSO, London.

Ritson, C. and Swinbank, A. (1991) The British Milk marketing Scheme – Implications of 1992. *Proceedings of the 25th EAAE Seminar, Braunschweig-Völkenrode*.

Ritson, C. and Williams, H. (1987) Reference Prices and the Marketing Mix for Fruit and Vegetables. *Journal of Food Marketing*, **3**, 1.

Rozin, P., Fallon, A. E. and Pelchat, M. L. (1986) Psychological Factors Influencing Food Choice. In C. Ritson, Gofton and McKenzie (eds), *The Food Consumer*, Wiley, Chichester.

Senauer, B., Asp, E. and Kinsey, J. (1991) *Food Trends and the Changing Consumer*, Eagan Press, St Paul.

Spencer, R. (1994) The milk shake-up turns sour over price-fixing fears. The *Daily Telegraph*, 25 August.

Food choice and the demand for food 3

Christopher Ritson and Richard Hutchins

The choice of food made by individuals and households is influenced by a large range of factors – including, for example, prices, incomes, tastes, religion, social attitudes, desire for convenience and so on. Much of this book is concerned with a detailed account of these various influences on food choice – what they are, and why they affect what we eat. It is, however, extremely difficult to say how important each individual factor is in influencing any particular food purchase decision. Indeed, arguably, there is no sense in attempting to explain whether, in an individual's decision to buy, say, a kilogram of tomatoes, he or she was most influenced by taste, relative price, income, nutritional knowledge or whatever.

For example, market research often concludes that price is relatively unimportant in food choice, when consumers are confronted with questions of the kind: 'How important were the following in your decision to purchase product X?' But when price goes up many consumers reduce or cease purchasing the product, and when price goes down consumption increases – and the food supplier discovers that price is, after all, rather important.

What we can do is attempt to explain why one household purchases a large quantity and another only a small quantity of tomatoes – by comparing differences in the households. An explanation might lie, for example, in income differences, social background, price advantage, a food allergy – or a host of other factors. We would, of course, be proceeding on the basis of inference, and it would be extremely dangerous to ascribe the different levels of tomato purchase entirely to a difference in, say, household income, even if we had failed to identify any major other differences between the two households being compared. There would almost certainly be other factors which were important but which could not be identified. We are on much safer ground when comparing *average* levels of consumption of a particular product between, say, high-income and low-income *groups* of households.

Similarly, it is possible to attempt to explain why average levels of consumption of different foods among the population as a whole change over time, by relating the changing patterns of food consumption to changes over time in the factors which we know influence food choice. A better understanding of the causes which change average levels of food consumption over time, and variations within the population, will be of great value to nutritionists, those involved in formulating food policies, and food marketers.

We do this by first identifying the set of factors which empirical research has shown influence consumer food choice and then relating these, statistically, to estimated total quantities purchased of individual foods over a specific period of time – which economists refer to as 'demand'. These factors can usually be divided into three broad groups: food prices; consumer incomes; and what economists describe as 'tastes and preferences' – roughly everything else, an amalgam of behavioural and attitudinal influences ranging from, for example, the number of single-person households to attitudes to diet and health.

Prices and incomes are sometimes referred to as the 'economic factors', and the rest as 'non-economic' (though there is, of course, often an economic component or influence within the other factors). This is also a useful distinction, not only because it indicates which discipline is likely to be most helpful in explaining the influence of a particular factor on patterns of food choice, but because it is usually possible to obtain independent, quantitative, estimates of the economic factors; whereas, in general, it is impossible (or at least very difficult) to do so with the other factors.

The ability to obtain numerical estimates of consumer incomes, and the prices paid for food products, contributes to our understanding of food choice in two important ways. First, it allows us to be able to predict how patterns of food consumption are likely to change in response to changes in food prices and consumer disposable incomes. Second, once we have obtained what we believe to be reliable estimates of the way patterns of food consumption respond to changes in prices and incomes, statistical analysis of the data collected over time allows us to identify that part of the past change in consumption which can be ascribed to changes in prices and incomes. The implication is that the remaining change, known sometimes as 'the underlying trend in demand', must be attributable to other factors – probably associated with more fundamental changes in consumer behaviour and attitudes with respect to food products. Thus, economic analysis of food demand can also contribute to our understanding of the extent to which attitudinal and behavioural factors are responsible for changing patterns of food consumption.

In this chapter, we first explore how patterns of food consumption are typically influenced by changes in food prices and consumer incomes. We then go on to discuss the concept of the *underlying* trend in demand and argue that, whereas some years ago, changes in prices and incomes had a profound effect on changing patterns of food consumption, increasingly it seems to be other factors which are responsible for the evolution of the modern diet. (This section is based on empirical work undertaken by the authors using British National Food Survey data.) Finally, having identified those foods which appear to have either strong positive or negative underlying trends in demand we explore some of the behavioural and attitudinal changes which appear to be responsible for these developments.

3.2
Food prices

We find that in almost all cases the quantity purchased of a food product is inversely related to price[1] – that is, when price goes up, quantity demanded goes down, and vice versa. But by how much? Economists conventionally use a measure of the responsiveness of quantity demanded to price changes known as *elasticity*. This is defined as:

$$\frac{\text{Proportionate[2] change in quantity demanded}}{\text{Proportionate change in price}}$$

If the proportionate change in quantity is greater than the proportionate change in price we say demand is *elastic* (i.e. responsive to price changes); and if it is less we call it *inelastic* (i.e. less responsive).

Table 3.1 gives estimates by Chesher and Rees of price elasticities of demand for the main food products and food product groups, using the delightfully named Almost Ideal Demand Model (Chesher and Rees, 1988). The data used are those collected by the British National Food Survey, and the table covers virtually all of the food items consumed in the UK.

The most obvious characteristic of Table 3.1 is that the majority of coefficients are *inelastic*: that is, they lie between 0 and −1.[3] In other words, for most main food products a proportionate change in price is associated with a less than proportionate change in the quantity demanded. Table 3.1 is consistent with one other normal characteristic – that the demand for a product will tend to be more elastic the better or more numerous are the substitutes for it. Thus products such as milk, cheese, eggs, butter, sugar, potatoes and bread – all of which lack good substitutes – are very price inelastic. This feature of demand means that the more we attempt to break food consumption down into individual product categories, the more price elastic the demand for each product is likely to become. It is for this reason that the demand for all carcass meat is less elastic than that for beef and veal, pork, or mutton and lamb,

[1]But see later for an exception.
[2]We use proportionate change, rather than just change, so that we can compare the responses of demand to price changes for different products.
[3]The signs are negative because of the inverse relationship – an increase in price causes a fall in demand.

Table 3.1 Estimates of price elasticities of demand for major food products and product groups in Britain (1977–1986)

Product	Price elasticity	
Milk and cream	−0.40	
Cheese	−0.26	
Carcass meat	−0.61	
Beef and veal		−0.70
Mutton and lamb		−1.45
Pork		−1.69
Other meat	−0.40	
Fish	−0.77	
Eggs*	−0.20	
All fats	−0.17	
Butter*	−0.40	
Margarine	−0.56	
Sugar and Preserves	−0.26	
All vegetables	−0.18	
Potatoes		−0.39
Fresh green vegetables		−0.79
Other fresh vegetables		−0.33
Processed vegetables		−0.66
All fruit	−0.51	
All bread	−0.09	
Cakes and biscuits	−0.58	
Breakfast cereals	−0.88	
Beverages	−0.62	
Ice cream	−1.08	

Source: Chesher and Rees (1988)
*Certain characteristics of the markets for butter and eggs during the 1980s made it impossible to obtain plausible estimates for these products. The coefficients quoted are taken from Ritson (1977), and relate to an earlier period.

Table 3.2 Estimates of price elasticities of demand for fresh fruits in Britain (1981–86)

Oranges	−1.29
Other citrus fruit	−1.12
Apples	−0.13
Pears	−1.59
Stone fruit	−2.25
Grapes	−1.49
Other soft fruit	−4.81
Bananas	−0.80
Other fresh fruit	−1.28
All Fresh Fruit	−0.43

Source: MAFF (1988)

taken separately. Similarly, the demand for all vegetables is less elastic than that for the individual product categories.

If we break down the food product categories further than in Table 3.1, then the products considered begin to be regarded by consumers as good substitutes for one another, and very elastic estimates result. This is shown in Table 3.2, where the category 'all fresh fruit' (itself a sub-category of 'all fruit' in Table 3.1) is broken down into nine sub-categories.

The responsiveness of the quantity demanded of one product (A) to a change in the price of another product (B) is known as cross-price elasticity of demand and is defined as:

$$\frac{\text{Proportionate change in quantity demanded of product A}}{\text{Proportionate change in price of product B}}$$

The cross-price elasticity of demand might be positive or negative – that is, an increase in the price of product B might cause the quantity demanded of product A either to increase or decrease, depending mainly on whether the products are complements or substitutes. Complements are goods which are customarily consumed together, for example gin and tonic, bread and jam, and bacon and eggs. An increase in the price of jam might therefore cause a decrease in the quantity of bread purchased (as well as a decrease in the quantity of jam itself).

Most food products are, to a greater or lesser extent, substitutes for one another. Table 3.3 gives calculations of cross-price elasticities for carcass meats in Britain. Reading down the first column of coefficients, an increase of 1 per cent in the average price of beef and veal would be expected to result in a decrease of 0.88 per cent in the average quantity of beef and veal bought, together with an increase of 0.90 per cent in average purchases of mutton and lamb, and increases of 1.10 per cent of bacon and ham and 0.57 per cent of broiler chicken.

The table indicates quite strong cross-price elasticities for the products

Table 3.3 Estimates of cross-price elasticities of demand for main meats in Britain (1977–1986)

| | Elasticity with respect to the price of | | | |
	Beef and veal	Mutton and lamb	Bacon and ham	Chicken
Beef and veal	−0.88	0.37	0.54	*
Mutton and lamb	0.90	−1.69	0.45	*
Bacon and ham	1.10	0.37	−1.23	0.36
Chicken	0.57	0.34	0.48	−0.56

Source: Chesher and Rees (1988)
*Estimates which are clearly not statistically significant have been excluded.

shown – suggesting that the various meats are good substitutes for one another.[4]

The concept of cross-price elasticity of demand provides a convenient basis for defining a food product. If, for example, all consumers regard Guernsey and Jersey tomatoes as perfect substitutes for one another, then, as far as market demand is concerned, they are a homogenous product ('Channel Island Tomatoes') and would possess infinite coefficients of cross-price elasticity. (The cheaper tomato would always be purchased in preference to the more expensive.) We might then expect to find quite high cross-price elasticity coefficients between Channel Islands and, say, Canary Islands tomatoes, and moderate coefficients between tomatoes and vegetables. For other food products we would expect positive, but very low, coefficients (unless we can find a product which is normally consumed jointly with tomatoes, in which case we would expect to find negative cross-price elasticity coefficients).

Earlier we suggested that the demand for food products would almost always be inversely related to price. There is in fact only one plausible exception to this, but it is a rather important one from the point of view of food marketing. It occurs when the price at which the product is offered for sale causes consumers' attitudes and perceptions towards the product to vary. There are two versions of this. First there is what is sometimes known as 'snob appeal' – where part of the value a particular consumer gets from something – say caviar or champagne – is displaying to others how much has been paid for it. In these circumstances, it is just possible (but we believe still rather unlikely) that a rise in price might cause *more* to be bought.[5]

The other version is much more plausible and can refer to everyday food products. It is where price is taken as an indicator of quality. Below a threshold price, consumers may become suspicious of the quality of a product – even though in reality there is no quality change when the price is reduced. Ritson (1977) refers to experiments suggesting that this applies to meat products. Another possibility would be famous estate-bottled wines, where too low a price might lead to doubts about authenticity.

Price as an indicator of quality is really a specific example of a more general phenomenon which has been the subject of a number of articles recently in the marketing literature (see, for example, Kalwami *et al.*, 1990; Thaler, 1985; Winer, 1986). This is the view that the relationship

---

[4]A somewhat similar table, published in Senauer *et al.* (1991), reports significant, though somewhat lower, meat carcass price elasticities for the USA. In contrast, the own-price elasticities which they show for the USA are broadly consistent with those in Table 3.1.

[5]Note we are not referring here to premium products – where a higher quality (or image) version of a particular product sells at a high price. What we describe is where the *same* product is offered at a higher price, with no other change in quality or information made available to customers.

between price and quantity purchased is influenced by a second relationship – that between the price of the product and some reference price level in the mind of the consumer. This reference level might be what the consumer believes is the minimum price for a 'safe' meat product, as in the above example; but there are various other possibilities which have been suggested, e.g. fairness, value for money, past prices, and so on. This view of the way consumption relates to prices should be of great interest to food manufacturers, because it implies that if the food firm can raise price expectations (by advertising or whatever) it might expect to sell more at prevailing prices (or the same amount but at a higher price).

3.3
Consumer income

Household income provides one of the most powerful explanations of differences in diet. It is, of course, not just a matter of the rich eating more than the poor (as might apply in a primitive society), but eating more of some things and less of others – as illustrated in Table 3.4. Similarly we find that diets vary between high and low income countries and that, as the average income in a country increases, so the pattern of food purchases changes.

We use the term *income elasticity of demand* to refer to the responsiveness of quantity demanded to a change in income, either for an individual household, or for the country as a whole, i.e. the responsiveness of national consumption to a change in the average level of income. Income elasticity is defined as:

$$\frac{\text{Proportionate change in quantity demanded}}{\text{Proportionate change in income}}$$

Table 3.4 A comparison of consumption of selected foods (per person per week) between high and low income British households in 1992[*]

| | Income (£ per week) | |
	>520	<140
Yoghurt and fromage frais (pt)	0.34	0.18
Cheese (oz)	5.18	4.05
Fresh fish (oz)	2.04	1.49
Butter (oz)	2.22	1.98
Margarine (oz)	2.14	3.36
Potatoes (oz)	21.35	41.39
Fruit juices (fl. oz)	11.49	6.69
White bread (oz)	5.34	16.65
Beer (cl)	26.71	15.18
Wine (cl)	34.67	5.89

Source: MAFF (1993)
[*]These are households without children.

Income elasticities can be:

- Elastic (more than 1) – if consumption increases more than proportionately with an increase in income.
- Inelastic (between 0 and 1) – where consumption increases less than proportionately with income.
- Negative – where consumption falls with income.

When the last of these applies, the food products are known as *inferior goods* – foods which are consumed at low incomes but are substituted to a greater or lesser extent by *normal* goods as people become better off.

Table 3.5 lists estimates of income elasticities of demand of various agricultural products in a selected group of countries. The coefficients are those which were used by the Food and Agricultural Organisation of the United Nations (FAO) when attempting to predict growth in demand for agricultural products between 1970 and 1980. The figures are therefore now a little dated, but they still provide an extremely valuable picture of the way demand for food products changes in response to growth in real income per head.

The most notable characteristic of Table 3.5 is that the majority of the coefficients are positive but less than one, indicating that an increase in income is associated with a less than proportionate increase in the demand for the product in question. This feature of food consumption was noted in the nineteenth century by the German statistician Ernst Engel and has since become widely known as 'Engel's law'. (Graphs plotting consumption against income are correspondingly known as 'Engel curves'.) Engel's law would usually be described as something like: 'the share of food in total household expenditure decreases with increasing income'. Table 3.6 shows that the 'law' still applies in Engel's homeland, and Table 3.7 provides corroborating data from the USA.

Returning to Table 3.6, there are a number of negative coefficients, indicating products which have on balance become inferior goods in the country concerned. This is only a common occurrence with cereals, but milk, butter, sugar and eggs all appear to be inferior goods in at least one country. (Many food products will, of course, be inferior goods for some members of the population but, for a negative income elasticity coefficient to emerge from analysis of national data, a sufficient proportion of the population must reduce their purchases of the product when their incomes rise so as to more than offset the action of those consumers for whom the product remains a normal good.) In general the highest coefficients are shown for meat, fish, fruit and vegetables; it is these products that consumers tend to substitute for staple foodstuffs, such as bread, potatoes and rice, when incomes rise.

Blandford (1986) has looked at food consumption patterns in the countries of the Organisation for Economic Cooperation and Development (OECD). By plotting income per head against food as a percentage of

Table 3.5 Income elasticities of demand for selected food items

Product	USA	Sweden	France	Australia	UK	Italy	Spain	Brazil	Kenya	India	Indonesia
Wheat	-0.3	-0.3	-0.4	-0.1	-0.2	-0.2	-0.3	0.4	0.8	0.5	1.0
Coarse grain	-0.1	-0.3	-0.1	0.0	-0.1	-0.4	-0.1	-0.3	0.4	-0.2	0.4
Sugar	0.1	0.0	0.3	-0.1	0.0	0.4	0.6	0.1	1.0	1.0	1.4
Vegetables	0.1	0.5	0.3	0.2	0.3	0.3	0.5	0.5	0.5	0.7	0.6
Fruit	0.2	0.6	0.5	0.7	0.5	0.6	0.7	0.5	1.0	0.8	0.8
Meat	0.2	0.2	0.4	0.1	0.2	0.7	0.7	0.5	1.0	1.2	1.3
Eggs	-0.1	0.1	0.2	0.0	0.0	0.5	0.6	0.6	1.0	1.0	1.2
Fish	0.3	0.3	0.6	0.3	0.3	0.4	0.7	0.5	0.8	1.5	1.0
Milk	-0.5	-0.2	0.1	-0.1	-0.1	0.3	0.5	0.6	0.8	0.8	2.0
Butter	-0.5	-0.2	0.2	0.0	0.0	0.4	0.5	1.1	0.9	0.6	n.a.
Coffee	0.0	0.3	0.5	0.8	0.8	1.0	1.0	0.1	1.0	0.4	0.2
GNP per head (1971, $)	5160	4240	3360	2870	2430	1860	1100	460	160	110	80

Source: Ritson (1977)

Table 3.6 The share of food* as a percentage of total consumer expenditure at different income levels, West Germany

	Low income	Middle income	High income
1965	50.2	40.0	28.5
1972	42.2	33.3	25.5
1980	33.5	28.1	22.0
1992	26.4	22.5	19.5

Source: Tangermann (1986), with supplementary data for 1992 provided by Tangermann.
*Including beverages, tobacco and restaurant meals.

total expenditure (Figure 3.1) he illustrates the applicability of Engel's law across countries, and by combining data relating to calorie consumption from three countries (Figure 3.2) he succeeds in producing a plausibly realistic Engel curve.

In more recent analysis of National Food Survey data, Chesher and Rees (1987) have produced some very interesting Engel curves using cross-sectional analysis of the household data – that is, looking at consumption of different households during the same year. Some examples of these which were published in the Annual Report of the National Food Survey for the main meats are shown in Figure 3.3.

All the curves demonstrate the expected shape. What is perhaps surprising is the extent to which the products become inferior goods at high income levels. Indeed 'other poultry' is the only one of some twenty meat product categories examined which displayed a positive response throughout the observed income range. However, all the Engel curves have a positive response at the average income of the households in the National Food Survey (marked on the curves).

When using data of this kind, an estimate of income elasticity would

Table 3.7 Household expenditure on food in relation to disposable income

Income group	Food expenditures as a percentage of disposable income
$5000–9999	28.2
$10 000–14 999	22.5
$15 000–19 999	18.4
$20 000–29 999	15.3
$30 000–39 999	13.3
$40 000–49 999	11.9
Over $50 000	8.5
All households	13.3

Source: Adapted from Senauer *et al.* (1991)

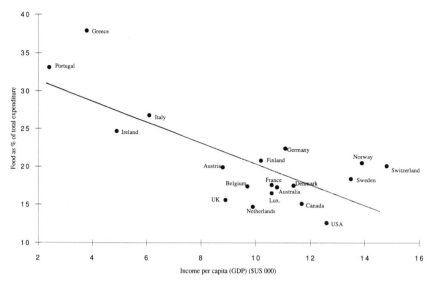

Figure 3.1 Relationship between income and food expenditure in the OECD in 1981 (*Source*: Adapted from Blandford, 1986)

usually be given for the average income. This indicates broadly the expected proportionate change in consumption of the product for a small change in the average income of consumers. (Strictly, it indicates how the consumer with the average income will respond.) Thus none of the meats shown in Figure 3.3 are inferior goods in the sense that they display negative income elasticities at the average income.

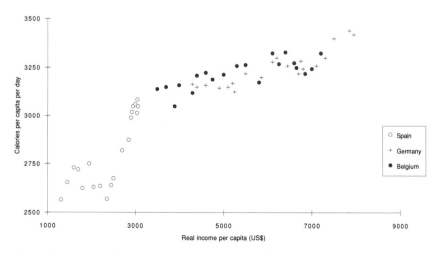

Figure 3.2 Relationship between food consumption and income in selected OECD countries for 1960–1980. (*Source*: Adapted from Blandford, 1986)

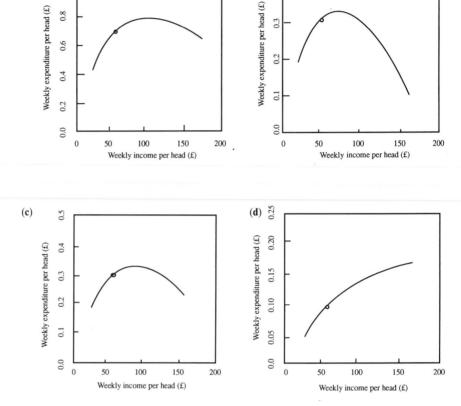

Figure 3.3 Engel curves for (a) beef; (b) pork; (c) broiler/chicken; (d) other poultry. (*Source*: Adapted from Ritson, 1988)

A distinction is sometimes made between 'income elasticity of quantity demanded' and 'income elasticity of expenditure'. Strictly speaking, if a product is precisely defined these measures should be identical since, if other factors remain the same, including the price and specifications of the product, a certain percentage increase in the quantity purchased must result in the same percentage increase in expenditure on the product. However, when an income elasticity coefficient is calculated for a broader commodity group, the measures will tend to diverge as there is a tendency for consumers to switch to higher quality (and higher priced) varieties within the broad food groups when their incomes rise; in other words, the income elasticity of quantity demanded tends to be higher for higher priced varieties, and thus the income elasticity of expenditure for the product group will exceed the income elasticity of quantity demanded.

The UK Ministry of Agriculture estimates income elasticities of both quantity (in terms of weight) and expenditure using data from its survey

Table 3.8 Income elasticities of quantity demanded and expenditure for major food products (UK, 1986)

	Income elasticity of quantity demanded	Income elasticity of expenditure
Milk	−0.11	−0.10
Cheese	0.37	0.46
Carcass meat	0.27	0.35
Fish	0.05	0.19
Eggs	−0.21	−0.08
Butter	0.18	0.20
Sugar	−0.41	−0.37
Potatoes	−0.32	−0.20
Green vegetables	0.17	0.43
Bread	−0.17	−0.04
Beverages	−0.10	0.12

Source: MAFF (1986)

of household food expenditure. Some of the estimates made for 1986 are shown in Table 3.8 (and this does indicate a number of products which have become inferior goods at average income levels).

It will be seen that the coefficients for expenditure are consistently, though not substantially, higher than those for quantity. The coefficients are closest for those products for which there is little scope for switching to higher priced varieties, such as liquid milk, butter and sugar. The difference between coefficients of income elasticity of quantity and expenditure therefore represents a useful indication of the extent to which consumers are purchasing better qualities and more processed products as their incomes rise; the difference is sometimes called the 'quality elasticity'.

3.4 The underlying trend in demand

In the previous two sections, we have shown how price and income changes can influence the pattern of diets and how we can obtain quantitative estimates of the strength and direction of these effects. Using these estimates it is possible to identify the extent to which past changes in dietary patterns are the consequence of changes in prices and income, and how much of the changes have been caused by something else. Ritson and Hutchins (1991) have studied the evolution of food consumption in the UK since the Second World War and concluded that, whereas much of the change which occurred in the post-war period can be ascribed to income growth and price developments, since about 1980 other factors have come to predominate.

In order to illustrate this point, Figure 3.4 provides a very simplified interpretation of the factors influencing changing patterns of food consumption over the past 50 years. We have divided the period into five

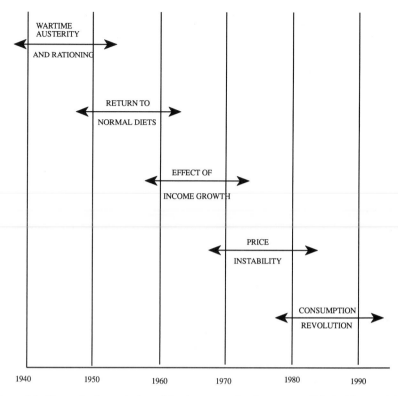

WARTIME
AUSTERITY
AND RATIONING

RETURN TO
NORMAL DIETS

EFFECT OF
INCOME GROWTH

PRICE
INSTABILITY

CONSUMPTION
REVOLUTION

1940   1950   1960   1970   1980   1990

Figure 3.4 Phases in the evolution of food consumption in post-war Britain (*Source*: Adapted from Ritson and Hutchins, 1991)

overlapping phases noting, for each phase, what we believe to have been the dominant factor influencing developments in food choice.

The first period has been labelled 'wartime austerity and rationing', in which individual food choice was largely imposed by availability. During the second phase, the end to rationing and more plentiful supplies allow British households to return to what would then have been regarded as 'normal' diets, under the constraints of prevailing incomes and prices.

By the mid 1950s, the post-war rise in living standards is under way. The effect for many households is to lift the income constraint gradually over a period of 20 years or so, allowing consumers to move their diets in what they would regard as a preferred direction. Table 3.9, which is based on contemporary demand analysis of National Food Survey data, gives examples of the main foods which would have been expected to experience increasing consumption as a consequence of rising personal incomes in the UK. In contrast, for some ('inferior') foods, rising incomes mean that less is purchased, and some of the main products which would then have fallen into that category are also listed in Table

Table 3.9 Food products for which income growth probably contributed to a significant (a) rise and (b) fall in average levels of consumption

(a) Rise in average levels of consumption

Cheese	Beef	Salad vegetables	Fresh fruit	Rice
Canned salmon	Pork	Salad oils	Chocolate biscuits	Coffee
Shellfish	Chicken	Frozen vegetables	Brown and wholemeal bread	Ice cream

(b) Fall in average levels of consumption

Canned meat	Margarine	Potatoes	Tea
Sausages	Lard	Dried pulses	White bread
Herrings	Canned milk puddings	Canned vegetables	Oatmeal products

Source: Ritson and Hutchins (1991)

3.9. In some cases, the effects of income growth were probably quite substantial, leading perhaps to changes of 20–25 per cent from about 1955 to 1975.

During the 1970s, however, prices became much more important as an influence on patterns of food consumption. A number of factors – for example, the world commodities price boom; the adoption of the EEC's Common Agricultural Policy; the drought of 1975/76; and the food subsidy programme – caused extreme volatility in the retail prices of many food products. (An example of the effect of price instability on the consumption of butter and margarine during this period is given in Figure 2.4.)

By the end of the 1970s, the volatility of prices had subsided; world commodity prices had stabilised and British food prices had absorbed fully the affects of adopting the Common Agricultural Policy. For most foods, prices then fell gently in real terms during the 1980s. Meanwhile the pace of income growth slackened, and the strength of the relationship between income and consumption seemed to lessen. Yet for many food products, either the previous change in consumption is reversed, or the pace of change accelerates. This last phase we have characterised as the 'Consumption Revolution'.

Figures A1–12 in the Appendix to this chapter show the evolution in average consumption (per person per week) of the main categories into which the NFS data is grouped. (1960 was chosen as the starting date to eliminate the early post-war food shortage as an influence on consumption.)

Before interpreting these diagrams, it is necessary to add a word of caution; what is shown are changes in purchases by households, in physical quantities, not all food consumed. Meals eaten outside the home are excluded (though not 'take-aways' eaten in the home). Also, in some cases, the form of the product has changed over time. For example, in the case of meat and vegetables, an increased proportion will have been purchased in a prepared and trimmed form – therefore reducing purchases by weight, but not necessarily amount consumed.

Nevertheless, the changes have been significant. During the 1980s, of the major food groups, only cheese, fish and fruit have seen a significant increase. In the case of milk and cream, fat and oils, eggs and beverages, relative stability at first has been succeeded by decline. Overall, meat consumption rose at first, but had declined back to its 1960 level by 1989. Purchases of sugar and cereal-based products have declined throughout, and vegetable consumption has remained relatively constant.

In many cases, a watershed appears to occur around 1975–1980. Stable or rising consumption is replaced by decline – or the pace of previous decline has accelerated. With fish and fruit, decline and stability, respectively, are replaced by growth.

Within the broad food groups there have been even more marked changes. The dramatic fall in purchases of full fat milk since 1980 has been offset to some extent by a rise in low fat milk. All the carcass meats

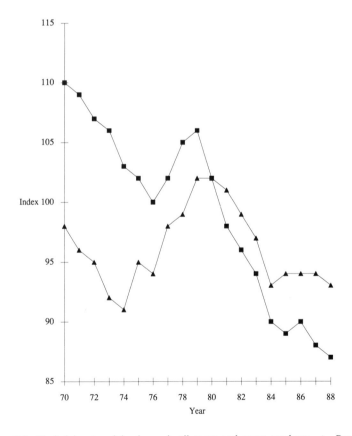

Figure 3.5 Underlying trend in demand: all meat and meat products. ▲, Purchases; ■, demand.

have suffered reduced purchases since 1980 – consumption of mutton and lamb has been in decline for some time. Increased consumption of poultry has offset this.

The breakdown for fish indicates that this product sector has experienced rather different conditions. The fall in overall consumption was mainly caused by the rapid decline in fresh fish sales between 1968 and 1975. Since then fresh fish sales have stabilised, and the growth in the frozen fish market has pulled up total fish consumption.

Butter consumption has plummeted since the mid-1970s and consumption of margarine has also eventually joined the decline. Within the vegetable sector, the growth has been in the 'other' category – within which the main products are carrots, onions, tomatoes, cucumbers and mushrooms. The growth in total fruit consumption has been partly accounted for by the rapid development of the juice market during the 1980s.

For beverages, the modest growth in coffee sales has failed to offset the substantial fall for tea. Finally, within the cereals market, the fall in white

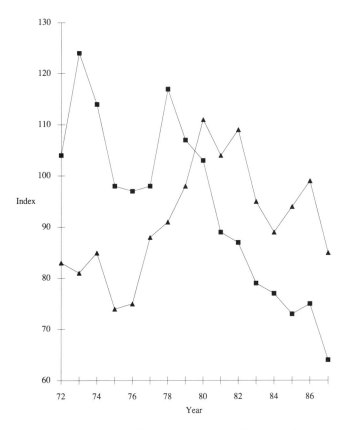

Figure 3.6   Underlying trend in demand: pork. ▲, Purchases; ■, demand.

bread sales has in no way been compensated for by the growth for brown and wholemeal bread – and, in fact, sales of wholemeal bread peaked in 1986. Otherwise, it is breakfast cereals which have shown sustained growth.

This then is a picture of how actual purchases of the food products shown in the Appendix have changed since 1960. What we now do is compare these changes with the underlying trend in demand. Arguably, this is a much better guide to successful marketing than actual consumption. In effect, it shows what *would have happened* to consumption if prices and incomes had remained constant in real terms. It can therefore be used to project future levels of consumption at constant real prices and incomes, revealing long-term changes in behaviour and attitudes.

Figures 3.5–3.8 show quantity purchased and the demand trend for four products by way of example. The figures are in the form of index numbers – that is, they are expressed as a percentage of the average values during a base period. It should be emphasised therefore that the

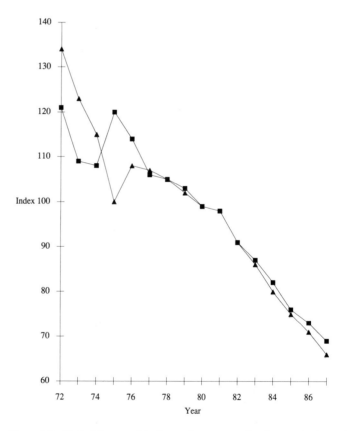

Figure 3.7 Underlying trend in demand: sugar. ▲, Purchases; ■, demand.

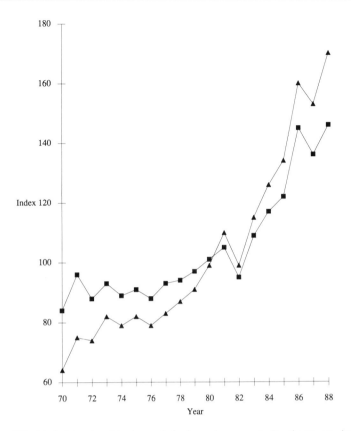

Figure 3.8 Underlying trend in demand: fresh mushrooms. ▲, Purchases; ■, demand.

graphs show relative trends – the absolute differences between the lines are not important (and data after 1988 is not available).

It can be seen that the trend in demand can be stronger or weaker than the trend in purchases; and in the interesting case of pork, for a period, the positive effect of declining real price more than offsets a negative trend in demand. In other words, contrary to belief at the time, pork was not becoming more popular, just cheaper!

These (and similar graphs) are also often consistent with the observation made earlier – that the period 1975–1980 represents a watershed. The underlying trend in demand begins to dominate the way patterns of food consumption in the UK are changing, succeeding a period in which much of the change in purchases was a result of price changes.

The long-term changes in behaviour and attitudes which underlie trends in demand are unlikely to cause average levels of food consumption of different foods to fluctuate erratically – except in exceptional cases such as the recent 'food scares', which have affected purchases of eggs and beef. Thus the year-to-year fluctuations in the demand trends displayed in

Figures 3.5–3.8 are most probably the consequence of a mixture of sample error, and a failure to identify accurately the effect of price and income changes.

In order to overcome this problem we have fitted simple linear trends to 155 foods, or food groups, for which this kind of data is available.[6] We have then drawn up a 'league table', ranking the products from those with the highest underlying trend in demand to those with the lowest. The 'top twenty' products are listed in Table 3.10.

The list makes interesting reading. Two features of 'star' products stand out – they seem to be associated either with convenience or 'healthy eating'. In both cases there is sometimes also an underlying element of novelty and variety.

The bottom twenty products are shown in Table 3.11.

To some extent the list of products with the strongest adverse underlying demand trends are a mirror image of the star products. In addition, however, it is clear that part of the growth in frozen and other convenience products is at the expense of canned products.

Table 3.10   Underlying trend in demand: top twenty

Product	Annual % change in demand
'Other' fresh green vegetables (e.g. spinach, broccoli)	+29.0
Wholemeal and wholewheat bread	+18.2
Frozen chips and other frozen convenience potato products	+13.1
All 'other' fats (e.g. low-fat spreads)	+11.6
Frozen convenience cereal foods (e.g. pastries and pizzas)	+11.2
'Other' vegetable products (e.g. salads, coleslaw, pies, ready meals)	+8.8
Fruit juices	+7.6
Crisps and other potato products, not frozen	+7.2
'Other' fresh fruit (e.g. melons, pineapples and exotics)	+6.9
Shellfish	+6.4
Instant milk	+5.4
Frozen convenience meat and meat products (e.g. ready meals, pies, sliced roast meat)	+5.2
Cooked poultry	+4.6
Fresh fat fish, excluding herrings (e.g. salmon, trout)	+4.6
Rice	+4.4
Breakfast cereals	+4.3
Miscellaneous fresh vegetables (e.g. courgettes, peppers, celery, bean sprouts, aubergines)	+3.9
'Other' bread (e.g. pitta, ciabatta)	+3.9
Spreads and dressings	+3.7
Mushrooms	+3.5

Source: Extracted from Hutchins (1994)

[6]A fuller discussion of this work is provided in Hutchins (1994).

Table 3.11 Underlying trend in demand: bottom twenty

Product	Annual % change in demand
Fresh white fish, unfilleted	−22.0
Fresh peas	−16.2
Processed fat fish unfilleted	−14.7
Soft fresh fruit, other than grapes	−14.5
Instant potato	−9.7
Offals, other than liver	−8.5
Baby foods, canned and bottled	−8.1
Canned peaches, pears and pineapples	−7.1
Canned potatoes	−6.8
Brussels sprouts	−6.8
Canned salmon	−6.3
Cakes and pastries	−5.6
Canned meat	−6.3
Liver	−5.2
Rhubarb	−5.2
Cream	−5.0
Uncooked sausages	−4.9
Marmalade	−4.9
Mutton and Lamb	−4.9
Butter	−4.8

Source: Extracted from Hutchins (1994)

3.5 Conclusion – what underlies the underlying trend in demand?

It is beyond the scope of this chapter to proceed to explore the factors which are causing such strong positive and negative adverse trends in demand, identified by eliminating those effects which can be ascribed to changes in prices and incomes. It is, however, instructive to attempt a brief categorisation of these factors as a prelude to some of the following chapters in this book. One useful distinction is that between factors relating to behaviour and structure of households, and factors relating to preferences of individual food consumers. Among the former would be influences such as:

1. the growth of single-person and no-children households;
2. the increasing proportion of women who are out at work;
3. changes in meal patterns with a trend towards lighter meals (see Marshall, chapter 11).
4. the impact of the above on habit formation by young people, making it more possible for new positive trends to develop, as young consumers carry these habits with them as they get older.

All of the above relate to time and convenience. For example, a food diary study carried out by Newcastle University a few years ago (Gofton and Marshall, 1989) produced what seemed to us the astonishing conclusion that 94 per cent of meals occupied less than 10 minutes of preparation time and 51 per cent no preparation time; and 93 per cent of meals less than 20 minutes cooking time and 61 per cent no cooking time.

Turning to the second category, it is also possible to identify four main influences:

1. Changing attitudes to diet, health and nutrition are the most talked about cause, and there is no question that this has been important. It is interesting to note that the watershed at which the underlying trend in demand begins to predominate as the force causing changes in the UK diet more or less coincides with the time at which the issue of the relationship between diet and health began to become high profile within the media.

2. Another interesting feature of individual food attitudes is probably a shift from quantity to quality. Average incomes have reached a level where, for most households, sufficient quantity, even of high priced foods, has no longer been an issue. This has affected meat probably more than anything else where, until quite recently, a large portion of meat was regarded as a luxury.

3. There is also an increasing element in consumer demand which reflects the relationship between food production and the environment. This, of course, expresses itself at its extreme in the demand for organic products, but also for products which guarantee some kind of acceptable welfare environment in terms of production.

4. Finally, and as already mentioned, there is an increasing demand for novelty and variety in food purchases which is no more than a food-related manifestation of the general development in patterns of consumer demand.

References   Blandford, D. (1986) The Food People Eat. In C. Ritson, L. Gofton and J. Mackenzie (eds), *The Food Consumer*, Wiley, Chichester.

Chesher, A. and Rees, H. (1987) *Food Expenditure Income Relationships: The National Food Survey*, unpublished report to the Committee of the National Food Survey, MAFF, London.

Chesher, A. and Rees, H. (1988) *Food Expenditure – Price Relationships: The National Food Survey 1977–1986*, unpublished report to the Committee of the National Food Survey, MAFF, London.

Gofton, L. R. and Marshall, D. (1989) *A Comprehensive Scientific Study of the Behavioural Variables Affecting Acceptability of Fish Products as a Basis for Determining Options in Fish Product Research and Development at Torry Research Station*, unpublished report to MAFF, London.

Hutchins, R. (1994) Changing Patterns of Tastes and Preferences for Food in Great Britain. *PhD Thesis*, University of Newcastle upon Tyne.

Kalwami, M. U., Yim, K., Runne, H. J. and Sugita, Y. (1990) A Price Expectations Model of Customer Brand Choice. *Journal of Marketing Research*, Vol. XXVII, August.

MAFF (1982, 1988) *Household Food Consumption and Expenditure: Annual Report of the National Food Survey*, HMSO, London.

MAFF (1986) The National Food Survey Supplementary Tables (Unpublished).

MAFF (1993) *National Food Survey Compendium of Results 1992*, London.

Ritson, C. (1977) *Agricultural Economics, Principles and Policy*, Blackwell, Oxford.

Ritson, C. (1988) Special Study on Meat and Meat Products. Section 3 of *Household Food Consumption and Expenditure, Annual Report of the National Food Survey*, HMSO, London.

Ritson, C. and Hutchins, R. (1991) The Consumption Revolution. In J. M. Slater (ed.), *Fifty Years of the National Food Survey*, HMSO, London.

Senauer, B., Asp, E. and Kinsey, J. (1991) Food Trends and the Changing Consumer, Eagan Press, St Paul.

Tangermann, S. (1986) Economic Factors Influencing Food Choice. In C. Ritson, L. Gofton and J. Mackenzie (eds), *The Food Consumer*, Wiley, Chichester.

Thaler, R. (1985) Mental Accounting and Consumer Choice. *Marketing Science*, **4** (Summer).

Winer, R. J. (1986) A Reference Price Model of Brand Choice for Frequently Purchased Products. *Journal of Consumer Research*, **13** (September).

Food consumption trends in the UK   **Appendix**

Source: Calculated from National Food Survey data.

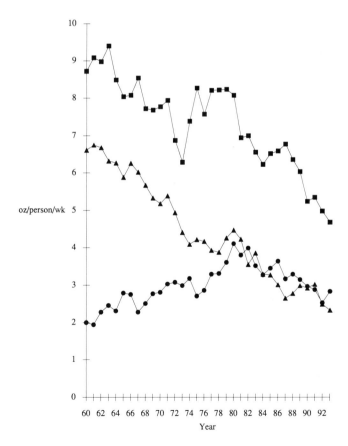

Figure A1   Carcass meat purchases (1960–1993). ■, Beef and veal; ▲, mutton and lamb; ●, pork.

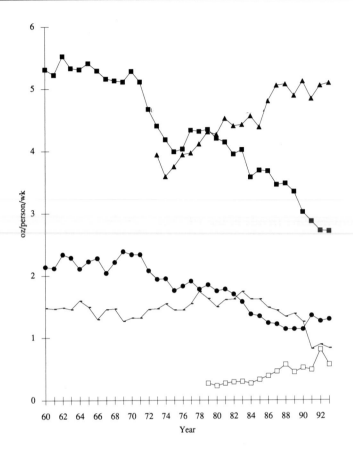

Figure A2   Meat purchases (1960–1993). □, Cooked poultry; ▲, uncooked broiler/chicken; ■, bacon and ham uncooked; –, beef sausage uncooked; ●, pork sausage uncooked.

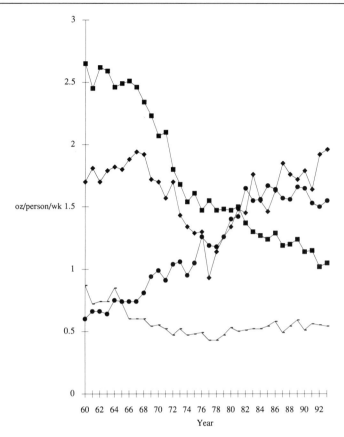

Figure A3   Fish purchases (1960–1993). ■, Fresh; ●, frozen; –, processed; ◆, prepared.

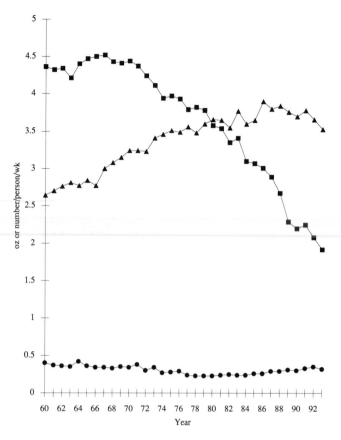

Figure A4   Cheese and egg purchases (1960–1993). ▲, Natural cheese; ●, processed cheese; ■, eggs.

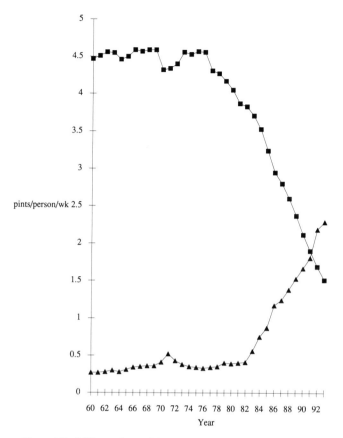

Figure A5 Milk purchases (1960–1993). ■, Full fat milk; ▲, other.

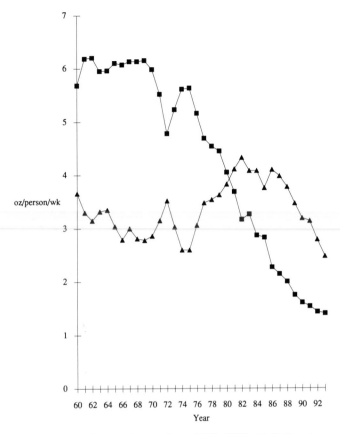

Figure A6   Butter and margarine purchases (1960–1993). ■, Butter; ▲, margarine.

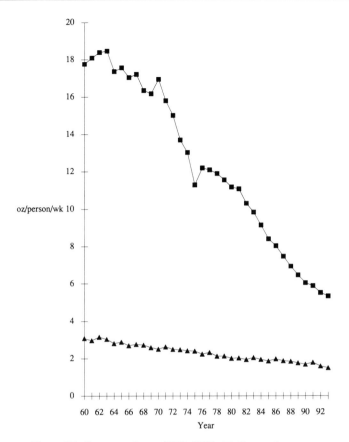

Figure A7 Sugar purchases (1960–1993). ■, Sugar; ▲, preserves.

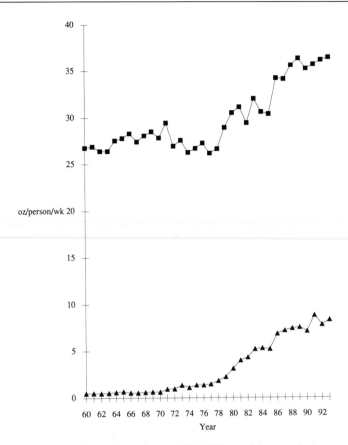

Figure A8   Fruit purchases (1960–1993). ▲, Juices; ■, fruit.

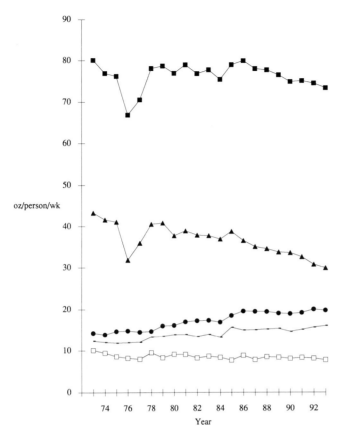

Figure A9   Vegetable purchases (1973–1993). □, Fresh green vegetables; –, other fresh vegetables; ●, processed vegetables; ■, all vegetables; ▲, fresh potatoes.

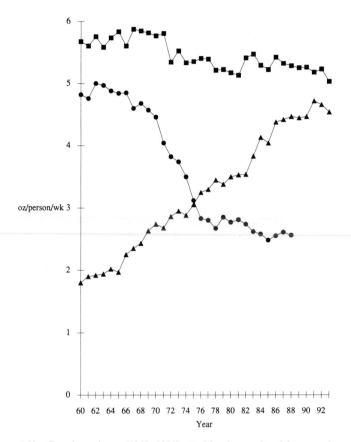

Figure A10   Cereal purchases (1960–1993). ■, Biscuits; ▲, breakfast cereals; ●, cakes.

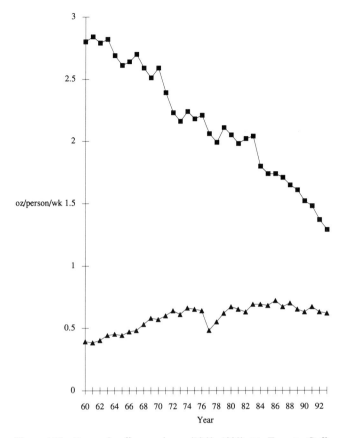

Figure A11   Tea and coffee purchases (1960–1993). ■, Tea; ▲, Coffee.

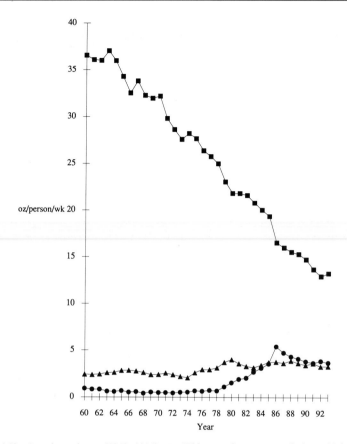

Figure A12 Bread purchases (1960–1993). ■, White; ▲, brown; ●, wholemeal/wheat.

Food retailing and the consumer 4

John Dawson

Food retailing represents the penultimate stage of the food chain from agricultural producer to individual consumer. The decisions by retailers on which products to sell circumscribes the choices which the consumer can make in respect of the what, where and how of food purchasing. A substantial part of the consumer's food purchase experience is determined by the retailer. Whilst the decisions on the product range, location of sales point and sales methods of retailers reflect the perceived needs of consumers, they also reflect the strategies and policies of the food retailers as profit-seeking organisations. This situation is equally the case for publicly quoted supermarket operators, for family-based market traders, or any other type of retailer – all these retailers have to make a profit whilst meeting consumer demands. The demise of the subsidised, non-profit, food retailing operations in former Communist countries only serves to emphasise the importance of this balance between responding to consumer demands and generating a profit.

The relationship between consumers and retailers is complex, and particularly so in the food sector (Ray, 1994; Sack, 1993; Trail, 1989). Whilst retailers respond to the culture of consumers and provide goods which are expected in a way which is acceptable, nonetheless the retailers also extend the horizons of consumers by presenting them with new products sold in new ways. Retailers, therefore, are both reactive and proactive agents in the process of consumer choice. One of the measures of success for retailers is the extent to which they can generate consumer interest and hence consumer satisfaction. This is particularly so in the wealthier economies of the USA, Japan, Australia and Europe where the choices available to consumers are very wide, competition amongst food retailers is intense, and food represents a declining proportion of household spending.

The aim of this chapter is to consider the approaches of retailers, in these wealthier economies, to the issues of food choice by consumers. The chapter explores four main themes. First, there is a trend towards retailers seeking to take more control and have influence over the decisions being made by others, including consumers, in the food chain. It is apparent that the structure of the food retailing industry is subject to some common influences and trends in all developed economies and thus consumers in different countries are presented with increasingly similar choices. This commonality is a second theme. Thirdly, the sales methods, including store location, used by retailers are changing in similar ways across many countries. This provides a third theme. Finally, the approaches to management used by different retailers indicate a convergence of strategies used by food retailers, but the basis of competition is to seek differences in the choices of product, price, promotional message and place to shop which the retailers offer to consumers. This dichotomy between common approaches to the consumer and the search for corporate differentiation provides the final theme.

4.1
Changes in the distribution channel for food retailers

Figure 4.1 indicates the main channels of distribution, to the final consumer, for food in the UK in 1985, the latest year for which such a chart is available (Slater, 1987). The calculation of the values of the flows is very difficult and the numbers shown on the figure must be read cautiously. Figure 4.1, however, indicates the variety of distribution channels which exist in the food industry and places in context the role of retailers within the total food chain. The relationships amongst the various agencies shown in this channel have evolved rapidly in recent years: a feature not limited to the UK. Corporate and governmental policies have influenced and directed the way these channels of distribution for food have evolved.

There is an ongoing shift in the balance of power within the food chain. The shift is towards greater power being in the hands of, and being used by, retailers. This power can be made manifest either in the response to or direction of consumer choice. This shift in power is driven by corporate policies and is apparent in several ways:

- the growth of vertical marketing systems;
- the increase of administered marketing channels;
- the use of new types of channel power.

Government policy, particularly on price control, competition control and the granting of development permission, also influences the relative balance of power of retailer, supplier and consumer.

The growth of vertical marketing systems is a widely recognised feature of consumer goods markets. In the vertical marketing system the various agencies, food producers, food processors, wholesalers, retailers, etc. no longer limit their activity to their traditional functions but take responsibility for functions and activities previously undertaken by others. By extending their functions, firms change their power relationships and the ability to respond to and influence consumer choice is changed. So wholesalers, for example, became involved in retailing through voluntary

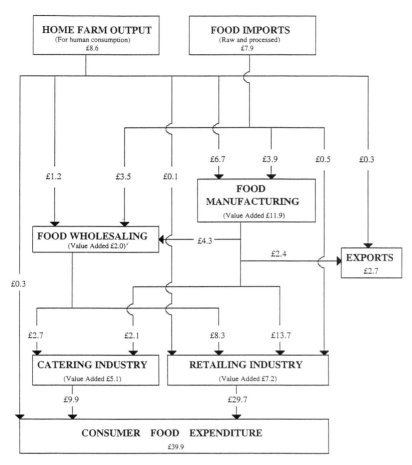

Figure 4.1 Main channels of food distribution in the UK in £ billion (1985). (*Source*: Slater, 1987)

chains, franchising, warehouse clubs or even direct store ownership; and the same wholesalers become involved in food processing through the development of distributor brand products. Similarly, retailers and processors extend their activities into new areas, even food producers move into craft-type processing, wholesaling and farm shops. The various agencies retain their core activity but seek to extend their activity, in these vertical marketing systems, with direct ownership, joint ventures, franchise arrangements, and other organisational relationships of a non-transactional form. The moves of brewery companies into the operation of off-licences, with clear consumer consequences for prices and product ranges, provides an example of this trend in the UK. In the food sector, not only in the UK (Kaas, 1993), the steady shift away from marketing systems coordinated by market transactions has been a feature of the last 20 years with vertical marketing systems now well developed and stable features of food distribution.

Administered relationships are a mechanism for management of the distribution channel in which one agent becomes the coordinator for the channel and manages the flows (product, finance, ownership, information, risk, etc.) in order to support a particular strategy or achieve specific objectives. Ownership structures in the channel often remain independent but activity is coordinated, not always with contracts, such that costs are taken out of the whole supply chain and the benefits are shared by the various agents in the chain. The emergence of strong and stable administered relationships in the distribution channel is also related to the decline of transactionally coordinated marketing systems (Carlisle and Parker, 1989; Dawson and Shaw, 1989).

Within administered channels the relationships between agents may be long lasting and stable. For example, an administered channel relationship may exist between a food processor and a retailer for the supply of either a manufacturer branded product or a retailer branded item. A retailer, for example, may act as the coordinator for an arrangement with a supplier which defines the supply terms, determines pack characteristics, coordinates the logistics, agrees the promotional plan, etc. for a manufacturer brand item (Shaw, 1994). In respect of retail brands the relationship of Marks and Spencer with its suppliers illustrates the use of administered channels. Marks and Spencer define the composition of the product and ingredient quality (even their source) for the food processors. The objective of the administered arrangement is to seek economies of coordination such that cost is removed from the channel, for example through minimising stock holding and transport costs, reducing paperwork with EDI, reducing the need for market searching by the retailer, etc. (Senker, 1986). The basis of this arrangement may be in place for many years with details, including price changes, being renegotiated as an ongoing process. The presence of these administered channels has increased substantially in the food sector over the last 20 years. They have become mechanisms through which horizontal competitiveness, not least for the retailer, is achieved through vertical coordination (Dawson and Shaw, 1990).

The consumer consequences of the extension of administered relationships in distribution channels are several, including the extension of retail brand product ranges, increasingly high in-stock positions in stores, more rapidly changing product ranges and more complex promotional offers. The opportunities and benefits of administered relationships tend to be greater for larger firms than for small ones and a further consumer consequence is the resulting increase in market concentration.

The third of the interrelated changes in the distribution channel for food, critically important to retailers and consumers, is the shift in the locus of power in the marketing system. The power of retailers has

increased relative to the power of others in the distribution channel
(Grant, 1987; Segal-Horn and McGee, 1989). Power in this context is the
extent to which one agent can influence the decisions of other independent
agents. The growth in retailer power has been caused by several factors,
some of which are covered in more detail later in the chapter, notably:

- the increase in size of retail firms such that they can exert considerable
  power in their buying activities;
- the increase in number of products such that there is considerable pressure on shelf-space in stores;
- the decrease in number of retailers and increase in market concentration such that market access by food suppliers becomes limited;
- the improved quality of information obtained by retailers from point-
  of-sale data collection methods;
- the growth of retailer branding and its ready acceptance by the consumer;
- improved management methods by retailers, particularly in store
  operations such that tighter control is exerted over store display and
  customer communication.

The activity of the three main food retailers in the UK: J. Sainsbury,
Tesco and Safeway, illustrate the growth of retailer power in respect of
all these factors (Knox and White, 1991).

The three changes, described above, in the distribution channel for
food are interrelated. The increase in power gives retailers a stronger base
from which to act as coordinators of administered channels and to extend
the range of their own activities within a vertical marketing system and in
so doing become stronger influences on consumer choice. The food processors, in reaction to their loss of power, have sought to administer their
relationships with their suppliers and to explore shifts of function in a
vertical marketing system. Wholesalers who have found their traditional
power base of smaller retailers declining have sought to secure their existence through vertical marketing systems. In all these changes there is not
only the facility for retailers to become more responsive to consumer
demands but also the possibility of retailers increasing their influence over
consumers. The remainder of this chapter focuses on the activity of retailers, but it is important to appreciate that retailers operate alongside
other profit-seeking agents in the overall food distribution channel shown
in Figure 4.1.

An important agent, external to the direct channel of food distribution,
is government. Government influence in food retailing is in three main
areas: control of prices, control of competition and control of store
establishment. Other governmental controls on product quality, store
opening hours, labour legislation, corporate taxation, etc. all affect food
retailers but mainly at an operational level compared with the controls

over prices, competition and establishment which have greater strategic impact.

Resale price maintenance has been a formative influence on food retailing in the UK in the second half of the twentieth century (Pickering, 1966; Yamey, 1966). The presence of Resale Price Maintenance (RPM) until 1964 limited the extent to which retailers could extend their power in the distribution channel. Its abolition provided a stimulant to the growth of new types of competition in food retailing (McClelland, 1963, 1967) and new types of discount food supermarket entrepreneur. For many years, through the high inflation of the 1970s, small traders lobbied, unsuccessfully, for the reintroduction of RPM. The freedom of competition resulting from the removal of price controls enabled an early start to the restructuring of British food retailing compared with some other European countries.

The control of competition through monopoly and merger legislation and through regulations on sales practices has influenced the nature of the restructuring. Whilst in food retailing no major merger has been refused on the basis of the legislation in place, the presence of the legislation acts as a restraint on predatory activity of large firms (Monopolies and Merger Commission, 1981, 1983). The rate at which concentration has occurred has been considerable but quite possibly would have been even greater had not legislation been in place. The control nonetheless is relatively light in the UK compared with other European countries, notably Germany. Regulation of sales practice, however, allows suppliers to refuse to supply retailers with a product if the product is then sold at retail at a price below its purchase price. Loss leading is an acceptable practice but has to have the tacit agreement of the supplier. For large manufacturers with strong brands this can be a powerful factor in their ability to retain some power in the supply chain but for smaller manufacturers in a commercial relationship with large retailers the control is difficult to apply in practice (Wrigley, 1993a).

Whilst legislation on price control and merger control have been important but shadowy influences on the strategies of food retailers, the control over new store development, through land-use planning legislation, is constantly spotlighted as affecting retail strategy, competition and consumer choice (Davies, 1984; Guy, 1988, 1994a). The requirement for all new store development to obtain explicit permission acts as a strong control over excessive store development and in effect acts as a spatial control on the intensity of competition. As retailers have sought permission to develop larger and larger stores, so this legislation has become a more important and controversial method of controlling competition, particularly for developments outside established shopping areas, for example at out-of-town or edge-of-town locations. Government policy has vacillated over the years between being against such development and granting few permissions to being relaxed about new stores of this type and granting permissions relatively freely (Davies, 1986). The network of stores is cumulative so when relative freedom is followed by strict control this gives a strong competitive advantage, locally monopolistic in some cases, to recently developed stores already in operation (Kleinwort Benson, 1991). This can have an influence on consumer choice, with for example one town having only one or two large food stores whilst another town of similar market character might have double the number of large stores. The side effects of the changes in government policy in this area can be a considerable influence on the way consumer demands for food are met by the retailers.

Food retailing is an industry sector in which there are considerable economies of scale and of scope. Realisation of the power of these cost economies has resulted in substantial restructuring of the sector (McClelland, 1990). It has also led to the shift from atomistic competition with a large number of small independent decision makers to oligopolistic competition in a market with a high level of market concentration (Akehurst, 1984). This change in market structure has affected consumer choice. Whilst, from the consumer viewpoint the range of shops to choose from has decreased, the range of products to choose from within a shop has increased (Gardner and Sheppard, 1987).

**4.2
The size and structure
of food retailing**

Within the UK there has been a substantial decrease in the number of firms in food retailing, due mainly, but not exclusively, to the net closure of small firms. The extent of the decrease is shown in Table 4.1. This pattern is seen in other European countries. Declines have been considerable over the last 15 years in Northern Europe and now are substantial in Mediterranean countries (Eurostat, 1993). There are many economic, social and political reasons for the decline of small retail firms in the food sector but it seems to be a trend which is unlikely to be reversed (Dawson and Kirby, 1979). Small firms do not have the resources to source product as effectively as large ones and in consequence are unable to match the selling prices of larger firms. Small firms also do not have access to the capital needed to install new information technologies and to make the investments in store equipment necessary to provide a range of products comparable to that of the stores of large firms. In these smaller firms capital is locked into the physical structure of the shop and is not easy to release for use as working capital. The physical condition of the shops of these smaller firms often is relatively poor for their economic purpose although fixed operating costs are relatively high. In general it also appears that managerial expertise is lower in the smaller firms and there are no government incentives or tax advantages for these small firms in food retailing. The death rate of small firms in food retailing is high.

Nonetheless, many of these small firms continue to survive. Routes to survival lie in either being part of a larger buying group, for example the

Table 4.1 Changes in the structure of food retailing in the UK

Number of businesses	1971[1]	1980	1984	1990	1992
Total	162300	90475	81680	65169	60119
of which					
large grocers[2]		114	98	78	71
other grocers	95500	43396	34053	21489	18557
butchers	22800	16613	16295	13137	12149
greengrocers	28300	14380	15119	11815	10622
fishmongers	4900	2411	2638	2444	2122
bakers	6500	6277	5523	5177	4006
Number operating					
1 outlet	142800	81835	74422	59238	55416
2–9 outlets	18950	8309	7031	5757	4554
10–99 outlets	482	289	196	159	133
100 and over outlets	55	42	31	15	16

Retail sales	1980	1985	1990	1994[3]
Sales value £bn				
food shops	21.66	31.69	45.34	56.7
grocers	16.17	24.62	37.01	49.0
Sales volume index				
all food shops	100	110.6	128.4	143.3
large grocers	100	125	160	190
small grocers	100	73	66	57
butchers	100	97	80	65
greengrocers	100	108	104	92
fishmongers	100	104	98	80
bakers	100	98	102	101

Sources: Business Monitors SDA25 and SDM28; Census of Distribution 1971; Corporate Intelligence Group 1994.
[1]Figures for 1971 are on a different basis and are not entirely compatible. Attempts have been made to maximise comparability to later figures.
[2]As defined in SDA25.
[3]Estimate.

Texaco buying group, or joining a wholesaler-sponsored voluntary chain with group marketing and support activity, for example Spar, or being in the specialist food trades rather than in general grocery, although in recent years specialists also have suffered severe competition from the large grocers. Such firms provide an important element in consumer choice of shop as the shops are often close to residential areas, are convenient to use on foot, and may provide consumer services such as home delivery. The loss of local small shops reduces the consumer choice of shop and has had a disproportionate reduction in choice for the less mobile sections of the community. The local small shop traditionally has served the very young, the old and the less mobile consumer groups who benefit most from the convenience factor. The element of convenience is important to the survival of such firms in Mediterranean countries where the high population density of towns and the form of the urban fabric makes it not only difficult for large firms to develop suitable sites for new stores but relatively large numbers of people live within a few minutes walk of the shop. Nonetheless, even in these alternative social environments small firms operating general grocery stores are decreasing rapidly and the same routes to survival are appropriate as in the UK.

The benefits to the food retailer of large organisational scale are considerable. Benefits exist in buying, advertising, retail brand development, information systems, store development, cost of capital, manager motivation and development, cashflow management, and many other retail business functions. Within the UK there is a long history of medium-sized companies being taken over by larger companies in order to obtain greater returns to operating scale and to release cost economies of organisational scale. In the UK, the growth of Allied Suppliers (Mathias, 1967) took this form but in the end they were taken over to create the basis of the Argyll Group which in turn took over several companies, not least the UK operation of Safeway. The history of Tesco also includes several take-overs (Powell, 1991), the most recent of which was William Low, a long established Scottish grocer which itself had acquired companies in its 125 year history. The same process is seen throughout Europe as food retailers seek organisational scale economies.

These large firms benefit from economies of replication in which new stores are cloned with some of the expenses of development being at marginal cost, obtaining leverage from investments already made and recovered from earlier store developments. The nature of retailing makes it possible for a firm to have many branches each serving a different spatial group of consumers. Not only are the firms large but they operate through many branches. Expansion is possible until the whole market is covered. Although the size of the four (J. Sainsbury, Tesco, Safeway and ASDA) main food retailers operating grocery superstores in the UK is considerable, they are all some way off complete market coverage. J. Sainsbury is particularly weak in Scotland, Tesco has relatively low

Table 4.2 Retailer Share (%) of grocery[1] trade by region of UK (1992)

Region	Company			
	J. Sainsbury	Tesco	Safeway[2]	ASDA
London	36.4	22.4	10.1	8.2
South	30.5	29.2	8.7	6.3
Anglia	30.7	27.0	4.2	5.1
SW and Wales	12.4	23.6	6.8	9.9
Midlands	20.0	12.1	7.5	10.2
Lancashire	11.3	12.7	4.1	14.3
Yorkshire	12.6	9.5	4.3	15.1
Tyne-Tees	9.3	4.5	14.7	13.8
Scotland	3.5	8.5	15.6	16.1
All GB	20.5	17.4	8.2	10.7

Source: Taylor Nelson AGB (1993)
[1]Grocery is defined as 73 product categories including ambient, frozen, health and beauty and non-food products.
[2]Excludes other Argyll Group formats.

market coverage in North East England and Scotland and Safeway is underrepresented in Lancashire, Yorkshire and East Anglia. ASDA, the fourth largest firm, similarly has a far from complete national coverage of stores, being weak in Southern England and East Anglia. Table 4.2 illustrates this pattern of regional strength and weaknesses.

The result of this is that for many consumers choice of full-line grocery superstore becomes limited with strong duopolies sometimes being established in local spatial markets. But these duopolies are constrained by other types of grocery and food retailers, for example convenience stores, small supermarkets, discount stores, greengrocers, etc., each with its own form of operation. Therefore consumer choice of store, and what that implies for choice of price, product quality, convenience, and ambience, may in reality be substantial although each of these other types of store provide less product choice for the consumer than the full-line superstore.

The implications of the growth of very large firms are complex and numerous. Some of the effects in respect of change in the distribution channel have already been mentioned. A major implication is the concentration of sales and of investment (Baden-Fuller, 1986). Table 4.3 shows the trend in market concentration in the UK. In 1992 the largest 10 food companies (excluding off-licences) accounted for 58.1% of the sales of all food retailers (Central Statistical Office, 1994). If grocery products only are considered then the three largest companies accounted for 50.6% of sales of grocery products (Taylor Nelson AGB, 1993) and the five largest 67.5%. Even higher levels of market concentration are present in some product areas. In respect of investment then, in 1992, 51.3% of the net fixed capital investment in all British retailing was accounted for by the 11 food retailers with sales of over £500 million (Central Statistical

Table 4.3 Trends in market concentration of food retailers in the UK

	CR5	CR10	Percentage of grocery sales by five largest firms
1973			29.4
1976			31.7
1980			45.5
1982			49.1
1984	28.2	41.1	51.4
1986	34.6	47.5	57.8
1988	40.7	52.8	63.1
1990	41.5	55.3	65.2
1992	43.5	58.1	67.5

Source: Central Statistical Office SDA25 and SDO25; Taylor Nelson AGB (1993)

Office, 1994). This contrasts with 23.1% for the 11 largest food retailers in 1984 (Central Statistical Office, 1986).

The very large firms in the food retail market in general operate large supermarkets and superstores. It is important to distinguish, however, between large firms and large stores. The returns to scale in firm and establishment in food retailing are very different (Shaw *et al.*, 1989) and are greatest at firm level. There are therefore some large firms operating extensive chains of smaller stores. Examples are: Shell with 1200 shops in the convenience store sector; Whitbread with over 1600 shops in the off-licence sector; and Greggs with almost 500 shops in bakery retailing. The growth of the large retailers is not limited to the general grocery sector.

Multiple retailers have been a feature of food retailing in the UK for over 100 years (Jefferys, 1954). Substantial chains were already present in the early years of this century but they accounted for a very small part of the market. The transformation in the structure of the sector more recently, particularly in the UK since the abolition of Resale Price Maintenance on food in 1964, is characterised by the large multiple retailers becoming the dominant form of retailing and the small firms and consumer cooperative societies losing considerable market power. The scale economies associated with centralised operation, especially central buying, have given the large food retailers a sustainable and reinforcable competitive advantage over other small firms.

4.3
Food retailing methods

In most western countries, year on year, fewer firms are involved in food retailing. There are also fewer food shops (Eurostat, 1993). This is despite small increases in sales volumes and value. In parallel with the change in industry structure there have been significant changes in operational practice. The decrease in shop numbers has been mainly the result of small shop closures by small firms but there also has been a rationalisation of

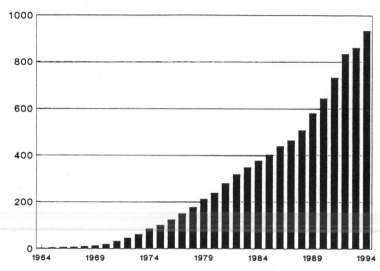

Figure 4.2 Cumulative number of grocery superstores in the UK. (*Source*: Institute of Grocery Distribution)

shop numbers by the larger firms either as a result of consolidation after take-over activity or the closure of less profitable shops in favour of fewer openings of more profitable shops,

Two notable trends have been the search for new economies of scale in store operation and the exploitation of existing economies of scope in the larger supermarkets and food superstores. Figure 4.2 shows the increase in number of grocery superstores in the UK. The increase has been substantial (Davies and Sparks, 1989) and the proportion of sales accounted for by superstores is considerable, being over 30% of grocery sales by 1990. A large superstore of 35 000 square feet might have sales in excess of £50 million (1994 prices) per year. The economies of scale sought in large food stores come mainly from costs of labour whereby labour cost as a percentage of sales can be reduced through use of part-time low-cost labour for standardised operations and only a small number of more expensive managers are necessary. The economies of scope result from the ability to sell related product ranges such that consumers can use the shop to purchase all their needs for food and everyday household goods. Logistics costs are also reduced as vertical marketing systems become effective. The result, from a viewpoint of consumer choice, is that whilst product choice is increased, for example in 1994 J. Sainsbury superstores had 17 varieties of fresh pasta in the range and 40 varieties of potato, the consumers carry additional costs of travel to store, undertake their own product selection and become responsible for more transport and storage costs on the product within the total food distribution system. Undertaking a large volume of purchases every week or two weeks passes to the consumer some stock-holding and financing costs compared with shopping every few days for items as they are needed.

The large food superstore represents one of several retail formats for the sale of food. Changes both in consumer demand and in behaviour patterns have resulted in an overall consumer requirement for differentiated stores. At the same time retailers have sought different ways to achieve cost economies and to obtain competitive advantage. Food stores have become positioned in consumers' minds on the dimensions of price and range. Against a conventional supermarket with which consumers have become familiar over 20 years or more, other types of store have been developed (Figure 4.3).

The food market has become more segmented with more choice of type of food shop now available to the consumer (Burt and Sparks, 1994). Thus whilst a standard supermarket may have around 8000 product lines, a food superstore may have 18 000 different products, a convenience store 1500 and a limited range 'hard' discount store less than 1000 different products (McGrath, 1994). In respect of price differentials a 'hard' discount store may be 10–15% lower in price than the standard supermarket, whilst specialist stores may be able to charge in excess of a 10% price premium. Price differentials, however, are difficult to establish as the product ranges of the different types of store are quite different and only a limited number of the same products would be common to the several store types. On competitive grounds these few common products may be identically priced across all types of shop. Price also reflects not just the physical product which is bought but also the services available to the consumer. These vary considerably across the store types. Range issues are also more complicated than might appear from Figure 4.3; the food superstore may contain specialist food sections which are similar to

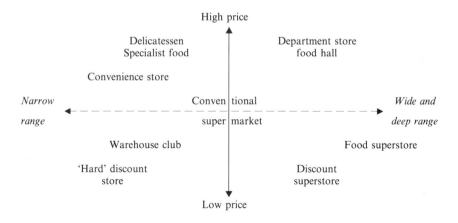

Figure 4.3 Market positions for food formats.

standalone specialist stores, having for example a service butcher, delicatessen and bakery within the large superstore.

The trends in recent years in the UK have been towards clearly defined store formats and towards greater variety of non-price competition. Simple comparisons of price for standard shopping baskets now have little meaning (Yarrow, 1992). Promotional techniques, for example multi-buy, in which the consumer gets volume discounts, link-save, in which discounts are provided on particular combinations of products, and shopper loyalty schemes, in which discounts are related to total purchase volume, make prices variable depending on the quantity or combination of items bought. Firm-specific and even store-specific promotions generate considerable price variation from week to week for a single product. As a result the number of known price items (KPIs) to the consumer is small. The previously used methods of price promoting these items in order to build shopper traffic is less widely used (Ehrenberg *et al.*, 1994). Shopper loyalty is now more likely to be based on availability of the appropriate product variety, shopping convenience, additional services, and personal rewards for frequent shoppers than on price, except for the limited line discount stores which attempt to attract shoppers almost wholly on price – although even here convenience of shopping and of location are starting to be used increasingly as shopper attractions.

Changes in logistics have been important in enabling the development of different types of store (McKinnon, 1985; Sparks, 1986). Improvements in logistics systems have reduced the time of the supply cycle so that fewer items of a product are necessary in-store, with back-up stock held at the distribution centre, but stock service levels are higher with fewer out-of-stocks for the shopper. The shortening of the supply cycle also results in fresh products, for example fruit and vegetables, spending less time in the supply chain between producer and store. With improvements in handling facilities in distribution centres these same products can arrive in the store not only more quickly but also in better condition (Smith and Sparks, 1993). These improvements in logistics have been implemented particularly by the larger firms, which again is a factor providing them with competitive advantage.

The main strategy of the firms in food retailing has been to seek to maximise market share through the development of a particular retail format. The efforts of Kwik Save to become the dominant firm in the discount sector, of Iceland to be the dominant firm in freezer centres, of Tesco and J. Sainsbury to be the major players in food superstores, of ASDA in hypermarkets, of Shell and BP in forecourt convenience stores, Thresher and Victoria Wine in off-licences, etc. all illustrate this trend. This common strategy has been a notable driver of change in food retailing. Each firm has concentrated on a particular form of retailing, even to the extent that firms which operate more than one format have divested their interest in a particular format. For example, the Argyll group which operated three formats of Safeway (food superstores), Presto (super-markets) and Lo-Cost (discount stores) has sold their interest in Lo-Cost and has broken up the chain. The focus on a single type of food retailing has proved to be a successful strategy.

In the early 1990s, however, a number of firms have reacted against this common trend and have pursued diversification strategies with moves into other types of food retailing. Tesco, J. Sainsbury and Safeway have sought to develop medium-sized supermarkets in city centres, operating them with a product mix aimed at the central city worker. J. Sainsbury and Tesco have also begun to develop convenience stores. This shift marks a change in strategic direction. Previously when firms were operating more than one format this was the consequence of acquisition of a firm, not all of the stores of which were transferable into the preferred format of the acquiring firm. The situation in Argyll is more complex owing to its history in that the firm grew rapidly through a series of acquisitions, one of which (Safeway) provided the critical mass of a pre-ferred format (food superstore/large supermarket). The earlier preferred format (Presto supermarket) had to be assimilated into Safeway or retained as a separate chain – many were changed into Safeway but the supermarket chain was retained with the smaller units being operated as the discount stores which were later divested. This shift in strategy on the part of the major companies is, as yet, in its early stages (Duke, 1991; Wrigley, 1994).

4.4 Management issues

The changing structure of the retail sector in general and the changing operational methods of firms in the sector provide only a partial explana-tion of the way that food retailing has changed in the UK. The third factor of change is the reorientation of a variety of managerial processes, not least those of marketing.

The expansion of retail brands has been a feature of food retailing management. There are implications of this expansion for consumer choice and for retailer strategies. Retail brands are products which carry the name of the retailer, for example J. Sainsbury, or a name which is owned by the retailer – this latter form is sometimes more strictly termed 'retail label' or 'retail sub-brand', for example 'Novon' – a J. Sainsbury sub-brand for detergent. In some less strict terminologies the whole con-cept is called 'own brand' or 'own label'. A form of retail brand which has no identification of the product other than product description, for example baked beans, is termed a 'generic' (McGoldrick, 1984).

Retail brands in the food sector have a very long history in the UK with some of the early multiple grocers also being importers, notably of tea, and establishing their own brands, although perhaps not brands in the sense of today (Mathias, 1967). The more recent explosion of retail brand activity in food owes much to the price deregulation of the aboli-

Table 4.4   Development of retail brands in the grocery market in the UK

Shares of sales by retailer

| Year | Retailer | | | | All grocers |
	J Sainsbury	Tesco	ASDA	Safeway	
1983	56.2	34.2	6.3	34.5	27.1
1986	55.8	36.7	18.2	34.7	28.9
1988	55.1	36.4	30.0	35.8	29.6
1990	53.4	39.4	30.6	33.1	30.5
1992	53.1	41.8	29.3	35.2	31.7

Shares of sales by selected product groups

| Year | Product group | | | | |
	Dairy	Paper	Frozen	Cleaners	Pet food
1983	35.3	34.6	29.8	21.3	10.4
1986	37.5	38.3	38.6	21.4	8.6
1988	41.2	39.9	38.6	22.4	6.9
1990	42.3	44.2	43.8	20.7	6.8
1992	45.6	44.3	43.2	22.0	7.7

Source: Taylor Nelson AGB (1994)

tion of RPM. Whilst the initial tactical response was a surge in price competition and the extension of product ranges to bring to the consumers lower-priced products, the more strategic response of retailers has been to move towards administered marketing channels in which the retailer has more market power (de Chernatony, 1989; Davies *et al.*, 1986; Leahy, 1987, 1994). Table 4.4 shows the extent of development of retail brands in the grocery market and the contribution they make to the sales of the major grocery retailers. Typical characteristics of markets in which there is a high level of retail brand penetration are (Buck, 1993):

- surplus manufacturing capacity;
- commodity status;
- low technical barriers to entry;
- little product differentiation;
- low levels of manufacturer investment.

It is these market areas where retail brand development has been in place longest and where it has developed most. Nonetheless, even markets where these characteristics do not hold may be susceptible to retail brand development, for example the cola market where retail brand extension in 1994 has been a notable feature with several retailers copying Coca-Cola product ranges through their own retail brands.

The use of retail branding in the food sector has become increasingly sophisticated. It is not unusual for a single product to have three, occasionally four, different retail brand items each with a different position.

For example, in a superstore a product (blackcurrant jam) might have a main retail brand item competing with the major manufacturer brand; a second item aimed at a premium position and price (perhaps in this case containing more fruit); a low price discount variant of the item, still with the retail brand, aimed at competing with discount retailers' offers; and a special version of the product aimed at a very particular consumer group (in the case of jam, probably a low sugar variant). Retail brands now constitute an essential component of product ranges in food retailers. For some retailers, most notably Marks and Spencer, the retail brand is the only brand present in the store. Retail brands are not limited to the general grocery retailers but are used by the specialist retailers as a mechanism for differentiation, for example in the off-licence sector where retail brands are strong in wine products.

Within the evolution of the idea of the retail brand in food products it is possible to identify four stages of evolution of the concept (Laaksonen and Reynolds, 1994). These stages are shown in Table 4.5. The early forms of retail brands are low cost, and low quality, and may be copies of existing brand products. These are the generic products or simple

Table 4.5 The types of retail brands in the food sector

Factor	Type 1	Type 2	Type 3	Type 4
Branding form	Generic	Unsupported retail brand	Supported retail brand	Segmented and sub-brands
Strategy	Generic	Low price copy	Copy of major brands	Added-value
Objective	Improve margin	(a) Improve margin (b) Reduce manufacturer power	(a) Improve margin (b) Reduce manufacturer power (c) Extend assortment (d) Build image	(a) Improve margin (b) Extend assortment power (c) Build image (d) Create differentiation (e) Improve customer loyalty
Product	Commodities and basic products	Basic products with large volume sales	Major sale items	(a) Niche products (b) Special category products
Technology	Simple process	Copy of market leader	Comparable main brand	New technology processes
Quality	Low	Medium and below market leader	Comparable to market leader	Same or better than market leader
Price position	20% below main brand	10–15% below	2–5% below	Premium
Consumer motivation	Low price	Low price	Value for money	Different product
Supplier	Unspecialised	Specialised and has own brands	Specialised in retail brands	International specialist in retail brands

Source: Based on Laaksonen and Reynolds (1994)

copies which are identified by the name of the retailer. The simple copying of manufacturer brand products characterises first and second generation retail brands. From the consumer viewpoint such products are substitutes for the established and well known manufacturer brands against which these retail brands have a price discount. Third generation retail brands expand product ranges. Whilst still at a price discount they are close to the brand leader in quality and image. The fourth generation of retail brands includes added-value products, the aim of which is to differentiate the retailer from its competitors. These products are often not at a significant price discount and may be competing with premium brands of manufacturers. Such products are used extensively to build customer loyalty (Corstjens and Corstjens, 1995).

Retailers have expanded their retail brand range to derive profitability, range extension, quality control and build consumer loyalty (Simmons and Meredith, 1984; Mårtenson, 1992; Laaksonen, 1994). The margin which retailers are able to obtain on retail brands is higher than on manufacturer brands. This is because of lower advertising costs and, for copied products, lower research and development costs. The range extension is achieved through the retailer deciding which products to put in the range and then obtaining the precise product, to a specification, to build that range. In retail brands it is usual for the retailer to take some responsibility for quality control, even in production processes, because the product carries the name of the retailer (Shaw *et al.*, 1992a). If there is a difficulty with product quality, from a consumer perspective, then the problem is immediately transferred to being a problem with the retailer because of the common name of product and retailer. With retail brands it becomes possible for retailers to become closely involved with quality issues and so to ensure high levels of consumer satisfaction. The fourth benefit of retail brands is the opportunity they provide to build consumer loyalty. A retail brand product is unique to the retailer which has developed it. The range of retail brands can be used to attract and retain consumers.

The implications of retail brands for consumer choice suggest a theoretical widening of choice but a practical reduction of choice and of the capacity to make direct and meaningful price comparisons. As individual retailers develop their own retail brands so overall product choice can be extended but in reality is constrained as store choice and product choice become closely interdependent. Whilst in theory it is possible for a consumer to visit different stores and to 'cherry pick' retail brands from different stores, in reality the inconvenience of this procedure raises consumer search and shopping costs to an unacceptable level. Consumers, therefore, in selecting a store also select the product range from which they will make a choice. Because the same product items, as retail brands, are not available in the stores of competing retailers, direct price comparison is impossible. Nonetheless price comparison, in effect, is made of main item retail brand of one retailer against main item retail brand of another. It is likely that in the mind of the consumer this price becomes the surrogate of the perceived price level of the whole store. The easy price comparisons of identical manufacturer brand products across different stores now have little meaning. Retail brand development has made fully evaluative choice more difficult but has simplified in-store choices for the shopper.

A second major area of change in management processes in food retailing is in the use of information. The facility to collect item level data at point of sale has enabled a wide range of new management and control processes to be introduced into food retailing firms. Knowing what food is sold, when it is sold, from which store, in what combination of products has provided a potential capability to manage stores and firms more effectively for the consumer and more profitably for the firm. The changes and benefits of the management of this information are:

- more effective shelf and store layouts which reduce search time for customers;
- closer matching of checkout labour to shopper flows;
- improved effectiveness in buying by the retailer with consequential potential for price competitiveness at the store;
- more capability to have new types of promotion, such as link-save, consumer loyalty programmes, coupon printing at point of sale in response to the specific purchase pattern of a shopper, etc.;
- improved stock management with fewer out-of-stock situations and product ranges more suited to consumer demand.

These changes, and others associated with new communication technologies and with advances in analytical techniques, when implemented successfully, improve the levels of customer service provided by food retailers which potentially increase retail sales of the firm and reduce the operating costs of the firm. There are therefore improvements to both customer choice and retailer profitability. The successful implementation of the techniques, however, is not easy in what has traditionally been a very low-tech industry. The collection of the data at point of sale is easy but the transformation of the data into useful managerial information and the change of managerial process consequent on this information being used are much more difficult tasks (Quarmby, 1993) – tasks more easily undertaken by large organisations than by small firms.

A third aspect of the transformation of management in food retailing is in the growing internationalisation of retailer activity. The ranges of food products regularly stocked in superstores and supermarkets reflects the more international food eating habits of consumers. It is difficult, however, to establish whether the retailer is leading this trend to a more international diet or responding to it (Shaw *et al.*, 1992b). In many cases the items in the range of ethnic foods carry higher margins for the retailer than the more mainstream foods. These ethnic foods are ideal for the creation of retail brand added value recipe dishes which carry even higher margins for the retailer. The consumer, however, seeks this variety and difference of product and so the retailers respond with a steady stream of new products in the ranges. In 1994, J. Sainsbury had wine from 21 countries in its range. The variety and sourcing of fruit and vegetable ranges also illustrates this increasingly international aspect of product ranges. Improvements in information flows between retailers and primary producers, within administered channel relationships, and improved logistics make possible the sourcing of fresh products from Southern Europe such that the products can be moved from field to supermarket within 60 hours. Similarly improvements in air freight increase the accessibility of suppliers of tropical fruits and vegetables to the buyers in UK supermarkets. The high quality of foreign products and the wide range of exotic fruit and vegetables in UK superstores has been one of the important factors in increasing the market share of these stores in competition with the small traditional greengrocer (Shaw *et al.*, 1994).

The internationalisation of food retailing is also evident in the moves by British retailers towards international investment in the operation of stores in other countries (Wrigley, 1989, 1993b) and the moves into the UK of European food retailers (Duke, 1993; Burt, 1991). Food retailers have been active over the last 20 years in extending their activity to store operation in other countries. Moves by UK retailers have been modest (Thompson, 1992) compared with companies based elsewhere in Europe, for example in France with Promodès and Carrefour, in Germany with ALDI, and in the Netherlands with Ahold. The gradual acquisition of Shaws in North East USA by J. Sainsbury and the more recent acquisition of 50% minus 1 of the voting shares of Food Giant have made J. Sainsbury one of the top 10 food retailers in the USA ranked by sales volume. Purchases in France and Hungary by Tesco point to increasing international interest, perhaps stimulated by doubts about the long-term market expansion opportunities in the UK (Guy, 1994b; Wrigley, 1996). Not all such activity has been successful with some acquisitions, for example by Iceland, resulting in subsequent divestment and retreat from international store operation.

An alternative form of international activity is seen in the fourth notable area of managerial change in recent years, namely the increased activity in strategic and tactical alliances amongst food retailers. Three of the forms these alliances take are international buying consortia, international marketing groups and international joint ventures. There is considerable activity also in other forms of national alliance amongst food retailers, particularly joint marketing and buying alliances and co-development alliances. In each case firms are attempting to increase their own competencies by combining with other firms who in some cases are competitors

(Dawson and Shaw, 1992; Hughes, 1994; MacNeary and Shriver, 1991; Robinson and Clarke-Hill, 1995).

International buying consortia and marketing groups in the food sector have a substantial history with the international buying group involving consumer cooperatives being established in 1918. The international activities of Spar, the voluntary chain, have over 40 years history. More recently there has been a flurry of activity in this area with several European buying groups established each with a single member from several European countries. One of the largest is Associated Marketing Services (AMS), which contains Safeway from the UK. AMS has arrangements for joint purchasing agreements, for members to use each other's retail brands, for the development of a common retail brand on selected products across several members, and for sharing of logistics services. There are also international links for national buying groups. EMD is a group which combines the buying power of Selex and Markant buying groups, comprised of medium sized retailers, in several European countries and includes the buying group NISA from the UK. The recent joint agreements involving Metro from Europe, Ito Yokado from Japan and Wal-Mart from USA indicate the potential emergence of global alliances of this type. These groups illustrate attempts to obtain even more powerful scale economies than those available to individual members.

Alliances between food retailers for joint shopping centre development represent a different type of strategic alliances. In these cases the alliance has the aim of creating new food retailing floorspace in situations where joint activity has a higher probability of gaining the necessary permissions for new store establishment than has the activity of a single firm. Other benefits, for example the sharing of expertise in the store development process and in the potential for cross shopping by consumer (Davies, 1993), also accrue to these alliances. The presence of such alliances does not limit the intensity of competition between the alliance members in respect of day-to-day trading. Within the UK there are several examples of this type of alliance involving Marks and Spencer on the one hand and Tesco, Safeway or J. Sainsbury on the other. The creation of shopping schemes containing a food superstore and a Marks and Spencer store, with a substantial food offer, enhances consumer choice of product within the one shopping trip.

The successful introduction of new managerial processes has been instrumental in generating changes in the structure of food retailing and in operational methods. The nature of the change has not been unidirectional, with new managerial approaches also being the result of changes in the external environment of the firm. Such external changes occur in the structure of the industry and in distribution systems, with the emergence of new food, communication, material handling and information technologies, and result from changes in government policy on food retailing. The management of food retailing, and in consequence the choices offered to the food shopper, is a function of the complex interaction of factors external to the firm with factors of the firm's internal strategy and tactics.

4.5 Conclusion

Table 4.6 summarises the changes in food retailing and indicates some of the explanatory factors of the changes. Key trends from a consumer point of view are:

- fewer out-of-stock situations in stores;
- fewer local stores;
- more product variety in stores but less effective choice of store;
- a wider range of price and quality offers presented to the consumer;
- more variety of promotional offer for the consumer to evaluate;
- a rise in base product quality;
- an increase in the proportion of distribution channel cost carried by the consumer;
- more new products available for consumer evaluation;
- fewer seasonal constraints on product availability;
- more similarity of choice offered to consumers in different countries;
- increasing influence of retailers on consumer choice.

Food retailers have a substantial influence on the food choices made by consumers. This is not a new phenomenon. The early development of the consumer cooperative stores in the 1840s had as their explicit aim the influence on and extension of the food choice of consumers (Potter, 1930). The success of consumer cooperatives through the nineteenth century extended the food choices available to many millions of consumers in urban and also rural areas of the UK. Emerging food chains at the turn of the century introduced branding as a means of confirming food quality and so influencing food choice (Mathias, 1967). The small corner shop of the 1920s and 1930s suburbia was limited in the products it could stock by the size of store and the counter service it provided. The retailer chose the product range from the wholesaler lists and in so doing circumscribed consumer product choice, but the large number of these shops gave the consumer a wide choice of shop (Smith, 1948). Food rationing in the 1940s was implemented through food retailers, again with implications for consumer choice – reducing both choice of product and choice of store (Hall, 1949). The strong development of supermarkets and superstores and the abolition of RPM removed some of the space and price constraint which were limiting consumer choice but the consequential change in industry structure reduced the opportunities for choice of store whilst increasing those for choice of product (Tanburn, 1981). The increased concentration in food retailing and the clear emergence of segmentation of the food market in the last decade has reduced the opportunities for choice of retailer but the different store types operated by the

Table 4.6 Summary of major changes in food retailing

Change	Main related factor	Secondary related factor
Decrease in store network	Fewer independent retailers	Poor management Age of owners and lack of succession Difficulty in obtaining capital Many independent retailers operate a small store which has become below minimum economic size Higher cost of obtaining supplies, a particular cause in, but not exclusive to, rural locations Store location no longer appropriate
	Increase in productivity of existing stores	Widespread adoption of self-service Use of information technology Better stock management, e.g. fewer out-of-stocks Corporate and contractual chain store management methods
	Economic/financial benefits of large stores	Scale economies in buying and labour Improvements in logistics Larger transaction value of consumers
	Consumer preference for larger stores	Storage of food at home improved More mobile consumer Larger stores give more choice
	Alternative uses for stores	More profitable uses as offices and service retailing (e.g. financial services), particularly in town centres
Increase in number and market share of large stores, e.g. supermarkets and food superstores	Economies of scope of product range	Consumer preference for one-stop shopping Larger range in bigger stores Specialist buyers in retailers More new products launched
	Economies of scale of store operation	Returns to scale of labour cost, logistics costs, energy cost, etc. IT benefits
	Large companies capable of undertaking development	High profits from food retailing Perceived as safe sector for investment Specialist management becomes experienced in large store operation
	Availability of sites	New road construction, particularly urban ring roads Government policy on rules of establishment

Table 4.6 Continued

Change	Main related factor	Secondary related factor
Increase in market concentration (increase in minimum efficient firm size relative to market size)	Organisational scale economics	Access to capital finance Returns to scale in buying, marketing, merchandising, IT, staff training, logistics More merger and take-over activity
	Rising entry barriers	Higher sunk costs with large stores Easier and better sites are developed first Local 'monopolies' of large stores
	Cost economies of channel coordination	Reduction in stock in supply chain Reduced transactions costs through longer term and administered relationships Increased use of IT and EDI methods Cost sharing of activities between retailer and manufacturer, e.g. product development
Increase in activities of buying groups, franchises and marketing alliances	Firms seek scale economies through joint buying	Seek to improve competitive position of firms Opportunities for range extension
	Small firm reaction to increase in market power of larger firms	Extension of joint buying first into other marketing activities and subsequently into a wide range of business support functions
	Large firms seek international links	Extend retail label coverage Information exchange Cost savings on logistics sharing Benefits from joint buying schemes
Increase in share of distribution costs borne by consumer	Widespread use of self-service	Retailer wish to reduce labour costs Manufacturer wish for better display of product
	Less dense store network	Acceptance of longer trips to store Lesser ability to make interstore price comparisons Use of personal transport for shopping trips
	Consumers' trade-off of choice, time and convenience	Higher consumer value placed on choice Higher levels of consumer mobility
	Out-of-town location of new stores	Lower unit land costs for large sites Land availability

Table 4.6 Continued

Change	Main related factor	Secondary related factor
Growth of retail brands	Provide product variety and extend range	Consumer seeks wider choice
	Higher gross margin for retailers than on manufacturer brands	Low development costs with retail brands copying manufacturer brands Production from firms with excess capacity Low advertising and promotion costs
	Build store and company loyalty	Major companies seek new ways of differentiation
	Retail brands move from being low price basic products to higher value innovations	More margin available in innovatory products Limited scope for more basic products
	Obtain economies of supply chain coordination	Important to have product in stock
Adoption and expansion of use of new information and communication technologies	Requirement to have better control of costs	Change in nature of costs Increases in costs Increased product variety and more SKUs generate higher stock costs New techniques of cost management available, e.g. DPP
	Requirement to expand sales	Increase number of shoppers Increase purchase volume/shopper More variety of promotional methods
	Need to build customer loyalty	Demand for consumer choice More intense competition New loyalty building methods, e.g. frequent shopper programmes
Product diversification	Need to fill large stores with products	Space costs are significant element in retail cost structure
	Retailers pursue policies of differentiation and competition based on product variety	Consumers seek wider choice Retailers seek to avoid direct price comparisons and competition on price
	New technologies create new products and packaging	New cooking methods in kitchen New storage requirements of consumers

various large firms has started to increase the opportunities for choice of shop without diminishing the choice of product. Food retailers undoubtedly continue to have a significant influence on consumer culture and food choice – an influence which has increased in intensity in recent years. Food retailers have become more responsive to consumer needs and provide the types of food demanded by the consumer yet the retailers still continue to lead and direct food choice through their decisions on store location, product ranges and new product development.

Acknowledgement   Dr Donald Harris provided valuable responses to earlier versions of this chapter.

References   Akehurst, G. (1984) Checkout: The analysis of oligopolistic behaviour in the UK retail grocery market. *Service Industries Journal*, **4**(2), 198–242.

Baden-Fuller, C. W. F. (1986) Rising concentration in the UK grocery trade. In K. Tucker and C. W. F. Baden-Fuller (eds), *Firms and Markets*, Croom Helm, Beckenham, pp. 63–82.

Buck, S. (1993) Own label and branded goods in fmcg markets: an assessment of the facts, the trends and the future. *Journal of Brand Management*, **1**(1), 14–21.

Burt, S. L. (1991) Trends in the internationalization of grocery retailing: the European experience. *International Review of Retail, Distribution and Consumer Research*, **1**(4), 487–515.

Burt, S. L. and Sparks, L. (1994) Structural change in grocery retailing in Great Britain. *International Review of Retail, Distribution and Consumer Research*, **4**(2), 195–217.

Carlisle, J. A. and Parker, R. C. (1989) *Beyond Negotiation*, Wiley, Chichester.

Central Statistical Office (1986) Retailing 1984. *Business Monitor SDO25*, HMSO, London.

Central Statistical Office (1994) Retailing 1992. *Business Monitor SDA25*, HMSO, London.

de Chernatony, L. (1989) Branding in an era of retailer dominance. *International Journal of Advertising*, **8**(3), 245–260.

Corporate Intelligence Group (1994) Food Shops. *Research Report 62*, 251–259.

Corstjens, J. and Corstjens, M. (1995) *Store Wars*, Wiley, Chichester.

Davies, B. K., Gilligan, C. and Sutton, C. (1986) The development of own label product strategies in grocery and DIY retailing in the United Kingdom. *International Journal of Retailing*, **1**(1), 6–19.

Davies, B. K. and Sparks, L. (1989) The development of superstore retailing in Great Britain, 1960–1986. *Transactions of Institute of British Geographers*, **14**(1), 74–89.

Davies, G. (1993) Patterns in cross shopping for groceries and their implications for co-operation in retail location. *British Journal of Management*, **4**, 91–101.

Davies, R. L. (1984) *Retail and Commercial Planning*, Croom Helm, Beckenham.

Davies, R. L. (1986) Retail planning in disarray. *The Planner*, **72**(7), 20–22.

Dawson, J. A. and Kirby, D. A. (1979) *Small Scale Retailing in the UK*, Saxon House, Farnborough.

Dawson, J. A. and Shaw, S. A. (1989) The move to administered vertical marketing systems by British retailers. *European Journal of Marketing*, **23**(7), 42–52.

Dawson, J. A. and Shaw, S. A. (1990) The changing character of retailer–supplier relationships. In J. Fernie (ed.), *Retail Distribution Management*, Kogan Page, London. pp. 19–39.

Dawson, J. A. and Shaw, S. A. (1992) Interfirm alliances in the retail sector: evolutionary, strategic and tactical issues in their creation and management. University of Edinburgh, Department of Business Studies, Working Paper, 92/7.

Duke, R. C. (1991) Post saturation competition in UK supermarket retailing. *Journal of Marketing Management*, **7**(1), 63–75.

Duke, R. C. (1993) European new entry into UK grocery retailing. *International Journal of Retail and Distribution Management*, **21**(1), 35–39.

Ehrenberg, A. S. C., Hammond, K. and Goodhardt, G. J. (1994) The after-effects of price-related consumer promotions. *Journal of Advertising Research*, **34**(4), 11–21.

Eurostat (1993) *Retailing in the European Single Market*, Commission of the European Communities, Brussels.

Gardner, C. and Sheppard, J. (1987) *Consuming Passion: The Rise of Consumer Culture*, Unwin Hyman, London.

Grant, R. M. (1987) Manufacturer–retailer relations: the shifting balance of power. In G. Johnson (ed.), *Business Strategy and Retailing*, Wiley, Chichester. pp. 43–58.

Guy, C. M. (1988) Retail planning policy and large grocery store development. *Land Development Studies*, **5**, 31–45.

Guy, C. M. (1994a) *The Retail Development Process*, Routledge, London.

Guy, C. M. (1994b) Grocery store saturation: has it yet arrived? *International Journal of Retail and Distribution Management*, **22**(1), 3–11.

Hall, M. (1949) *Distributive Trading*, Hutchinson University Library, London.

Hughes, D. (ed.) (1994) *Breaking with Tradition: Building partnerships and alliances in the European food industry*, Wye College Press, Ashford.

Jefferys, J. B. (1954) *Retail Trading in Britain, 1850–1950.* Cambridge University Press, Cambridge.

Kaas, K. P. (1993) Symbiotic relationships between producers and retailers in the German food market? *Journal of Institutional and Theoretical Economics*, **149**(4), 741–747.

Kleinwort Benson (1991) *Food Retailing: Survival of the fittest*, Kleinwort Benson Securities, London.

Knox, S. D. and White, H. F. M. (1991) Retail buyers and their fresh produce suppliers: a power or dependency scenario in the UK? *European Journal of Marketing*, **25**(1), 40–52.

Laaksonen, H. (1994) *Own Brands in Food Retailing across Europe*, Oxford Institute of Retail Management, Oxford.

Laaksonen, H. and Reynolds, J. (1994) Own branding in food retailing across Europe. *Journal of Brand Management*, **2**(1), 37–46.

Leahy, T. (1987) Branding – the retailers viewpoint. In J. M. Murphy (ed.), *Branding: A key marketing tool*, Macmillan, London. pp. 116–124.

Leahy, T. (1994) The emergence of retailer brand power. In P. Stobart (ed.), *Brand Power*, Macmillan, Basingstoke. pp. 121–136.

McClelland, W. G. (1963) *Studies in Retailing*, Basil Blackwell, Oxford.

McClelland, W. G. (1967) *Costs and Competition in Retailing*, Macmillan, London.

McClelland, W. G. (1990) Economies of scale in British food retailing. In C. M. Moir and J. A. Dawson (eds), *Competition and Markets*, Macmillan, Basingstoke. pp. 119–140.

McGoldrick, P. (1984) Grocery generics: An extension of the private label concept. *European Journal of Marketing*, **18**(1), 5–24.

McGrath, M. (1994) *Grocery Retailing 1994*, Institute of Grocery Distribution, Watford.

McKinnon, A. (1985) The distribution systems of supermarket chains. *Service Industries Journal*, **5**(2), 226–238.

MacNeary, T. and Shriver, D. (1991) *Food retailing alliances: Strategic implications*, Corporate Intelligence Group, London.

Mathias, P. (1967) *Retailing Revolution*, Longmans, London.

Monopolies and Merger Commission (1981) *Discounts to Retailers, HC311*, HMSO, London.

Monopolies and Merger Commission (1983) *Linfood Holdings Plc and Fitch Lovell Plc: a report on the proposed merger, Cmnd 8874*, HMSO, London.

Mártenson, R. (1992) *The Future Role of Brands on the European Grocery Market*, Söderburg Research Institute of Commerce, Gothenburg.

Pickering, J. F. (1966) *Resale Price Maintenance in Practice*, George Allen & Unwin, London.

Potter, B. (1930) *The Co-operative Movement in Great Britain*, George Allen & Unwin, London.

Powell, D. (1991) *Counter Revolution: The Tesco Story*, Grafton Books, London.

Quarmby, D. (1993) Food retailing: maintaining a competitive edge with a quality approach. Paper to the Manchester Statistical Society, 15 December 1993. 12 pp.

Ray, D. (1994) Who is driving the food industry? Wye College, Food Industry Perspectives, Discussion Paper, 2.

Robinson, T. and Clarke-Hill, T. (1995) International alliances in European retailing. *International Review of Retail, Distribution and Consumer Research*, **5**(2), 167–184.

Sack, R. D. (1993) *Place, Modernity and the Consumer's World*, Johns Hopkins University Press, Baltimore.

Segal-Horn, S. and McGee, J. (1989) Strategies to cope with retailer buying power. In L. Pellegrini and S. K. Reddy (eds), *Retail and Marketing Channels*, Routledge, London. pp. 24–28.

Senker, J. M. (1986) Technological co-operation between manufacturers and retailers to meet market demand. *Food Marketing*, **2**(3), 88–100.

Shaw, S. A. (1994) Competitiveness, relationships and the Strathclyde University Food Project. *Journal of Marketing Management*, **10**(5), 391–407.

Shaw, S. A., Dawson, J. A. and Blair, L. M. A. (1992a) The sourcing of retailer brand products by a UK retailer. *Journal of Marketing Management*, **8**(2), 127–146.

Shaw, S. A., Dawson, J. A. and Blair, L. M. A. (1992b) Imported foods in a British supermarket chain. *International Review of Retail, Distribution and Consumer Research*, **2**(1), 35–57.

Shaw, S. A., Gibbs, J. and Gray, V. (1994) *The Strathclyde Wholesale Markets Study*, University of Strathclyde, Glasgow.

Shaw, S. A., Nisbet, D. and Dawson, J. A. (1989) Economies of scale in UK supermarkets: some preliminary findings. *International Journal of Retailing*, **4**(5), 12–26.

Simmons, M. and Meredith, B. (1984) Own label profile and purpose. *Journal of the Market Research Society*, **26**(1), 3–27.

Slater, J. M. (1987) The food sector in the UK. Paper to *Conference on Competition Policy in the Food Industries, University of Reading, September 1987*.

Smith, D. L. G. and Sparks, L. (1993) The transformation of physical distribution in retailing. *International Review of Retail, Distribution and Consumer Research*, **3**(1), 35–64.

Smith, H. (1948) *Retail Distribution*, Second edition, Oxford University Press, London.

Sparks, L. (1986) The changing structure of distribution in retail companies. *Transactions of Institute of British Geographers*, **11**(2), 147–154.

Tanburn, J. (1981) *Food distribution: Its impact on marketing in the 1980's*, Central Council for Agricultural and Horticultural Co-operation, London.

Taylor Nelson AGB (1993) *Share of Grocery Trade, 1973–1992*, Taylor Nelson AGB, London.

Taylor Nelson AGB (1994) *Packaged Grocery Private Label, 1983–1992*, Taylor Nelson AGB, London.

Thompson, K. (1992) The serpent in the supermarket's paradise. *European Management Journal*, **10**(1), 112–118.

Trail, B. (ed.) (1989) *Prospects for the European Food System*, Elsevier Applied Science, London.

Wrigley, N. (1989) The lure of the USA: further reflections on the internationalization of British grocery retailing capital. *Environment and Planning*, **A21**, 283–288.

Wrigley, N. (1993a) Abuses of market power? Further reflections on UK food retailing and the regulatory state. *Environment and Planning*, **A25**, 1545–1552.

Wrigley, N. (1993b) Retail concentration and the internationalization of British grocery retailing. In R. D. F. Bromley and C. J. Thomas (eds), *Retail Change: Contemporary issues*, UCL Press, London. pp. 41–68.

Wrigley, N. (1994) After the store wars. *Journal of Retailing and Consumer Services*, **1**(1), 5–20.

Yamey, B. (ed.) (1966) *Resale Price Maintenance*, Weidenfeld and Nicholson, London.

Yarrow, S. (1992) Are UK supermarket prices competitive? *Consumer Policy Review*, **2**(4), 218–226.

Food and nutrition: helping the consumer understand 5

Annie S. Anderson, Kathryn Milburn, and Michael Lean

During the 20th century, the impact of nutrition on health has been of major concern in public health medicine throughout the world. However, by the 1939–45 war a comprehensive food policy incorporating food rationing, mass nutrition education and dietary supplements illustrated that an adequate and nutritious food supply could be provided and consumed by the entire nation, with documented health benefits in terms of reduced heart disease and reduced diabetes.

Post-war food rationing was, however, changed within decades by modern agriculture, food technology, new food industries and an expanding retail network into a food provision which supplied ample quantities of foods previously restricted. Sugar and its many forms in beverages, confectionery and manufactured foods and fat in a range of meat products, baked goods and pastries provided the luxuries denied to a generation of war-weary adults and children. Such foods could be produced, transported and stored cheaply, and the agriculture industry was directed towards reducing food costs.

By the 1970s, the health effects of a high-sugar, high-fat diet, low in fibre-rich carbohydrates, fruit and vegetables, were being questioned. The increase in chronic diseases such as cardiovascular disease, cerebrovascular disease, diabetes mellitus, cancer, bowel disorders, obesity and dental caries and their epidemiological links with diet led several leading nutritionists to address problems of malnutrition relating to overeating rather than undereating. In the USA, the McGovern report (Senate Committee on Nutrition and Human Needs, 1977) identified specific nutrition targets for the general population. This approach was revolutionary compared with the previous decade when as Southgate (1992) reports: 'nutrition messages in the western world seemed to call for "moderation" and for foods to be "eaten regularly" with certain foods (e.g. confectionery) being eaten "infrequently" rather than in specific amounts.'

Six years later in the United Kingdom the National Advisory Committee on Nutrition Education (NACNE) (Health Education Council, 1983) produced a discussion paper *Proposals for Nutritional Guidelines for Health Education in Britain*. Within this paper specific nutrition targets and goals for the population were presented, in a style to be followed by future WHO (1990) and British (Department of Health, 1994b) reports (Table 5.1). These reports spell out, in detail, the current understanding on the relationships between sucrose and dental caries, dietary fibre and bowel disorders, fat and coronary heart disease, salt and blood pressure and energy intake and obesity. Such reports helped initiate the change in 'public attitude towards food and health from a minority fetish into a mass public concern' (Winkler, 1991). The NACNE report of 1983 was widely accepted by medical and nutritional scientists, who welcomed an activist group. Sadly, despite the consensus of expert opinion expressed in this publication, no official action was taken by government. The *Health of the Nation* White Paper (Secretary of State for Health, 1992) and the development of the Nutrition Task Force in 1992 represent the latest United Kingdom moves which do, however, indicate a greater willingness of government to initiate change to prevent ill health. Hitherto, there has been little improvement in the nation's diet, and perhaps over-

Table 5.1 NACNE (HEC, 1983) targets for the UK population and WHO (1990) and Department of Health (1994a) recommendations

*Energy	Composition	Short-term proposals (inc. energy from alcohol)	Long-term proposals (inc. energy from alcohol)	WHO (1990) Upper limit	Lower limit	DH (1994b)
Protein	energy %	11%	11%	10%	15%	not specified
Fat	energy %	34%	30%	15%	30%	35%
Saturated fat	energy %	15%	10%	0	10%	10%
Carbohydrate	energy %	50%	55%	55%	75%	50%
Sucrose	per head per year	34 kg	20 kg	0 (energy from sugar)	10% (energy from sugar)	not specified
Alcohol	energy %	5%	4%	not specified	not specified	not specified
Dietary fibre	g	25 g	30 g	16 g 24 g	24 g** 40 g	not specified
Salt	g	11 g	9 g	0	6 g	6 g

*Nutritionists have to consider the balance and relationships between nutrients in a food, or in the diet. They therefore express nutrients like protein, fat, etc., as a proportion of energy (calories).
**NSP = Non Starch Polysaccharides
(*Source:* Black and Rayner (1992).

reliance on health education which seems to have little influence on behaviour.

The diet and disease concerns of the British public relate principally to coronary heart disease (CHD), which is the major cause of premature death in the United Kingdom with Scotland having the highest rates of CHD in Western Europe. Raised serum cholesterol levels (specifically LDL cholesterol) are known to be associated with a diet high in saturated fat and are a recognised risk factor for coronary heart disease. It is also now recognised that the damaging LDL fraction is activated by oxidation from free radical action (such as those from smoking) and that vitamins A (beta carotene), C and E protect against vascular damage. It is estimated that around 75% of adult Scots have a high LDL cholesterol (and therefore total cholesterol) above the acceptable range (> 5.2 mmol). The WHO (1988) MONICA study showed that the whole Scottish population has the highest risk from elevated cholesterol in the western world and also that the Scots have lower levels of protective vitamin E and C than other western countries. Additionally, there is increasing evidence for causal relationships between the consumption of sodium with the level of blood pressure – a problem of particular concern when manufactured foods provide 65–85% of total dietary sodium (Department of Health, 1994b).

There is also increasing concern over the relationship between diet and cancer. This is summarised in the WHO report (1990) which concludes that a high intake of total fat, and in some cases saturated fat, is associated with an increased risk of cancers of the colon, prostate and breast. Diets high in plant food, especially green and yellow vegetables and citrus fruit, are associated with a lower occurrence of cancers of the lung, colon, oesophagus and stomach – probably through the preventive action of antioxidant vitamins A, C and E. Additionally, high intakes of alcohol have been clearly linked with cancer of the upper alimentary tract.

Obesity, another major health risk, is associated with the development of metabolic disorders (e.g. diabetes, gallstones, infertility and a number of cancers), mechanical disorders (e.g. back pain, breathlessness and psychological disorders such as depression). The MONICA study shows the incidence of overweight to range from 70% in Russian women to 40% in Scottish women (age 40–49) and from 70% in Portuguese men to 42% in Spanish men in the same age category (Scottish Office, 1993). The reasons for these high rates relate principally to a reduction in physical activity as private transport became more available (and households more mechanised) coupled with high energy intakes.

Although there is continuing debate about the level of dietary change required to reverse trends in coronary heart disease, cancer and other diet-related disorders, it is clear that the majority of national and international committees now agree on the need to promote a diet high in fibre-rich carbohydrates, fruits and vegetables and low in fat (especially saturated fat), sugar, salt and alcohol (Cannon, 1992; WHO, 1990). Both the World Health Organisation (1990) and the UK Department of Health (1994b) reinforce the need for major changes in the intake of these nutrients.

5.2
Dietary intake and
changes in the 1980s
and 1990s

Following the publication of the 1983 NACNE report, the Health Education Council, the Joint Advisory Council on Nutrition Education (JACNE) and local National Health Service (NHS) Food and Health policy groups started major public campaigns to encourage individuals to change dietary intake (Anderson and Lean, 1987). Despite such widespread dietary advice, it appears that dietary intake has changed little in the decade since 1983. The United Kingdom National Food Survey, which is the longest running continuous survey of household food consumption in the world, documents the food intake to households throughout the UK and is used to estimate changes in the nutrient intake of the population. In 1980 this survey estimated dietary fibre intakes at 12.0 g/day and in 1992 intake remained at this level. In 1983 the percentage of energy derived from fat was estimated at 42.6% and in 1992 this had decreased to 41.7%. However, the percentage of food energy derived from saturated fat had decreased from 18.7% in 1983 to 16.3% in 1992 (MAFF, 1994). Clearly, only saturated fat has seen a change in line with current recommendations, although even intakes of this nutrient is well in excess of the 10% recommended by national and international bodies (Table 5.1). Despite these statistics, there is a great deal of evidence to suggest that the public are responding to health messages which target specific foods as opposed to nutrients. For example, in many campaigns of the 1980s, four principal food selections were targeted. These were: 'use reduced fat milk instead of whole milk'; 'use a polyunsaturated spread instead of butter'; 'use brown and wholemeal bread instead of white bread'; and 'reduce sugar intake'. The National Food Survey has documented major changes in consumption of these foods (see also Ritson and Hutchins, chapter 3, for a discussion of the underlying trends in food consumption in the UK since the 1980s). For example, whole milk consumption has decreased from 4.2 oz per week in 1980 to 1.7 oz per week in 1992, butter consumption has decreased from 4.1 oz per week in 1980 to 2.8 oz in 1992, consumption of white (and other) breads has decreased from 29.6 oz per week in 1980 to 22.8 oz per week in 1992, and sugar consumption has fallen from 11.2 oz per week to 5.5 oz per week. These increases are also supported by retail data (NTC Publications, 1993). Again, however, it is clear that these changes are most notable in higher social class groups. Data from retail sales presented in *The Scottish Diet* (Scottish Office, 1993) report show that sales of salad, fruit, brown bread and low-fat milk were all higher in stores which served a higher proportion of social class I and II residents.

It is clear that despite apparently healthy food choices by some adults, the population is not achieving current dietary targets. The major problem in dietary education has been described as 'the triangular tug of war' (Stockley, 1991), that is the compromise that arises from recommendations which are scientifically accurate, which meet public health needs and are applicable for 'the man in the street'.

Stockley presents a convincing argument that one of the major problems for both consumers and professionals in understanding dietary advice is the language and form in which advice is presented, and emphasises this by defining and discussing the approaches to dietary advice which have been used. For example, the quantitative guidelines outlined in the NACNE report and subsequent publications (DHSS, 1984; USDA, 1985; WHO, 1990; Department of Health, 1991; Scottish Office, 1993) are designed for professional nutrition educators. In these reports, it is anticipated that health education will translate the guidelines into readily usable advice for individual consumers based on food rather than nutrients. Such guidelines can take a number of forms (as described by Stockley, 1991) although they tend to comprise *population nutrient goals* which represent long term aims or ideals and *population nutrient targets* which are time-defined strategies. This does not mean that nutritional science will have changed its recommendations after a short term, but more realistically it is known that people are unlikely to change their diet immediately and setting short-term targets permits revision of strategy for the forthcoming years which will help achieve the ultimate goals. These goals or targets are set out in strictly nutritional terms as relative proportions to be derived from the energy-giving nutrients.

An alternative to nutrition guidelines is to offer *nutrient consumption 'advice'*. This approach is usually intended for individuals and is often qualitative (e.g. avoid too much fat), although it has been presented in a quantitative manner. For example, the Committee on Medical Aspects of Food Policy (DHSS, 1984) recommended that individuals should reduce their total fat intake to less than 35% of energy from fat and 15% of energy from saturated fat. However, such advice is almost impossible for the consumer to understand and difficult even for dietitians to assess easily. Despite the development of many dietary self-assessment tools and quizzes, the consumer cannot easily and accurately measure their own nutrient intake and thus assess the adequacy of their food intake.

Some countries and regions have set *food goals and food targets for populations*. These are designed for planners and tend to be done on a large-scale basis (as in Norway) and have the advantage that people from all sectors can comprehend exactly what is being measured. The recently published reports on *The Scottish Diet* (Scottish Office, 1993) and *Nutritional Aspects of Cardiovascular Disease* (Department of Health, 1994b) use nutrient targets and dietary (food) targets. Thus specific details of key

5.3
Promoting a healthier diet

foods are clearly spelt out for the entire population. For example, *The Scottish Diet* report recommends that consumption of fresh fruit should double between 1993 and 2005.

An alternative approach is *food consumption advice*, where individuals are advised on specific levels of consumption of different foods. This approach enables an appropriate nutrient intake to be achieved, but once again returns to a rationing-type approach which is often considered too terse and didactic. For example, in Sweden high energy consumers are recommended 0.2 litres of low-fat milk and 280–350 grams of potato per day.

The most common approach to nutrition education used within the UK is to target individual knowledge of nutrition and provide enough information to make appropriate food choices. This approach has meant concentrating on information transfer rather than developing motivating or enabling strategies. Anderson *et al.* (1988) suggest that there are three important components to nutrition knowledge: understanding nutrition terms; understanding the theoretical principles of good nutrition; and understanding the practical application of these theoretical nutrition principles.

It appears in many instances that people do have a good knowledge about nutrition and the public are aware of the need to reduce consumption of saturated fats, sugar and salt (Marshall *et al.*, 1994). A survey conducted by MORI (1992) for the National Dairy Council among 1709 members of the general public showed that 95% were aware of the message to eat more fibre, 93% less sugar, 93% less fat, 87% less salt and 60% less fat. However, an understanding of the practical interpretation of these nutrients into food seems to cause more concern. From a list of food items, only 25% could identify starch sources, 14% saturated fat sources and 2% polyunsaturates.

Even in younger adults dietary knowledge appears to be reasonably good. In the West of Scotland Twenty-07 study, a cohort of around 1000 15-year-olds answered the question: 'If someone asked you whether the following foods were good or bad for them if they wanted to lose weight or remain slim while remaining healthy, what would you say?' A total of 99.5% described fresh fruit, 99.3% fresh vegetables and 93.4% fish as good in this context, while only 5.7% described pastries/pies, 0.7% chips and 0.4% sweets/lollies as good for the healthy weight watcher (Anderson *et al.*, 1994a). This suggests that, at the extremes of high calorie foods, adolescents were not ignorant about the current orthodoxy regarding a healthy diet.

However, if nutrition knowledge is considered in terms of food choice, it means that people need to be aware of what they are currently eating, which nutrients they should be eating in greater or lesser quantities and how to achieve those changes in terms of which foods to eat more or less of. In practice this means that people need to know what foods to buy, where to buy them and how to prepare and cook them, and can of course only help in dietary change in the presence of an enabling and facilitating environment (e.g. resources being available). Changes are necessary at all stages in the food provisioning process.

Current approaches to increasing nutritional knowledge have tended to focus on familiarising people with important *nutrition terms*, emphasising *reductions* in intake of certain nutrients and focusing on certain foods which should be *substituted* by 'healthier' variations or reducing intakes of certain foods. This approach has yet to succeed in achieving major dietary change.

It seems clear that for dietary advice to be effective it must be understood and hence must be able to provide a true practical understanding of appropriate food choice concerning issues that confront consumers on a daily basis. Food choice and dietary intake are influenced by all components of the food network (Figure 5.1) and it is clear that the 'meshes of a net are interconnected and a pull on one will affect the shape of all' (Hurren and Black, 1991).

To put a good understanding of nutrition into practice the consumer must be able to cope with issues such as quantities of food needed for maintenance of health and prevention of disease, changing media messages about food, the nutrient content of food, and the flourishing 'health' and dietary supplements market. Additionally, nutrition educators need to take account of lay perceptions and beliefs about nutrition and health before making advances in achieving a real consumer understanding of a healthy diet.

One reason why the British diet has not changed significantly in the last decade may relate to emphasis on qualitative changes in the con-

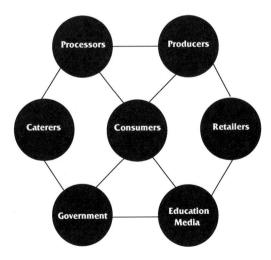

Figure 5.1   The food network. (*Source:* Hurren and Black, 1991)

sumption of certain foodstuffs with no guidance on quantitative change. For example, changing from white to wholemeal bread will only bring about a change in dietary fibre consumption if it is accompanied by a significant increase in the amount of bread eaten. Similarly, to decrease the percentage of energy derived from fat in the diet and increase carbohydrate it is necessary to decrease intake of foods which provide significant amounts of fat (e.g. whole milk, manufactured meat products and total spreading fat) and increase the amount of foods which provide significant amounts of carbohydrate (e.g. total bread, potatoes, and other cereals).

National Food Survey data shows a steady decrease in consumption of carbohydrate-rich foods since 1940. Energy intake which has been reduced by decreasing milk fat and spreading fat has been replaced by fat from other sources such as cakes, pastries and ice-cream (which are known to have increased in consumption between 1980 and 1992).

The failure to emphasise a corresponding increase in carbohydrate-rich foods when fats are reduced was illustrated by the dietitians study (Cole-Hamilton *et al.*, 1986). In this study dietitians and their partners were given detailed, personalised advice on decreasing fat, sugar and alcohol intakes to meet NACNE guidelines. However, the failure to advise corresponding increases in intakes of bread, potatoes and other cereals meant that many people lost weight, which in the long term may have deterred them from continuing with this type of dietary regimen. Additionally this approach pays little need to the lay perspective of consumers and priorities within families (see section 5.6).

To achieve a practical understanding of the specific amounts of nutrients required and amounts of food to be eaten in a healthy diet, knowledge about quantitative amounts of nutrients or foods is required. In a recent study of barriers and incentives to fruit and vegetable consumption in Scotland (Anderson *et al.*, 1994b), 82% of the respondents said they believed that fruit and vegetables were associated with health. However, amongst respondents with low fruit and vegetable intakes (< 2 portions per day) 55% said they felt they were eating the right amount of vegetables and 24% the right amount of fruit. In addition, this study reported that 81% of Scots said that being better informed on the number of fruit and vegetable portions which should be eaten would encourage consumption and 24% thought portion size information would be useful. Hence it appears, in this instance, that the consumer knows what nutrition information they want and recognises that this information is difficult to find.

The MORI survey (1992) reports that four in five respondents agree that much food and diet advice goes against what they were told when they were young (and this includes 20- to 30-year-olds). Examples of this sort of advice include the concept that 'sugar is needed for energy' and 'bread and potatoes are fattening'. An additional problem which the consumer has to face is the confusion over dietary advice. For example, the MORI report *Attitudes Toward Healthier Eating* (1992) shows that 72%

of the British public think that 'experts never agree what foods are good for you' compared with 16% who do not think this. This point is illustrated further by Rudat (1993) who quotes a respondent in this survey as saying: 'I think the experts keep changing their mind. An expert is an expert because he thinks he is; until someone comes along and says "I've got a better idea than you".' In conclusion Rudat suggests that the general public are 'aware of buzz words' but have 'little input from informed sources', 'lack actionable knowledge' and are 'sceptical of media portrayal' of dietary issues.

Perhaps it is a combination of some these issues concerning quantities, confusing messages and conflict with tradition which leads to complacency in dietary change. In a recent project on barrier and incentives to dietary change in Scotland (Anderson *et al.*, 1994b), 60% of Scots said that they did not wish to change their current diet. It appears that dietary beliefs, values and overall attitude against changing eating habits conspire to produce dietary inaction within Scottish culture. Similarly, in the MORI survey (1992) half of the respondents said they were used to the food they eat and did not want to change. Additionally, 42% agreed with the statement: 'I tend to ignore most food and diet advice and eat what I like'.

Motivating dietary change is neither straightforward nor instant, but addressing some of the existing beliefs and making current consensus on healthy eating more widely known and clearly understood may help remove some of the barriers to dietary change. It seems that increasing knowledge about nutrition may have little impact on improving people's understanding about current dietary recommendation unless specific information on all foods are available – such as would be achieved by nutrition labelling.

5.4 Nutrition labelling and nutritional claims

With the evolution of the food industry in the present century has come a vast increase in the range of available food products. In parallel, there has been increasing recognition within the food industry that the nutritional content of the diet, and thus of component foods, is important to consumers and therefore of value for marketing.

Historically, there have always been certain choices to be made by consumers – usually by the purchaser or gatherer (beef or mutton, cabbage or carrots, apples or pears). These choices were heavily influenced by availability as well as cost. Consumers now have to make decisions about brand or variety for virtually every item of food purchased. Large supermarkets may stock 30 different brands of breakfast cereal, 5–6 different types of milk to go with it (thus 180 different combinations). A cheese sandwich for lunch can involve deciding between 20–30 different types of bread or roll, 12 different types of cheddar cheese and up to 30 different types of spreading fat. Ready prepared foods or complete meals in tins and packets, e.g. a deep-frozen chicken pie for dinner, come with an almost infinite range of specifications. Many homes now routinely stock several varieties of common commodities and family members often use specific brands for health reasons, including slimming. Health claims such as 'low fat' can be powerful determinants of brand selection. But how can consumers make the best decisions?

There seem to be at least two approaches towards helping consumers. These approaches unfortunately have tended to become confused. The first is to educate and guide choices between broad 'food groups' which have general characteristics and which can be considered an equivalent in terms of function or role in the diet and in nutritional terms. Thus all breakfast cereals, bread, pasta and rice are made from seeds of cultivated grasses with similar contents (relatively high in protein and carbohydrates, dietary fibre, vitamin E, magnesium, etc.). These foods are also low in saturated fats and can be encouraged in one form or another to become the main part (i.e. most calories) for each meal of the day. On average, if a range of products is used, this nutrient contribution is fairly predictable.

There is still enormous mileage in the food group approach for dietary education, as evidenced by the current campaigns to eat five portions of fruit and vegetables daily which has been used in the USA, Australia and the UK, the Department of Health *National Food Selection Guide* (Department of Health, 1994a) and the new 'Plate Model', developed to show how a healthy diet can be created by simply placing foods in the correct positions on a standard plate (Figure 5.2) (Armstrong and Lean,

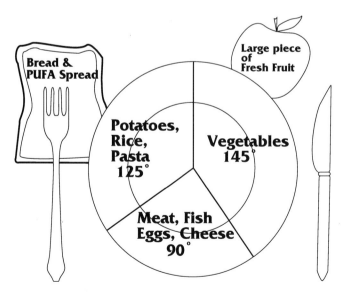

Figure 5.2 'Plate Model' for creating a healthy diet. (*Source:* Armstrong and Lean, 1993)

1993). However, problems immediately arise when it is realised that amongst the huge number of choices between individual foods, there are combinations which introduce undesirable nutritional consequences and consumers need to be able to select combinations which are desirable. For example, it would be very easy to eat our day's menu from the right combination of food groups, get the calories, fats, carbohydrates and protein right, but run into difficulties with sodium intake or saturated fat. Commercially prepared breads, butters, margarines, tinned and pre-packed foods all tend to be high (quite unnecessarily so) in salt, and pre-packed ready-to-eat meals are often high in saturated fats. The recently published COMA report, *Nutritional Aspects of Cardiovascular Disease* (Department of Health, 1994b), has attempted to describe recommended nutrients and food for groups, but individuals may still be easily confused. For these reasons, there is a need for consumers to have some guidance about which brands or varieties of foods to choose. Nutritional labelling is therefore the second angle on health promotion.

Many foods already have data on their composition detailed on the packaging. This is often used to support product marketing and promote a food as high in something (e.g. fibre) or low in something (e.g. fat). The information provided is expressed in terms of chemical composition (g per 100 g, or mg per 100 g), and sometimes as a proportion of the Recommended Daily Amount (RDA) of a nutrient (usually provided by 100 g of the food, or by a single unit such as one biscuit).

The need to be consistent within Europe, together with the problem of labelling on imported foods, have led to a new EC Nutrition Labelling

Table 5.2 Standard compositional data (g/100 g)[*]

	per 100 g	per serving (142 g)
Energy (kJ)	1120	1635
(kcal)	267	390
Protein (g)	9.8	14.0
Carbohydrate (g) of which	25.3	37.0
Sugars (g)	0.2	0.3
Fat (g) of which	13.7	20.0
Saturated fats (g)	5.5	7.8
Dietary fibre (g)	0.3	0.4
Sodium (g)	0.5	0.7

[*]The nutrients listed here are energy (calories) and the 'Big 7' nutrients required by law. The information is prepared in the chemical (g/100 g) terms used by food scientists, which are rather different from the terms used by nutritionists. The standard compositional data do not declare the amount of water in the food. (*Source:* Black and Rayner (1992)).

Directive which came into force in 1994. This requires standard chemical compositional data (per 100 g of food) for a limited number of nutrients (see Table 5.2) to be presented on foods on a voluntary basis, or when some nutrition or health claim (e.g. low in, high in) is being made. This form of food labelling is convenient for producers and manufacturers and familiar to the committees of the Ministry of Agriculture, Fisheries and Food (MAFF) in the UK. It allows very precise analysis and comparisons to be made by food scientists, but from the point of view of nutritionists, and of consumers, it has serious shortcomings.

Perhaps the simplest, but most pervasive, problem is that the current analytical labelling of food is expressed per 100 g weight of food. To work out what a meal contains, the consumer must weigh the food (or know what 100 g looks like). More seriously, it is not possible to make nutritional comparisons between foods or to decide in nutritional terms whether the food is higher or lower in a nutrient if the food contains an unknown and possibly variable amount of water. For nutritional comparisons, the nutrients have to be expressed per unit of nutrition (e.g. as % calories or per 1000 calories). It is possible, with knowledge of 'Atwater factors' (the calories per gram of different nutrients) and a calculator, to work out the nutritional compositions of foods from the information required by this EC Labelling Directive, but it is cumbersome. Well-conducted research from the Coronary Prevention Group (CPG) and Co-Operative Wholesale Society (CWS) has found that the average consumer is unable to understand or use the 'nutritional' information currently provided on foods (Black and Rayner, 1992).

Various attempts have been made to express the current information in a more comprehensive user-friendly format. Graphical representations are not currently permitted under the EC Directive, but will be considered in 1998. However, as argued above, even if the average consumer could grasp the current information with better presentation, a degree in nutrition would be needed to relate the chemical composition to health education advice on dietary recommendations.

One alternative to the 'per 100 g' criterion for nutritional labelling is to base the analytical nutrient content of a food on an average portion size. This approach, which has been adopted particularly in the USA, has the attraction that consumers can relate more easily to 'a portion'[1] than to 100 g. It is easier to add up nutrient consumed in a day in this way. Many nutrients, e.g. vitamins on breakfast cereal packets, are already described in these terms. Although, on the face of it, average portion sizes are attractive, they can be very misleading, and this approach is not suitable for a standard system of nutritional labelling.

[1]In the USA 'cupfuls' are frequently used as a unit measure, but 'portions' are still the main descriptor. It is interesting that even for foods such as fruit and vegetable a 'serving' is still described in terms of cupfuls (an average serving being equal to a half cup for most fresh or cooked vegetables).

The most serious shortfall of average portion sizes is of course that no individual actually eats the average portion size. We do not eat the same amount each time, or at different meals. There are major regional differences, and individuals in high risk groups for diet-related diseases often have very different portion sizes for socio-economic and cultural reasons. Finally, since our health education advice is attempting specifically to change people's portion sizes, it would be a nightmare to base nutrition labelling of foods on some statistical or hypothetical average portion size.

A modification of the average portion size system proposed from time to time is to base nutritional labelling on the contribution of a portion to overall nutrient requirements of individuals. As well as the problem over portion sizes, this approach would have to take account of the enormous variation in total requirements according to age, size, sex, activity, health, etc. This is a good exercise in nutritional science but inappropriate for food labelling.

The signs are now more optimistic that the UK Department of Health and Ministry of Agriculture, Fisheries and Food may be working more closely together, with the *Health of the Nation* (Secretary of State for Health, 1992) objectives which should represent common aims. It is hoped that health considerations may similarly begin to influence Food and Agriculture Divisions at a European level.

A system for nutritional labelling of foods which is consistent for all foods, links with health recommendations and forms a basis on which nutritional claims can be met was developed by the Coronary Prevention Group (CPG), and has been piloted by the Co-Operative Society (CWS) on its own brand products. This system (Table 5.3) does not represent the final word in nutritional labelling but it does offer consumers an approach which research has shown to be more accessible, better suited

Table 5.3 Nutrition bandings

Nutrient	Dietary target	Low	Med–low	Med–high	High
Protein	12.5% energy	<6.25	6.25–12.49	12.5–18.75	>18.75
Carbohydrate	57.5% energy	<28.75	28.75–57.49	57.5–86.25	>86.25
Sugars[1]	12% energy	<6.0	6.0–11.9	12.0–18.0	>18.0
Fat	30% energy	<15.0	15.0–29.9	30.0–45.0	>45.0
Saturated fats[2]	10% energy	<5.0	5.0–9.9	10.0–15.0	>15.0
Dietary fibre	30 g/10 MJ	<15.0	15.0–29.9	30.0–45.0	>45.0
Sodium	2 g/10 MJ	<1.0	1.0–1.9	2.0–3.0	>3.0

[1]Energy from sugar is part of the total energy from carbohydrate.
[2]Energy from saturated fat is part of the total energy from fat.
Source: Black and Rayner (1992).

to their needs in trying to choose foods which will be good for them, and helpful in evaluating health claims.

The basis of this system is that for major nutrients (and indeed for many minor nutrients) there are now well-established recommendations for intake. These recommendations represent the views of an overwhelming consensus of nutritional scientists and are extraordinarily consistent internationally. The recommended amounts are not intended to be followed slavishly by individuals, but represent goals for average intakes in whole populations. They represent a benchmark for assessment of the nutrients in different diets on meals, and then offer a point of reference for the nutrient contents of foods.

Nutritional recommendations for diets are expressed on an energy basis (per 1000 kcal or on % total calories) and this was chosen by CPG as a way to express the nutrients in foods (the CPG system, for example, would identify dried and cooked pasta as the same food – whereas the EC-MAFF g/100 g system would show radically different contents). Foods which have nutrient contents similar to the dietary recommendation or targets can be termed 'medium' in content. Arbitrarily, a content over 50% greater than the dietary recommendations is termed 'high' in that nutrient and a content below 50% under the dietary recommendation is 'low'.

This information is proposed as supplementary nutritional information which can be usefully added to the chemical data (g per 100 g) required by existing law. Variations are possible on this CPG system which use Dietary Reference Values (DRVs) (Department of Health, 1991), expressed per 10 MJ, as the target figure – and this system could be used for all the micro nutrients as well as the customary 'Big Seven'[2] nutrients. The CPG approach lends itself also to visual expression on bar charts or by stars to indicate the value of a particular food.

The CPG system for supplementary nutritional labelling has obvious educational value at a time when nutrition and health are of major public and political concern. It also offers a very simple objective system on which to base nutritional claims on foods. At present claims such as 'low in fat' on food labels can be made simply by comparison with some other (often unspecified) food which is even higher in fat. Thus certain brands of fromage frais have been labelled 'low fat' by comparison with cheddar cheese. Similarly, consumers certainly need to know that reduced fat cheddar cheese has less fat than conventional cheddar, but to label it 'low fat' demands some objective point of reference. Dietary recommendations, as employed by this CPG system, provide this basis.

---

[2]'Big Seven' nutrients are carbohydrate, protein, sugars, fat, saturated fat, dietary fibre, sodium.

To achieve a healthy diet the concerned consumer needs not only to understand nutrition but also the relevance of other food and health issues such as salmonella food poisoning, BSE, microwaves, irradiation, additives, colourings, pesticides and contaminants. All of these issues mean that food manufacturers have to address health seriously and this was illustrated by the fact that in the first half of 1989 40% of all new food products launched in the UK made one or more health claims. Food manufacturers now offer low-fat foods, low-salt products, salt substitutes, low alcohol drinks, as well as spreads high in polyunsaturates, artificial sweeteners, sugar-free products, and products with dietary fibre added.

Because of this concern with health, the response of the food industry means that they must develop market strategies which show their product in a 'healthful' light or develop new products which are genuinely helpful for consumers who wish to follow current dietary guidelines. A good example of how health claims have helped develop a range of food products but confuse the consumer is provided by the case of spreading fats.

In Europe during the 1960s butter was generally perceived as tasting good and margarine was perceived as a cheap substitute. In 1964 Flora margarine was introduced in the UK and promoted as high in polyunsaturates, and thus useful for preventing CHD. In 1968 Outline low-fat spread was introduced and promoted for slimmers (who wish to cut calories) and those concerned about heart disease (who wish to cut total fat). In 1978 St. Ivel Gold appeared and was promoted on health and taste. This was rapidly followed by a range of spreads such as Delight and Golden Churn which suggested a butter-like flavour but also had 'health' advantages. In 1983 Gold unsalted appeared and was thus promoted for those concerned about hypertension (who wished to cut down on salt). More recently 'Olio' has been developed from olive oil for those concerned about heart disease (who wish to increase their monounsaturated fat intake). It is widely believed that many households will in fact purchase more than one type of spread to meet the varying health demands and taste preferences within a family, therefore increasing overall sales of spreading fats. It is notable that, with respect to spreading fats, the most 'healthful' actions would in fact be to reduce total amount of fat spread (of whatever type), to stop using spread altogether or to use oils as a substitute, none of which will bring increases in sales to the margarine and butter industries. All of these actions would require a major education campaign to be effective.

However, in this market there are a number of margarine products being promoted on the basis of their low cholesterol content. Yet, in the UK there is no recommendation for intake of dietary cholesterol in the general population (this is because the main source of cholesterol is in fact from within the body).

5.5
Educating for healthy eating in the next century – coping with the health market

The annual expenditure on slimming foods[3] was estimated at £57.6 million in 1992 (Mintel, 1993) and growing. Dietary food supplements such as meal replacement have been around for a decade. Whilst it might be argued that slimming foods would be a self-limiting market (i.e. as people lost weight the demand for such products would diminish), it is clear that there is a continued market growth for these products.

Dietary supplements are also becoming increasingly marketed as an essential health product or as 'insurance' during periods of low food intake (e.g. slimming, children's food fads). The recent case of vitamins and intelligence showed how vulnerable the consumer is to apparent health claims. The Consumers Association (1992) recently reviewed the multivitamins market and noted that these were promoted in three main areas: when suffering from emotional stress; owing to extra physical activity (e.g. as in sportspeople); and to improve children's IQ. None of these areas have ever been scientifically shown to improve from vitamin supplementation. They also noted that a wide range of unnecessary ingredients are added to vitamin preparations, including other vitamins (apart from those stated on the label), minerals, artificial colourings and flavourings.

The *Dietary and Nutritional Survey of British Adults* (Gregory *et al.*, 1990) reported that average intakes of vitamins (from food) were well above recommended dietary allowances for all the age groups studied, yet within that sample 17% of women and 9% of men took dietary supplements. These supplements included multivitamins, cod and halibut liver oils, and vitamin C and B complex vitamins. Interestingly, informants who took dietary supplements had higher intakes of vitamins from food than informants who did not take supplements. It is clear that a large number of consumers are clearly attracted by taking extra vitamins in food (often for added value) and through supplements. It seems that a nutrient-dense diet which is high in vitamins and minerals and low in fat and sugar (as opposed to the nutrient-dilute or calorie-dense diet) needs to be promoted for all its dietary qualities and not just macro nutrients.

It is clear that the cost of 'healthful' versions of basic foods is higher (e.g. reduced-sugar products) and for some people the only route to a healthy diet is 'healthful' manufactured foods. In this instance the additional cost is likely to prohibit the adoption of a more healthy diet. These 'healthful' alternatives are not only promoted as more healthy but as an attractive, exciting and convenient part of modern lifestyles. Advertising is undoubtedly contributing to the successful promotion of certain products.[4] In the words of Tim Lefroy (chairman of The Alliance agency and member of the IPA advertising effectiveness committee): 'There is a close

[3]The slimming business is not only about food but is also about magazines. The three bimonthly slimming magazines had a combined circulation of 466 000 in 1994 (NTC).
[4]An advert for milk won the 1992 Award for Advertising Effectiveness.

correlation between the adverts popular with the public and those which are effective' (*Independent*, 1992). Adverts are also perceived as a source of information about nutrition. In the MORI survey (1992), 43% of respondents said they used television adverts as a source of nutrition information, 32% used adverts in women's magazines and 28% adverts in newspapers or magazines. An additional 15% said they derived nutrition information from materials produced by food companies. However, at the acquisition stage, external factors such as market prices, distribution, etc. have a major impact on what consumers choose to eat.

5.6 Food choice, eating and the lay perspective on health

Discussion of the consumer's perspective on domestic food provisioning has to take into account the wider socio-cultural context in which the 'consumer' exists. It is perhaps helpful, therefore, to acknowledge that the use of the term 'consumer' reflects only one aspect of any individual's behavioural and motivational repertoire. In this section therefore the value of understanding food choice and eating behaviour within one such wider context, that of the 'lay perspective' on health and illness, is suggested.

All societies have often complex belief systems about how the properties of various foodstuffs relate to the physiological and emotional health of those who consume them (Fieldhouse, 1986). Examination of such systems may be central to anthropological studies. Thus, for example, many Latin American cultures distinguish between 'hot' and 'cold' foods. This categorisation relates to basic beliefs that health is a temperate condition, and that disease results from an imbalance between 'hot' and 'cold'. Foodstuffs defined as 'hot' or 'cold' may play a major role in avoiding such health damaging imbalance. For example, menstruation is regarded as a 'warm' state, when foods defined as 'cold' should be avoided as they may cause cramps (Moloney, 1975). Similarly, in western society there is currently a distinction between 'junk' and 'real' foods. This has led, in one instance, to an ongoing debate about how the supposed lack of vital nutrients, or the addition of many artificial ingredients, in 'junk' food may lead to various levels of behavioural or emotional disturbance in young people.

These examples illustrate that debates about the relationship of food and eating behaviour to health and illness are common currency in all cultures. In western societies, however, the dominant discourse has been biomedical; and, particularly during the past decade, this discourse has increasingly focused on the identification of properties of certain foodstuffs as potential risk factors for illness and disease. As is discussed elsewhere in this chapter, marketing messages have played their own part in this discourse, either challenging or supporting the biomedical directives. Furthermore, public attention has been directed towards official reports on 'good diet' and the consequent exhortations to individuals to make changes in what they eat and drink in accordance with nutritionally sound directives (HEC, 1983; Scottish Office, 1993). However, even within the biomedical discourse the precise relationship between diet and health remains controversial (Research Unit in Health and Behavioural Change, 1989). Critics have also indicated that some of the dietary advice may be inappropriate for certain groups in the population because of its financial implications (Lang, 1984), its sexist assumptions (Hunt, 1985; Kerr and Charles, 1986) and its lack of appreciation that eating is a culturally based social experience (Fieldhouse, 1986).

Although the biomedical discourse has been dominant in attempting to define the relationship between food, eating and health, social scientific studies have begun to document the importance of a strong lay discourse which, it has been suggested, could be seen as a lay epidemiology of health and illness (Davison, 1991). Examination of this lay perspective shows how other social and culturally determined priorities may compete with the dominant discourse to legitimate and influence everyday behaviour (Backett, 1990). Furthermore, studies of the lay perspective on health have demonstrated the importance of seeing how aspects of everyday living provide the framework for understanding health-relevant attitudes and behaviours, and, indeed, whether or not a concern for health is always the relevant factor in such behaviour! In particular, qualitative work has challenged the tendency of quantitative surveys to detach particular items of health knowledge and behaviour (for example food choice) from the realities of daily social life where they are experienced, tested out, and given meaning (Backett, 1990).

Although the lay perspective is clearly influenced by the scientific biomedical discourse, the gap between the two is regularly shown in practice by the so-called problem of non-compliance. This has been a long-standing concern of preventive medicine (Dean, 1984) and, more recently in the field of health education, has been summed up by the term 'unhealthy lifestyles'. For the purposes of this present chapter this conundrum of the lack of fit between dietary knowledge and attitudes, and actual behaviours, may be formulated as follows. If certain everyday and individually controllable behaviours such as eating junk or fatty foods, or having a diet lacking in fruit and vegetables, are identified as harmful by both lay and professionals, and given that good health is a valued concept, why is it that many people persist in choosing nutritionally deficient foods for themselves and their families?

Here it is useful to make an analytical distinction between what is rational behaviour and what is reasonable behaviour (Backett and Davison, 1992). The difference between the two hinges on the acknowledgement that socio-cultural factors play a part in health-relevant decisions and action. The concept of rational behaviour derives from a logical, almost mathematical, model. For example, it suggests that, if A and B are true then C is the only logical outcome. Applied to the healthy lifestyles model a rational approach would suggest, being simplistic for the sake of example, that if high levels of blood cholesterol are implicated in ischaemic heart disease, then everyone should cut down on saturated fats. By contrast, the concept of reasonable behaviour is culturally based and asserts that the assessment of the normality and meaning of an action is relative. It suggests that if A and B are true then action C is a logical outcome in cultural context X, but in cultural context Y, action D is more appropriate. Therefore, to continue the simplistic example, change in dietary behaviours relevant to ischaemic heart disease may appear to be perfectly reasonable with Mediterranean or Japanese cultures but not, perhaps, within Scotland (Anderson *et al.*, 1994b) or to particular groups of the population in other western countries.

Qualitative research in the UK has shown, therefore, that health-relevant behaviours are the outcome of processes which may very well draw on *parts* of scientific rationality. Indeed, most British studies, whether quantitative or qualitative, indicate that most people are well aware of epidemiologically identified health risks (Blaxter, 1990). However, it also appears that, overall, there is no automatic long-term translation of knowledge of behavioural risks into modifications of personal lifestyle. For example, collaborative work on the data sets from three recent qualitative studies throughout the UK (Backett *et al.*, 1994) confirmed that most smokers, excessive drinkers, non-exercisers and 'bad' eaters (and, in one study, their children too) (Backett and Alexander, 1991) were fully aware that such behaviours were potentially health damaging. The studies also indicated that even those respondents who identified some personal hereditary illness risk factor varied considerably in making any of the behavioural changes, dietary or otherwise, which they also 'knew' might affect their chances of actually getting the illness. Few respondents, however, saw their lives as being guided by irrationality, and this was because potentially damaging behaviours were seen as 'reasonable' in a wider personal and social context.

To take a specific example, in a study of health in families, it was found that fulfilling perceived familiar obligations or achieving harmonious relationships were often accorded priority, or indeed defined as 'healthier' than, say, the achievement of physical fitness or the imposition of a 'healthy diet' on a reluctant family (Backett, 1992). This study also considered the social accomplishment of the 'family meal', which provided illustration of the many interactive factors underlying the usual 'data' about foods actually purchased and eaten. Leaving aside the practical constraints on meal preparation of economics, cooking ability and time availability, study respondents' accounts showed that the content of the 'family meal' was affected by the following factors: time schedules of different family members; attitudes towards appropriate socialisation of children; biographical experience of the eating preferences of family members; the importance attached to commensality as a forum for family communication; and pre- and postprandial activities of individuals and subgroups within the family. These data illustrate that behavioural outcomes are better understood as 'reasonable' responses to complex social situations, of which considerations of health and physiology are only one small part (McKenzie, 1986).

Another characteristic of the lay perspective on food choice, eating and health is confusion and scepticism about the validity of 'scientific knowledge'. The lay knowledge which is brought to bear on making sense of health-relevant issues is also much broader, and more multifaceted and pragmatic than is usually the case with the biomedically based information presented by official sources. For example, studies have shown that when considering the causes of good and bad health lay respondents regularly vacillated between ideas about individuals being able to control health and illness through personal behaviour and ideas about individuals being subject to factors over which they had little or no control. In this latter category, lay respondents drew on observation and experience of many factors such as heredity, upbringing, personal differences, factors in the social, natural and manmade environment, and forces such as luck, chance, randomness and personal destiny (Backett *et al.*, 1994). To be fair, such an appreciation of the complex interrelationships amongst causal and risk factors, and scepticism about the quality of scientific evidence (based in part on contradictory findings and conflicting advice in the media), echoes the uncertainties of many epidemiologists themselves.

What implications, then, did these characteristics of the lay perspective on health have for everyday decisions about food and eating? Carrying out 'reasonable' courses of action meant that it was acknowledged that, although supposedly health-damaging behaviours involved risks, they also involved benefits in terms of, for instance, wellbeing, social acceptability or pleasure. So-called risky behaviours might therefore be seen as 'life enhancing' even if they were not considered to be 'health enhancing'. Studies have indicated that, in practice, what constitutes a 'risk' may be viewed differently at different points in the life course. The important point is that, from the lay perspective, it was not the behaviours themselves which were necessarily viewed as reproachable, risky or unhealthy, but rather that they were viewed in that way in inappropriate socio-cultural contexts.

For example, lay respondents have indicated that young single adults were expected to be relatively free of ailments and that they would therefore not experience any great impetus to stick to official tenets of healthy eating. Moreover, the still-young body was defined, as in childhood, as being able to deal more efficiently with toxins and therefore to not show many ill effects of poor eating habits. Studies have indicated that, at this point in the life course, therefore, enjoyable though potentially health-damaging behaviours such as eating junk food, having irregular eating patterns or paying little attention to nutrition were variously viewed as 'life enhancing', or at least were tolerated since they were not expected to be permanent (Backett and Davison, 1992). Also, such behaviours were often seen to be balanced by parallel 'healthy' behaviours such as leading an active or sporty life. Thus, the process of evaluating risk from the lay perspective took place in a much broader landscape since each behaviour was assessed in terms of its socio-cultural and environmental context, not simply the physiological.

Furthermore, this lay process of risk assessment was found to be intimately connected with assessment of a repertoire of potential consequences. Studies have shown that, regardless of respondents' understanding of probability or long-term consequences of health-damaging behaviours, there was a strong tendency to pay attention to the short-term rather than long-term consequences (Backett et al., 1994). From the lay perspective, if a person looked all right, felt all right, was not suffering any immediate effects from being overweight, then there was less experienced pressure to change any personal health-relevant behaviour such as diet. Emphasis was also placed on the short-term disadvantages of certain so-called healthy behaviours. When deciding on reasonable courses of action, lay respondents took into account their observations that, for example, people may put on weight after stopping smoking, may have to have dietary supplements if they become vegetarian, and that children may opt to eat nothing at all rather than consume the amounts of vegetables which are currently regarded as necessary for good health.

From the lay perspective, making sense of the relationship between food, eating and health is a complex social accomplishment which is fraught with contradictions and confused messages. The main solution to this, as evidenced in both qualitative (Backett et al., 1994) and quantitative work (Blaxter, 1990), was to avoid excess and to aim for moderation and balancing out the 'good' and the 'bad' (Mullen, 1993). This was carried out both in general and specific terms. For instance, balance and health were connected in terms of general way of life. This meant that one part of life, or area of behaviour, should not become dominant at the expense of others. The ideal was for a satisfactory balance between, for example, work and family commitments, personal satisfaction and social obligations, having good health but not becoming obsessed with 'healthy living'.

Specifically, with regard to food and eating, lay respondents have regularly reported that they aimed to achieve a 'balanced diet' (Backett, 1992). A detailed analysis from one qualitative study indicated that this was not simply in terms of 'meat and two veg' but also, with varying degrees of scientific sophistication, in terms of their perception of a nutritionally balanced diet. As with many other studies, respondents were well able to speak in general terms about nutrition. However, within the context of daily life the achievement of a 'balanced diet' could be seen as an everyday social achievement. For example, respondents' reports of family eating patterns showed that this achievement of 'balance' involved interspersing so-called 'good' meals with 'bad' snacks and junk food; being aware of eating 'healthily' after a spell of inattention to dietary requirements; and trade-offs between foods perceived as 'junk' or 'convenience' with items considered to be nutritionally 'good'. It was also important to have a balanced view about eating and to see food as a pleasure as well as a bodily necessity.

A high public profile is currently given to the relationship between food choice, eating behaviour and health. This has been dominated by a biomedical discourse which has attempted to identify how the properties of certain foods may have implications for the development of illness and disease. However, as this section has demonstrated, a proper understanding of the links between knowledge, attitudes and practices necessitates the acknowledgement of the lay perspective. The preceding discussion of the lay perspective on health has highlighted how socio-cultural processes interact with health knowledge and attitudes to shape their translation into potential behaviour. It is clear, therefore, that from the lay perspective, the relationship between food choice, eating behaviour and health makes sense only against the background of daily social experience and the cultural meanings attached to health and illness.

5.7 Conclusions

In attempting to address why a good nutritional knowledge fails to result in the consumption of a healthy diet, it is clear that being well informed does not mean that people will act according to one particular set of beliefs in promoting health and preventing disease. Information may be disregarded, altered or even used to justify existing behaviour rather than stimulate behavioural change. The scope for improving existing nutritional knowledge (to provide a practical understanding of food choices required to achieve a healthy diet) is vast but dependent on all participants of the food network. Thus there is a general feeling amongst many nutritionist educators that nutritional knowledge functions only as a tool if and when individuals are ready to make changes (Parraga, 1990).

This is further illustrated by the lay beliefs of consumers, who show not only confusion over dietary issues but complacency and sometimes anger over mixed dietary advice which continually emphasises individual choice and lifestyle change. Priorities about food choice in real life are rarely made through dietary education and information but are determined by social and cultural norms and financial, physical and retail resources.

Ultimately, nutrition education is only one tool in assisting dietary change (which has to date been largely ineffective). Structural and collective approaches based on a government-directed food and health policy are long overdue to assist the members of the food network who strive for dietary change and the health of the nation. Opportunities abound for all parts of the food network, including producers, manufacturers and retailers – often labelled as the enemy by health educators.

References

Anderson, A. S. and Lean, M. E. J. (1987) Setting an example: food and health policy within the national health service. *Health Education Research*, **2**, 3, 275–285.

Anderson, A. S., Macintyre, S. and West, P. (1994a) Dietary patterns among adolescents in the west of Scotland. *Br. J. Nutr.*, **71**, 111–122.

Anderson, A. S., Lean, M. E. J., Foster, A. and Marshall, D. (1994b) The Chief Scientist Reports … Ripe for Change: fruit and vegetables in Scotland – current patterns and potential for change. *Health Bulletin*, **52**, 51–64.

Anderson, A. S., Umapathy, D., Palumbo, L. and Pearson, D. W. M. (1988) Nutrition knowledge assessed by a structured questionnaire in a group of medical inpatients. *J. Hum. Nutr. Diet.*, **1**, 39–46.

Armstrong, J. and Lean, M. E. J. (1993) The Plate Model for Dietary Education. *Proc. Nutr. Soc.* **52**, 1, 19A.

Backett, K. (1990) Studying health in Families: A Qualitative Approach. Chapter 3 in S. J. Cunningham-Burley and N. P. McKeganey (eds), *Readings in Medical Sociology*, Tavistock/Routledge, London and New York.

Backett, K. (1992) Taboos and excesses: lay health moralities in middle class families. *Sociology of Health and Illness*, **14**, 255–274.

Backett, K. and Alexander, H. (1991) Talking to young children about health: methods and findings. *Health Education Journal*, **60**, 434–438.

Backett, K. and Davison, C. (1992) Rational or reasonable? Perceptions of health at different stages of life. *Health Education Journal*, **51**, 2, 55–59.

Backett, K., Davison, C. and Mullen, K. (1994) Lay evaluation of health and healthy lifestyles: evidence from three studies. *British Journal of General Practice* **40**, June, 277–280.

Black, A. and Rayner, M. (1992) *Just Read the Label*, Coronary Prevention Group, HMSO, London.

Blaxter, M. (1990) *Health and Lifestyles*, Routledge, London.

Cannon, G. (1992) *Food and Health: the experts agree*, Consumers Association, London.

Cole-Hamilton, I., Gunner, K., Leverkus, C. and Starr, J. (1986) A study among dietitians and adult members of their households of the practicalities and implications of following proposed dietary guidelines for the U.K. *Hum. Nutr.: Appl. Nutr.*, **40A**, 365–389.

Consumers Association (1992) *Which multivitamins. Which way to Health*, August 1992, 128.

Davison, C. (1991) Lay epidemiology and the prevention paradox: the implications of coronary candidacy for health education. *Sociology of Health and Illness*, **13**, 1–19.

Dean, K. (1984) Influence of health beliefs on life styles: What do we know? *European Monographs in Health Education Research (127–150)*, Scottish Health Education Group, Edinburgh.

Department of Health (1991) Dietary Reference Values for Food energy and Nutrients for the United Kingdom. *Report on Health & Social Subjects No. 41*, HMSO, London.

Department of Health (1994a) *National Food Selection Guide*, HMSO, London.

Department of Health (1994b) *Nutritional Aspects of Cardiovascular Disease*, HMSO, London.

Department of Health and Social Security (1984) Diet and Cardiovascular Disease. *Report on health and Social Subjects 28*, HMSO, London.

Fieldhouse, P. (1986) *Food and Nutrition: Customs and Culture*, Croom Helm, London.

Gregory, J., Foster, K., Tyler, H. and Wiseman, M. (1990) *The Dietary and Nutritional Survey of British Adults*, HMSO, London.

Health Education Council–NACNE (1983) *Proposals for Nutrition Guidelines for Health Education in Britain*, Health Education Council, London.

Hunt, S. (1985) Below the breadline. *Community Outlook*, October, 19–21.

Hurren, C. and Black, A. (1991) *The Food Network*, Smith-Gordon, London.

Independent Newspaper, 19th November 1992.

Kerr, M. and Charles, N. (1986) Servers and providers: the distribution of food within the family. *Sociological Review*, **34**, 115–157.

Lang, T. (1984) *Jam Tomorrow?* Food Policy Unit, Manchester Polytechnic, Manchester.

McKenzie, J. (1986) An integrated approach – with special reference to the study of changing food habits in the United Kingdom. In C. Ritson, L. Gofton and J. McKenzie (eds), *The Food Consumer*, Wiley, London. pp. 155–170.

Marshall, D., Anderson, A., Lean, M. and Foster, A. (1994) Healthy Eating: Fruit and Vegetables in Scotland. *British Food Journal*, **96**, 7, 18–24.

Ministry of Agriculture, Fisheries and Food (1994) The British Diet: finding the Facts 1989–93. *Food Surveillance Paper*, MAFF.

Mintel (1993) *Slimming Foods Mintel Market Intelligence*, Mintel Publications, London.

Moloney, C. H. (1975) Systematic valence coding of Mexican 'hot–cold' food. *Ecology of Food Nutrition*, **4**, 67–74.

MORI (1992) *MORI research – attitudes to food, health and nutrition messages among consumers and health professionals*. MORI report.

Mullen, K. (1993) *A Healthy Balance: Glaswegian men talking about health, tobacco and alcohol*, Avebury, Aldershot.

NTC Publications (1993) *The Food Pocket Book*, NTC Publications Ltd, Henley-on-Thames.

NTC Publications (1994) *Marketing Pocket Book*, NTC Publication Ltd, Henley-on-Thames.

Parraga, I. M. (1990) Determinants of Food Consumption. *J. Am. Diet. Assoc.*, **90**, 5, 661–663.

Research Unit in Health and Behavioural Change (1989) *Changing the Public Health*, John Wiley and Sons, Chichester. Chapter 6.

Rudat, K. (1993) MORI research – attitudes to food, health and nutrition messages among consumers and health professionals. In *Getting the message across – nutrition and communication*, National Dairy Council.

Scottish Office (1993) *The Scottish Diet*. Report of a Working Party to the Chief Medical Officer, Scottish Office Home and Health Department, Edinburgh.

Secretary of State for Health (1992) *The Health of the Nation – A strategy for Health in England*, HMSO, London.

Senate Committee on Nutrition and Human Needs (1977) *Dietary goals for the United States*, US Government printing Office, Washington, DC.

Southgate D. A. T. (1992) Dietary advice: foods or nutrients. *Proc. Nutr. Soc.*, **51**, 47–53.

Stockley, L. (1991) Do the experts agree? A review of dietary goals and guidelines. In C. Hurren and A. Black (eds), *The Food Network*, Smith-Gordon, London.

USDA (1985) *Dietary Guidelines for Americans*, United States Department of Agriculture.

Winkler, J. (1991) The consumer's agenda for nutrition in the 1990s. In C. Hurren and A. Black (eds), *The Food Network*, Smith-Gordon, London.

World Health Organisation (1988) MONICA project. Geographical variation in the major risk factors of coronary heart disease in men and women aged 35–64. *Wld Hlth Statist. Quart.*, **41**, 115–140.

World Health Organisation (1990) Diet, nutrition and the prevention of chronic disease. *Technical report Series 797*, WHO, Geneva.

Preparation

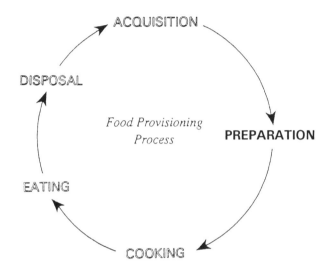

ACQUISITION

DISPOSAL

Food Provisioning Process

PREPARATION

EATING

COOKING

The omnivore's paradox 6
Nick Fiddes

As a member of the species *Homo sapiens sapiens*, I am an omnivore. This gives me considerable capacity for choice – for pleasure, or for survival. My fellow humans in the far North have lived on little but fish and other fruit of the sea; major populations have thrived on a diet of rice with minimal supplements to savour; yet others conjure a living from desert sands. Meanwhile, I myself enjoy the miracle – or is it mirage? – of a wider choice of foodstuffs than almost anyone who has ever lived.

It is dinner time. I consider my options: I could drive to the supermarket, to pace aisle after bewildering aisle of mouthwatering-looking packages, to select on a whim. I could drop in on my local shops, and pick up a few basics: those tasty new Egyptian potatoes, tomatoes from Holland, mushrooms, and an Israeli avocado for starters, perhaps, and maybe a pack of free-range bacon from the deli ... I should manage something with that lot.

Then again, since writing this is making me even hungrier, perhaps I should take the lazy option and get the Chinese take-away's excellent Chicken with Cashew Nuts ... or the pasta shop's Spirelli Christina? Or – since there's that chapter on the Omnivore's Paradox I have to make a start on tonight – maybe I should be really indulgent and phone out for pizza. It's up to me. I can have what I like. It's my choice.

Or is it? The choices available to me, and indeed my very propensity to value choice, are far from obvious, normal or natural. After all, why not economise by netting a couple of those pigeons that scuttle around my feet on the walk home? Or make a spicy ragout out of my neighbour's cat?

The answer, of course, is my socialisation – that conditioning from family, friends, teachers, peers, media and figures of authority which, over the course of my life's experiences, has taught me what is normal, natural, legal, decent, healthy and prestigious.

Having discounted at least two of my options, I am left to reflect on how quickly my choices are narrowed. Whenever I decide what nutritive (or even non-nutritive) substance to ingest, I do so in the context of a subtle set of influences and restrictions which govern my choice. This chapter will touch on a few of these factors. In so doing, it may shed light on why so many contemporary western consumers appear to be looking again at their own food choices.

But first it may help to establish a few principles. Even many leading nutritionists and other 'food professionals', let alone policy makers, remain inclined to ascribe explanations for human food habits to qualities inherent in the substances – or to putatively objective values such as the pursuit of health – rather than to the culturally defined political, environmental, and economic processes which condition and define these. Residual beliefs abound that good or bad 'tastes' are somehow absolute, inborn, or residing in particular foods. They must, however, be relinquished. Although human infants have an 'early and probably innate preference for sweet tastes', which may reflect the likelihood of the first denoting nutrition and the other possible poison, since 'most drugs and poisons have a bitter taste' (Cowart, 1981: 60), the variety of globally documented culinary likes and dislikes suggests that cultural factors condition taste far more than does flavour, or even safety.

How else should we explain the peculiar tastes that 'others' enjoy (be it grubs, chilli, or dog flesh), or should 'they' make sense of 'ours' (such as fungi, unfertilised hens' ova, or rotten and mouldy mammary secretions)? Farb and Armelagos (1980) suggest that we humans are unique in persisting, under cultural influence, with using an initially unpleasing substance until we enjoy it. And indeed, most of us can perhaps recall a food item which we 'grew into' liking. I personally used intensely to dislike the smell of coffee – until I tasted it. Even then, the first time, I did not actively enjoy it. I merely learned that it would not kill me, so that the next time I was again offered it in a situation where it would be socially prescribed as impolite to refuse, I could swallow it down to keep up appearances and not offend my host. Soon, I was addicted.

One unduly rationalistic interpretation of food choice is Paul Rozin's (e.g. 1976, 1982) description of a 'double bind' in which human beings find themselves, as omnivores – from which this chapter's title derives. Rozin argues that any new food represents both a potential danger and an opportunity for dietary diversification, and that this precipitates two contradictory urges: of neophilia (curiosity towards new foods, and aversion to monotony) and nephobia (culinary conservatism, or 'familiarity breeds content'). The human desire for variety is such that new foods can and must be sampled, although they might be repulsive or even dangerous. But this conflicts with the desire for familiarity or security, and these factors must be reconciled.

Biological advantages to both of these traits are indeed evident. Sticking to the tried-and-tested, for example, makes obvious sense in an environment of unknown risks. This is why rats are difficult to kill by poison (Rozin, 1976). (They are reluctant ever to consume unfamiliar foods; even if forced by hunger, rats will only initially sample new foods in small quantities, presumably to try for side-effects.) It is a mistake directly to equate the behaviour of any two species, not least rats and humans. But it seems reasonable that a precautionary principle must offer advantages to us, too.

Dietary adventurousness also offers biogenetic dividends. The diversity of potential nutrition from culinary Catholicism (and proclivity to boredom with sameness) represents a safeguard against shortage, with a motive continuously to identify alternative comestibles. It enables the species, or individual societies, to modify habits, such as under ecological stress, knowing that some subgroup has already beaten an experimental track. As the Irish potato famine of the 1840s showed all too starkly, a population that is overdependent upon a single staple is vulnerable to fluctuations of supply owing to climate, disease or politics (Woodham-Smith, 1987).

Biological factors clearly condition food choices. But transcending our biological conditioning can seem to be one of the human animal's few consistent characteristics. So, just as it would probably be unwise to explain an individual's preference for the music of Nirvana over Beethoven by the physical evolution of the human ear, to focus on the nutritional component of food choices is usually far from adequate. Many other considerations adhere, and anthropological experience consistently suggests that we are far more likely to tailor 'facts' to suit pressing social imperatives, than to allow ourselves to be moulded by mere materialism. Culture develops under its own momentum and with its own internal logic which may, or may not, include rationalistic concepts of nutrition, purity, or convenience. Our culturally constructed sense of taste empowers us to select from the range of ingestible substances available. But it subsumes and sometimes defies physiological reasoning.

It also belies the title of omnivore. We eat far more narrowly than that name implies, for we only perceive as food that which we deem food. Meat provides one example. As Vialles demonstrates for the culturally mediated transition of (certain parts of) certain animals into meat, we deploy a complex symbolic logic to arrive at the range of potential species we exclude from consideration, which greatly outnumber those which we will accept:

> We do not, after all, eat just any animals. Exclusive carnivores, particularly carrion-eaters, are not on the list. Aquatic animals are not looked upon as furnishing meat. Birds and rabbits do not fall within the category of so-called 'butcher's meat'. The flesh of animals seems in fact to be hierarchised according to the particular animal's habitat. Those that live in or beside water are not meat at all; those that can fly or live above or below ground appear to constitute an intermediate category; 'true' meat comes from four-footed animals, exclusive herbivores, domesticated ruminants – in other words, the animals with which in the most tangible sense we live on an equal footing, as it were, the animals of which we are able to consume the milk of reproductive females, or the flesh, but not the blood, of specimens that have been fattened up. (Vialles, 1994: p. 128)

In any case, Rozin's 'Omnivore's Paradox', too, reflects much more than biological imperatives. We are not merely digestive tracts which absorb at one end and excrete at the other. Nor are we rational economic decision makers, maximising our utility like a clutch of utilitarian economists. By choosing some substances and refusing others, we deploy consumption as an expressive process by which to define who we are, no less. A willingness to experiment, whilst retaining distinctive patterns in what we find acceptable to eat, is in no small measure, what allows us to be human. From the early spice trade, to the colonisation of the New World, to development of the contemporary European Union, to a vibrant market and spice bazaar in the Sahel (Stoller, 1989), the exchange of food has provided an incentive and framework for economic activity as well as for much of what we recognise as culture. Without the yearning to 'live to eat' as well as to 'eat to live', our time passed on this planet would be immeasurably poorer.

Most of our daily acts are communicative at least as much as they are materialistic, and eating more than most. We make full use of the capacity of foods to say things for us, to us, and about us. In Lévi-Strauss's (1970) memorable phrase, foods must be 'not only good[s] to eat, but also good[s] to think [with]'. We choose foods for more than their nutritional qualities, which may at least partly reflect the fact that eating:

- is one of the few means by which we permit transgression of the bodily membrane that contains us as discrete individuals, separate from the outside world;
- incorporates part of the animal, vegetable, and mineral universe into our own substance;
- keeps us alive.

Any one of these might make eating an exceptional activity. Taken together, it would be astonishing if food and its consumption were not profuse with symbolism that fed cultural discourse, the world over. And so it is. Eating together is the standard means by which hospitality is expressed, and community shared. Around the gathering, preparation and consumption of food, daily, annual and perennial rhythms are organised. Eating and food provision is near the centre of almost any celebration, such as when we enjoy a Carnival (from Carne = 'meat') or Fiesta (a feast), and in Harvest or Thanksgiving, for example, we devote entire festivals to their honour. It would be surprising to encounter any language not littered with testimony to food's wider cosmological significance.

But food is not just an aggregate compound. 'Food' is a malleable category comprising countless endlessly manipulable substances and traits. So intricate is its allusiveness, indeed, that in attempting to convey the complexity of iconography involved in Amazonian mythology pertaining to food, table manners, fire and other subjects, Lévi-Strauss was drawn to an extended musical analogy (e.g. 1970, 1973, 1978). The metaphor is appropriate, since infinitely patterned variations can be played on food's many themes (and a skilful chef, like a great composer, can orchestrate a gloriously stirring culinary symphony). In the western world, these themes include:

- biochemical categories (e.g. fats, vitamins, proteins);
- nutritional categories (e.g. fattening, body-building, healthy, poisonous);
- freshness categories (e.g. unripe, fresh, stale);
- transformative categories (e.g. raw, cooked, rotten);
- culinary categories (e.g. roast, marinated, raw);
- event categories (e.g. main meal, course, snack, mouthful);
- sensory categories (e.g. crunchy, subtle, pungent, filling);
- ethnic or geographical categories (e.g. French, sushi, Hawaiian);
- ethical categories (e.g. 'cruel' battery chicken, 'kinder' free-range eggs, 'fair trade' produce);
- emotional categories (e.g. comfort foods, pick-me-ups, aphrodisiacs);
- social categories (e.g. celebratory, hospitality, traditional, exotic, fashionable, women's, men's, children's).

This list is far from sufficient. It only alludes to the menu of analytical frameworks available to an individual in describing, and selecting, foodstuffs. A master chef will emphasise some, which would not be precisely the same as those of a hospital caterer. An office worker in a hurry for lunch would have different priorities. And an historian, picking over the scraps at the banquets of long-dead dynasties, would surely have other interests again. None is more true than any other.

6.2 The cultural meanings of foods

Every society has its own classificatory categories, many of which differ from 'ours': heating foods and cooling foods, yin and yang, sacred, profane, inauspicious. And, of course, even within – let alone beyond – the western orbit, new themes are perpetually encountered, or being invented, as familiar ones become redundant. Few in the west today would pay much heed to how acceptable a particular food was to their head of state's palate (whatever American broccoli-growers might have thought when George Bush revealed his dislike of that vegetable), although that distinction could once have been a decisive.

Over time, particular ingestibles develop their own socio-culinary identities. Meat, for example, derives much of its fabled nutritional and ideological potency from its ideal suitability as a culturally constructed statement of our capacity to control the unruly natural world (Fiddes, 1991). Even in France, the historic growth of meat consumption has been slowing, and this trend does not seem entirely attributable to economic factors. According to Fischler, public criticisms of meat there are rarely constructed in explicit or learned nutritional terms, but rather:

meat was perceived as a somehow special food, overconsumption of which was seen as self-indulgence or conspicuous consumption. The superfluous, it was felt, could easily become harmful, if not immoral, as was expressed in the oft cited: 'You don't really need meat at every meal.' (Fischler, 1986: 952)

The forces that have been moving this change come not simply from the deliberations of hard science, but from a public mood that is responding to a complex cultural dynamic in which science is but one factor. The rest is bound up with the food's social identity, which is modulated by people going about their daily lives – and subtly continues to develop with each and every transaction. Every recognised food has its own image, which will be only partially related to nutritional or sensory qualities. In much of the western world, our daily bread even today retains its traditional image as the honest basis of all eating, although brown bread has recently come to displace white's historic prestige, as technology has made the refined substance non-exclusive, and as fears of global and local ecological crisis have brought imagery of naturalness to rival or even eclipse that of modernity in popular appeal. 'Natural' foodstuffs, which formerly had to be 'tamed' by processes of acculturation to be fit for the table, have become the stuff of fashion, as industry has fallen from grace:

> Refined foods, which we were craving for until recently, are now rejected for nutritional (white sugar = 'empty calories') and/or symbolic reasons (refined foods are 'artificial', or 'dead'). White, the triumphant colour of the sixties (white sugar, white bread, white veal, white laboratory-like kitchens, white blouses in the supermarkets, etc.) is no longer cherished. The time has come for brown bread, brown sugar, grey flour, pink veal. 'Nouvelle Cuisine' prizes the raw over the cooked, the art of *selecting* highest quality foodstuffs over that of *processing* it. (Fischler, 1980: 946, italics in original)

Water (Illich, 1986) and salt (Visser, 1986) enjoy mythologies which have long made them much more than mere foods or drinks. On a shorter timescale, a single decade saw kiwi fruit cycle from unknown exotic, to fashionable garnish, to yuppie cliché, to mundane fruit. Caviar is sophisticated. Lentils are wholesome.

The meanings of particular foods, thus, are neither immutable nor homogenous. The broad-brush caricatures painted above do little justice to a social process of infinite complexity. The language of food is a discourse in which each of us enters as an individual, but whose grammar we learn from our parents and peers. And, like any symbol, what we eat can say very different things for us, and about us, at different times, and within different communities.

Beef, for example, has a singularly rich heritage of signification in the western world. The roast beef of Olde Englande and the American hamburger both testify to the identities which each nation is wont to project. It also records the traditional status of cattle flesh as the most prestigious of meats locally. Even in these areas, however, beef's reputation will vary widely. For example, in a growing and increasingly vociferous subgroup in each country, beef is more likely to evoke feelings of outrage at the 'inhumanity' of rearing sensitive creatures for slaughter and ingestion, or perhaps hostility towards a hamburger chain's allegedly venal ecological or employment practices. Most Hindus would regard consumption of the 'sacred cow' with not only distaste but abhorrence. And in the Shetland Islands of Scotland, meat from locally reared sheep is the traditional favourite, not beef (Fiddes, 1990).

No two individuals will conceive of any foodstuff in exactly the same way, owing to their own personal histories. Nonetheless, commonalities of 'taste' and 'tradition' develop and persist within families, localities, religions, ethnic groups, political affiliations, secular memberships, sporting societies or philosophical orientations. Indeed, most social groupings, at whatever scale, probably have characteristic food tastes. Whether we call it habit, familiarity or tradition, the power of that which we have eaten before to keep us consuming it in the future should not be under-estimated. By repeated exposure to 'right' foods, and admonishments with regard to 'wrong', communities teach their members what to enjoy, and what not to. In just this way, most young westerners have, until recently, been reared to regard meat as not only an acceptable food, but a special, manly, strength-endowing, prestigious, adult, complexly mythopoeic item. The Japanese have been learning to love it as 'western', and so preferable to their traditionally more fish-based cuisine. But most Hindus continue to avoid beef. Muslims and Jews are taught to abhor pork. Gypsies regard hedgehog as supremely edible, since its spikes guard against pollution, as well as for a host of other reasons (Okely, 1983: p. 101). Catholics (officially) eat fish on Fridays. And British environmental activists circulate in a society in which not only meat, but all animal products, are widely disparaged. Some Indians eat sweets at the beginning of the meal; the French before the final cheese course; and the British after cheese.

But likes and dislike develop by more than random whimsy. We consume food in patterned ways, which carry meanings that are consistent within their own cultural frame of reference, and which can be 'read' to reveal cultural trends as well as archetypes. In her classic essay *Deciphering a Meal*, Mary Douglas argued that if 'food is treated as a code, the messages it encodes will be found in the pattern of social relations being expressed' (Douglas, 1975: p. 249). A liking for familiar foods is not just a biological, or even a socially developed, preference for the familiar. It is also a strong statement of continued adherence to whatever shared values the particular food items, preparation methods and contexts of consumption are, tacitly at least, agreed to represent. As one conspiratorially inclined American sociologist puts it:

> In studies of social movement and the formation of sects and dissident groups, the role of food cannot be underestimated. In adhering to some dietary rules, what to eat, when to eat, or when not to eat, groups maintain control over their members. They also require members to deviate from the general population when they venture outside their group. This behavior is one of the most effective ways of assuring adherence to special group codes. (Back, 1977: p. 31–32)

He is correct, except in ascribing this only to groups with which he, personally, prefers not to identify. The obtaining and sharing of socially sanctioned food has long been recognised as a ubiquitously eloquent statement of shared ideology that expresses affiliation and solidarity, within dominant as well as dissident cultures. Over a century ago, W. Robertson Smith observed that 'those who eat and drink together are by this very act tied to one another by a bond of friendship and mutual obligation' (1889: p. 247). Radcliffe-Brown thought the getting of food 'by far the most important social activity' for Andaman Islanders (1922: p. 227). And Darlington (1969) proposed commensality as the most important basis of human associations. And food, said Barthes (1975), is a system of communication, a body of images, a protocol of usages, situations and behaviour. What these writers share is an appreciation of the central function that eating performs in any community's discourse and self-definition.

6.3
Continuity and change

In his 1978 work, Rozin suggests that every culture in the world has its own characteristic flavour principles (such as curry, garlic, lemon grass, chilli or oregano), repeatedly occurring in basic food dishes; he suggests these may help to reassure by providing familiarity, and so promote acceptance of new foods. Twentieth century western society has been no exception. The products which supplied the modern supermarket were routinely constructed from fillers, fats, sugars, colorants and MSG: a list of regulation ingredients that provided a reassuringly familiar basis on which to engineer a few unthreatening variations which avowedly evoked the cuisines of the world, the fruits of nature or whatever the latest marketing stratagem may have been. Standardised fabrication ensured that the safely bland uniformity which characterised so much industrial food – largely stripped of local distinction, seasonal variety, batch variability, species flavour or individual culinary creativity – became a watchword in its own right, whilst permutations of presentation aimed to gratify the omnivore's quest for novelty. Together they provided an iconic framework by which millions of consumers could recognise the unselfconsciously modern society to which they felt they belonged. The irony of the omnivore's paradox was that the commercial food industry, over time, metamorphosed from being simply a trafficker of already validated meanings to developing an identity all of its own, which became the dominant icon in modern western food signification.

The concept of a 'convenience' food, for example, became so familiar to large sectors of the population that anything else stood to represent unwelcome, old-fashioned drudgery – almost independently of factors such as the number of minutes a given dish takes to prepare. Commerce largely succeeded in arrogating unto itself the mythology of ease, of hygiene and of modernity, and did so with such consummate thoroughness that many hardly dared stray beyond its benign assurances. We became affronted by suppliers' deficiencies which once would have been the legitimate subject of elementary domestic precautions – such as cracking eggs into a cup to check their freshness, or cooking poultry thoroughly to ensure its safety (see Gofton's discussion of convenience, chapter 7).

The basic human urge of culinary conservatism described by Rozin was so abundantly usurped that the strange became the familiar. We preferred not local fare, but that from far-off fields. We no longer ate that which was in season, but we manufactured seasons to produce what we desired to eat. Our role models for what is normal and natural to eat were no longer our parents and peers, but celluloid celebrities staring out from the silver screen. According to Fischler (1980), the modern western world has been suffering a biocultural crisis as a result of the body's wisdom being overridden by an overabundance of external signals, such as commercial advertising, which have been short-circuiting the internal biological signals which otherwise would regulate eating patterns. He is perhaps misleading only in seeming to imply that the process was unfortunate and unforeseen. For, to an extent at least, it was the consummation of a coherent agenda towards clear commercial aims.

Our taste for the familiar is, therefore, much more than a rattish proclivity to avoid the unfamiliar in search of security. It is more than our becoming comfortable with certain sensory experiences. It says far more about us as culturally constructed, and constrained, centres of consciousness than as biological beings. It is nothing less than an ongoing opportunity to affirm our identity as individuals, as affiliates and affines of our own communities. Industrial culture is just one identity that is at our disposal to don. It achieved almost uniform popularity in the 20th century, in keeping with the corporatist spirit of the times. But there can be no certainty that this cultural hegemony will last forever, at least with such ubiquity; indeed, it will endure only so long as – individually or as a society – it suits us to do so.

Just as shared consumption can be taken to imply shared values of some sort, refusal to partake of particular items can indicate our distinction from others. A rebellious teenager might adopt a vegetarian diet, or a dissident sect its own prescribed manna, partly at least to set themselves apart from their 'parental' authorities. And so too, where we see societal-scale rejection of particular foods, or fragmentation of familiar eating habits, we may validly expect to be able to read into this process that profound social changes may also be afoot. Indeed, although George Orwell's statement that 'it could be plausibly argued that changes of diet are more important than changes of dynasty or even of religion' (1937: p. 82) may have been meant merely materialistically, it is also more broadly apposite.

One illustration of the capacity for food to reflect deeply rooted social processes is the differentiation which has been evident, particularly in the late 20th century food market, which reflects the fragmentation of what was previously a more homogeneous society. Almost any survey of food habits prior to this historical period would have shown local peers to share broadly similar tastes and perceptions of the edible. A European peasant of the Middle Ages, for example, would have enjoyed a narrowness of diet that today would appear to be 'dismal monotony' (Tannahill, 1988: p. 184), and even then contrasted sharply with courtly menus. But that peasant would have known, and seldom doubted, their place in society and the menu that status dictated.

But today, since industrial preservation, mechanisation, retailing and transport permanently altered the provisioning landscape (Goody, 1982: p. 154), and increasingly bolstered by rapid communications, 'postmodern' multiculturalism is becoming the norm. It is no longer new to observe that traditional institutions, cultural forms and scales of values have been subject to disintegrative processes (e.g. Béjin, 1976). Indeed, this is but one articulation of a transnational trend towards traditional orientations being jettisoned in favour of uncertainty. Beck (1992), for example, perceives that over the past decade or so there has been a striking transformation in the nature of people's social identity, and that this is caused by fundamental changes in the organisation and culture of contemporary societies, particularly through the shift from an industrial to a global 'risk' society. His thesis of reflexive modernisation describes industrial society's current cultural shift, involving a radical transformation of politics and modern subjectivity, beginning to undercut its own foundations, and eroding the antithesis between nature and society, with the resulting 'transformation of threats to nature from culture into threats to the social, economic and political order'.

Beck argues that the process by which this is occurring is bypassing formal structures of political debates and governments. Sexual, social, political and dietary assumptions have all undergone examination, such that traditional normative constraints wield considerably less influence than previously. And indeed it is daily more evident that careers, politics and even family life are becoming characterised more by serial infidelity than by lifelong security. One manifestation of this deeply rooted shift is the increasingly evident trend for large sections of the populace to asseverate their own nutritional wisdom, selecting from the cogent opinions of their 'experts' only those views which accord with a culturally defined agenda. It is as if we are learning to regard the pronouncements of western science not as necessarily false, but rather, just as anthropology has taught us to do when interpreting the cosmologies of 'other' cultures on distant continents, as 'simultaneously real, social and narrated' (Latour, 1993: p. 7).

Amid this deconstruction of traditional social forms, the security of a single prescribed culinary identity is increasingly difficult to maintain. We find ourselves cast adrift in a sea of potential affiliations from which we are free to pick'n'mix – and to eat accordingly. We can represent ourselves through a diet appropriate to an African-American environmentalist, or a Bolivian Jain. Tandoori chicken pizza jostles for space on the supermarket shelves alongside 'fat-free' ice-cream and vegetarian bacon. We can select soya yoghurt, steak and fries, nachos, or bouillabaisse. What we choose depends upon what sort of person we wish to represent ourselves as.

Such anarchy can threaten anxiety for the consumer, as old certainties evaporate. For the food industry, on the other hand, the dissolution of traditional modes bestows an ideological void, a latent demand which the well-organised marketing operation seems ideally-positioned to satisfy (see Buisson, chapter 8). This quintessentially metacultural process throughout the western orbit, of moving towards an individualistic decentralisation of choice and legitimisation of hedonism and away from the 'fordist' food provision model of the post-war years, appears to offer a golden opportunity for market segmentation and specialisation. Gofton (1989), for example, describes how this has presented the industry with an opportunity to differentiate 'value-added' products to appeal to niche sectors of the emerging market – the 'health-conscious', the 'gourmet', the 'ethical' or 'green' consumer, and so on. Indeed it presents, at least in theory, the chance effectively to redefine the consumer, and to develop a set of models by which demand as well as supply might be more advantageously managed.

However, overreliance on this pliant-consumer archetype would be a mistake. Human creativity is eternally certain to outflank such mechanistic conceptions. The impression that our tastes in food are becoming so eclectic that little order survives amid the chaos is superficial for, on closer examination, clear trends and themes do emerge – a point made repeatedly throughout this book.

6.4 Natural as cultural

One such matrix is the manner in which 'nature' has permeated the realm of food provision, which may yet subvert prognoses based entirely on historic circumstance. Having previously touched on the nature of contemporary culture, it is time briefly to consider the rapidly modifying culture of nature. For nature, as they say, is not what it used to be. The academic jury is still out on the extent to which non-western societies share a concept of nature that is comparable to ours. It is indeed arguable whether terms such as nature or natural are valid at all, outwith the western orbit, since they are contingent upon a binary distinction between humanity and the rest of nature that is alien to most other societies' thinking, and which may above all reflect the alienation from a direct experience of personhood that has been prevalent in our recent history (Ingold, 1986).

In the west, at least, we have tended to think of nature as largely unproblematic, conceptually if not ecologically. But, clearly, no such 'thing' actually exists. As a discrete entity, 'nature' is, and always has been, a model which has existed nowhere but in our own minds. There can be nothing which is absolutely natural, or unnatural, other than what we deem so to be. It is an invention that western thinkers have been content to deploy over the years with a variety of motives and constructions (e.g. Gruen and Jamieson, 1994), but generally as a residual category to contain that which is not human, and to serve particular political purposes.

In the main, it has existed as a function of our desire to draw a philosophical distinction between 'ourselves' and the non-human sphere, and to isolate a resource that can be exploited with moral impunity. Thus, when Francis Bacon wrote that 'man by the fall fell at the same time from his state of innocency and from his dominion over creation. Both of these losses however can even in this life be in some part repaired' (*The New Organon, Aphorisms*, quoted in Clarke, 1993: p. 85), or when Descartes claimed to have observed that the body 'is of its nature always divisible; mind is wholly indivisible ... I understand myself to be a single and complete thing' (*Principles of Philosophy*, quoted in Clarke, 1993: p. 88), thereby laying the foundations for the empirical scientific ethos that pervaded the modern consciousness, each was an example of:

> the presupposition in our thinking, of western thought in general, and of our philosophy in particular: that the world is made up of a plurality of discrete individual substances: the world has been viewed, since classical times, as an array of individual objects which are logically mutually independent but bound in a web of causal ties. (Mathews, 1991: p. 7–8)

This conception of the universe's essential nature became so endemic in modern western thought that we are still seldom, if ever, truly able to escape it. In the arena of food, a separation of nature from human culture has been as central to environmentalist vegetarians as to the most devout carnivore. The meat-eating ethos places conspicuous emphasis of the value of 'human' domination of that other thing we call 'nature'. But the ecologist issuing earnest appeals for society to respect 'nature' has equally tended to be primed to perceive a thing called nature that can be revered.

As the product of human creativity, nature can as easily be re-invented. Indeed, it *must* continuously be re-invented, in order to endure. Thus, the nature that surrounds us today is a very different nature to that which Bacon and Newton, or indeed Rousseau, inhabited. Whilst many of our economically oriented institutions still seek to train the young to perceive 'it' as a resource, a technological challenge, and a scientific adventure, a quiet sea-change has been sweeping through the population at large, such that nature has been metamorphosing into a vulnerable giant, and repository of traditional values and virtues. In fact, Fischler's observations about modern society must be qualified by Latour's (1993) contention that 'we have never been modern'. For, as Latour explains, the unified cosmological image long ago broken by Hobbes and Boyle was never, in fact, as utterly destroyed as the official record had it. For many years we dutifully dubbed ourselves modern, paid lip-service to scientific dogmas through enactments of appropriate rituals that to this day litter our daily lives. But we sustained this performance only by practising a customary public denial of much that we know gives our lives value. Behind the façade of modernity, we continued to rejoice in diverse appreciations of the 'other half' of our nature(s) – the non-techno-scientific, culturally constructed reality.

Latour's observations do not emerge from an intellectual vacuum. Rather, his contribution reflects a continuing grassroots debate, in which many assert not only a right but also a requirement to blur the rigid boundary between orthodox 'rational' science and subjective values. A diffuse cosmology embracing ethnical, environmental and, indeed spiritual, orientations informs an increasingly influential meat-wary disposition, and not only amongst committed vegetarians.

In its entirety, this world view, which typically values diversity over uniformity, creativity over authority, spontaneity over predictability, empathy over control, or animism over utilitarianism, remains marginalised among a (growing) handful of subcultures, whose dissidence can lightly be dismissed as meaningless or misguided. But few remain entirely oblivious today to such conflicts of loyalty in their own lives – perhaps between 'commercial imperatives', material indulgences, and ethical and environmental considerations. At a crucial level, this represents a discourse about cosmologies, which are ultimately incompatible.

But the conception of reality characteristically held by those with most conventional political power is handicapped in countering this movement, by being constrained from recognising its negotiability, and therefore vulnerability. Adherents to the corporate conception of nature have been taught to assume its tenets to be so normal and natural that they are often poorly prepared to join in appropriate conceptual battle, being used to working within the system, rather than working to defend it:

> Power through market mechanisms and intercollated corporations is widely dispersed. No one person or small cadre is in charge of the course of the economy or human development, nor can they be. Corporations collectively have created a language of getting things done. This discourse has not been designed consciously, but on inspection its most salient feature is that it is meant to persuade for material and social effectiveness. New members are socialized to this language and its uses. (Rose, 1994: 96)

The food industry's response to the changing cultural agenda has been to make concerted efforts to project and protect an image of 'naturalness', even while its processes have been scaling new heights of technological sophistication:

> Technological changes in agri-food production, now enhanced by biotechnology and genetic engineering, have manipulated biological processes, appropriating some and substituting others, redefining and refashioning nature as a source of profit and capital accu-

mulation. As the food we consume has become more processed it has been presented as more 'natural' by the food industry. (Goodman and Redclift, 1991: p. 250)

Indeed, so enthusiastic has been this effort that governments have been obliged to consider guidelines for the uses of such terminology. Butchers' promotional material features healthy, happy chickens or pigs, set against a background of green rolling landscapes – not the production-line homes in which most such animals actually pass their days. The fashion for 'artificially' strong colours that characterised a period of self-consciously modern sensibility has waned in favour of gentler hues that the public is more willing to associate with its rural idyll. Chemical companies have even been permitted legally to term their compounds 'natural' or 'nature-identical', so long as they proximally mimic extracts from substances found in the wild.

This repositioning by the food industry has achieved notable success in enabling it to accommodate, to date, the majority of the increasingly ecologically minded public. Still, it seems, only the most fundamentalist of buyers insist on fresh, seasonal, organic, unprocessed, local produce supplied with minimal packaging by small-scale enterprises, as most 'deep green' analysts would advise for sustainable ecological (and personal) health. The values inherent in industrial food manufacturing, and the discourse thereby conducted, still appear to hold more general sway. Yet, despite all this, there are good reasons to doubt whether events will necessarily always continue in this direction, at least with such ubiquity.

For, in spite of its attempted appropriation of the discourse of naturality, the food industry has set itself irreducibly in the corporatist and quasi-scientific camp that aims to control and subdue natural diversity, and so opposes itself to the emergent anti-authoritarian ecological paradigm. As industry has achieved ever greater control over the processes of food production and supply, the modern global ecosystem has become so specialised, so internationalised, that entire bioregions are devoted to largely monocultural production (e.g. the corn belts of America, or Third World nations' primary cash crops for export), for global distribution. The industrialisation of eating took another step forward in the UK when the British government recently decreed that cooking should be taught to schoolchildren only under the rubric of Food Technology, as an option of Design and Technology. And in deliberately fostering the decontextualising and deskilling of domestic consumption (including through such devices as the coordinated promotion of microwave ovens and ready-to-eat meals), the food establishment has sought to configure itself as a quasi-substitute parent, a sort of paternalistic mothering figure, guaranteeing the consumer's comfort and security, but simultaneously setting itself apart from their lived experience.

Through the increasing marginalisation, by virtual commercial invisibility, of local, seasonal, and familial modes of food production and preparation, consumers have been counselled to deposit their faith in insipid dishes flashed with an apparently authoritative 'low fat' hallmark, regardless of – or even, perhaps, because of – their expense. It is quite normal, today, for people to prefer Coca-Cola to milk, or margarine to butter, and countless millions feel happier with the 'vetted' contents of tins than with fresh fruit and vegetables. Indeed, George Orwell remarked on this tendency already prior to the Second World War, when he noted that the 'number of people who *prefer* tinned peas and tinned fish to real peas and real fish must be increasing every year, and plenty of people who could afford to have real milk in their tea much sooner have tinned milk' (Orwell, 1937: p. 89; italics in original). So successful has the industry been in selling its message of safely sanitised fare that it has been scarcely surprising to hear of someone rejecting the flavour of an extensively reared rare breed of chicken as 'wrong', because it tastes of more than water and fishmeal.

The amorphous entity of commercial production articulates an ersatz breast on which the trusting consumer is entreated to suckle. He or she is invited to experience this as an ever-increasing range of potential foodstuffs, subject only to relatively minor seasonal fluctuations in market price ('flattened out' by the major retailers through a system of subsidies and oncosts in the interests of price stability and consumer confidence). And with the majority this has, to date, proven popular. But such power carries a burden of responsibility, of a sort that is not easily fulfilled. Each lapse in implicit promises of vigilant guardianship brings not only fleeting disappointment, but also inflicts some small fissure in the edifice of technological prowess. Fears regarding the abuse of that responsibility can motivate concern for not only the immediate issue, but can generate deeper and persistent doubts.

Moreover, the paradoxical price of this variety has been homogenisation in terms of loss of local identity, as one supermarket (or McDonald's) increasingly resembles another, whether in Maryland, Moscow or Manchester, and as the consumer is literally distanced ever further from involvement in, or even awareness of, the processes of production. We have been losing our local roots. It is anxiety stemming from this loss of identity which Fischler (1980) identifies as the wellspring of modern society's growing demand for symbols of nature in what it eats. He suggests that industrial culture presents us with novelties galore: as we browse the supermarket bins emblazoned with special introductory offers, we seldom find ourselves in danger of enduring the excessive monotony from which our urge to experiment is equipped to spare us. However, that basic familiarity and continuity which stems from personal contact with the context of production, required to give us confidence in what we are eating, is less secure. As Symons has observed, where 'we might once have herded our own goats or bent over our own paddy-field, different people now sow the wheat, build the tractor, design the ball-bearing assembly, teach the engineers...' (1993: p. xi). In three distinct stages, he suggests, first:

> it was agriculture that was industrialised; then, late in the nineteenth century, food proces-
> sing and retailing; and finally, since the Second World War, cooking. That is, the house-
> hold has successively lost its own self-sufficient garden, its cellar or pantry for storage, and
> now its kitchen. (Symons, 1993: p. xi)

But whether desirable or not, it is questionable if a global industry –
however sophisticatedly it targets its promotions at ethnic, gender, age, or
ideologically-oriented subgroups – can ever adequately substitute for the
eloquent iconography of locally produced and historically situated cuisines with which small-scale human groups have, historically, been used
to associating, or realise an appropriate degree of sensitivity to serendipi-
tous interests. Certainly, if current trends continue, and the amorphous
caricature of an undiscriminated and undiscriminating 'modern' proletar-
iat upon which the nineteenth and twentieth century food industry fed for
its development continues to disintegrate into a kaleidoscopic diversity of
groups and allegiances, then the monolithic producer faces a severe challenge to imbue the variety of relevance that is likely to be demanded, at
least to maintain its 'market share'.

This debate partly underscores the confusing, and indeed confused,
message which consumers seem to be sending through their purchasing,
patronising both health food shops and burger bars in ever greater numbers, and demanding both hygiene and naturality in tandem. It is subject
to this discourse, directly or indirectly, that categories of pure and polluting foods have been undergoing continuing reappraisal. Tempting though
it may be to dismiss such developments as food fashion, food preferences
are conveying a coherent if paradoxical message. Western society is
reconsidering its very identity and trajectory, but rather than replacing
one previously hegemonic ideology with a new one, we are (so far at
least) fragmenting into a society of many parts and various viewpoints, as
if the more energetically to conduct our deliberation.

6.5
Crises of confidence

The world is changing, right before our eyes. What we take for granted is
not what it was ten years ago, let alone fifty, or two hundred. Nature is
not what it was, and its imponderables are not what they were. It is not
that we know ourselves to be a part of nature, and feel secure in that
knowledge. It is not that we know ourselves to be apart from nature, and
live confidently on that basis. It is both, and neither. The separation of
culture and environment has been impressed on us, but we know it does
not make sense, and we are perplexed by its contradictions.

Meat provides a vivid example of this incongruity in operation. For
willingly, or at least accepting, in some period of the past, doubtless
linked with urbanisation, we allowed ourselves to become dissociated
from the processes of agricultural production. We consigned that function
to specialist professionals. Once upon a time these were contrived as

jovial, hardy, trusty farmers – and to this day where farming appears in our public fora, more often than not the system depicted more resembles an Arcadian image of the past than that in which most modern beasts actually spend their days. But now we are beginning to blink and open our eyes to find, almost in the manner of Dr Jekyll, that our farms are run by biotechnologists, industrial chemists and process engineers, shut away behind that very high-walled techno-industrial encampment that many have been learning to distrust. The bleating calves we left in farmers' care have become efficiency-oriented production units. The manger is run by the manager.

The two images do not fit: a fact that can make us uncomfortable indeed – at least when we have to confront it. We prefer to avoid distasteful and discomfiting realities, or inconsistencies in our philosophy, so we conspire with our media guardians to avoid confronting such issues. A recent study of British children's television, for instance, found that although animals appeared at a very high rate in both fictional and non-fictional programmes, the issue of meat consumption was almost entirely avoided, which was interpreted as an expression of adult society's discomfort with the paradox of advocating kindness to animals on one hand, but the acceptability of meat eating on the other:

> The task that society has set itself is to tell children that mammals are human-like and sentient and it is wrong to hurt them, but that it is, nevertheless, OK to eat them. And the consequence, it seems, is that the subject of meat (especially mammal) production and consumption is almost completely avoided on British children's television. (Paul, 1994: 6)

Such contrivances cannot, however, be sustained always and forever. Probably the most common single catalyst for individuals who have become vegetarian is a 'moment of realisation', when some incident has unavoidably brought to the attention the fact that meat – previously perhaps an unproblematic meal ingredient – derives from the flesh of once-living creatures' bodies. Meat is perhaps a singular example. But what is true for the meat industry, as the epitome and touchstone of the culinary conjunction of human industry with the wider world, could become equally true for the entire edifice of food manufacturing and supply if the fashion for a less alienated relationship with the natural world continues to diffuse more widely beyond its 'alternative' ghetto.

And diffusing it is. For instance, a consistent thread runs through the many and various health concerns and 'food scares' of recent years. These imply rejection not only of particular production practices, at which, from time to time, scientific or complementary epidemiologies point an accusatory finger. Clearly implicit too is fear of intrinsic non-naturalness in our constitutional settlement, as reified in eating habits. One example was the notorious British television 'sausage programme' (*World in Action*, 7 October 1985) which gave consumers a vivid insight into the production of Mechanically Recovered Meat, after which sales of sausages and other processed meat products plunged overnight, never fully to recover. Above all, it was the artificiality of the mechanical process whereby heavy-duty industrial equipment is fed with parts of animals that most chefs would regard as beneath consideration, to be processed and disgorged as a pink slurry for concoction into human foodstuffs such as pies and patés, that offended viewers' sensibilities.

Or consider the various 'health scares' that intermittently preoccupy the media. At first sight the issues are diverse and unrelated: listeria; salmonella; irradiation; BST; BSE; cook/chill; additives; hormones; antibiotics; pesticides; unhygienic slaughterhouses – the list seems endless. Yet each has a common factor: the threat is perceived to originate in some aspect of 'unnatural', or technological, food supply. This is not to say that there may be no physical basis for the concerns; technological interference in natural processes has indeed bequeathed an extensive repertoire of previously unimagined threats. These are the issues in which our reductionist society, that demands isolated causal relationships for scientific and political credibility, conducts its discourse. But viewed as a whole they express more. Just as the consumption of convenience food can symbolise participation in modern consumer society, so such health fears metaphorically signal public loss of confidence in the excesses of a food production and supply system that, by and large, had hitherto been trusted.

Or take, for example, the place of fat in contemporary consciousness. We are told that most of us consume too much of it, although fatness was once thought (like meat) to indicate prosperity. Today, however, it is regarded – by the western middle classes at least – as a sign of cloth, indiscipline, and decadence. And since nature is thought to be a state where the fittest survive, so fat has come to signify the unfitness of unnatural culture.

That a single television documentary concerning the methods of industrial food production now has the power to catalyse enormous numbers to reconsider their purchasing choices is the consequence of a long-developing ambivalent relationship between the public and its providers. Indeed, the food industry in general, and the meat industry in particular, may have reached a point where it is inflicting serious long-term damage upon itself for the sake of short-termist gain. By being unwilling or unable to recognise that progressive public concerns may be more than misguided in their suspicion of the prevalent idiom in corporate food supply, a legacy of mistrust is being established which could be difficult to disperse. As Hirschman (1970) has pointed out, consumers of any product denied a legitimate 'voice' are left only with the option of 'exit' from the relationship, after which 'loyalty' may be long in re-establishing.

The 'vegetarian', who disavows not only the substance of meat – who, so to speak has 'exited' from allegiance to the intrinsic ideology of meat-eating – is less prone to the paradox that is inherent in the omnivore's conception. For, their sharing a traditional understanding of nature separate from culture notwithstanding, the moral threshold implicit in the vegetarian's dietary stand bequeaths a philosophical vantage point from which to disown at will the ideological baggage which the omnivore must carry regardless. Having taken one step off the carousel, others become easier, if not inevitable. It is no coincidence that meat-abstinence correlates with a tendency to prefer fresh and unprocessed produce, as well as various other 'unindustrial' indicators. So just as numbers of self-defined vegetarians have been steadily rising, the products of the industrial food business, which traditionally met with widespread public acceptance, indeed favour, are being greeted with growing diffidence by a rising proportion of the populace.

The buying habits of a generation speak of still diffuse but potentially damning disillusionment with the great white hope of scientific civilisation that once promised so much, permeating the attitudes of millions. The excitement has gone. The spell has been broken. As we look around at the brave new world that once so enticingly beckoned, we are hard pressed fully to appreciate the triumphs of technology, whilst new reports of ecological and human disasters wash over us with every news report, whose ultimate cause a growing consensus places in the court of the Dominant Ethos. The shock of the new is that the new is no longer shocking, nor even particularly alluring. It sometimes seems like nothing so much as an unstoppable juggernaut, trampling everything of human value in its path. We yearn for the old days, when everything seemed so simple.

We have, therefore, travelled almost a full circle. Novelty and variety has, by a curious twist of logic, become the one thing that is stable and familiar. Amid the ubiquitous profusion of tastes and experiences, the novelty which opinion-formers increasingly demand is that which ironically is characterised by non-modernity: by naturalistic icons of 'organic' production, 'free-range' farming, and 'unadulterated' processing. The new is old, the old new. So here is the final paradox that afflicts the western omnivore today. It is not merely that we find how to conceptualise our victuals ever more problematic, as Fischler has argued. Rather, the naturality around which our culinary aspirations have become peculiarly fixated is subject to an intractable dilemma, precipitated by a combination of contemporary circumstances.

Commanding processes of industrialisation have come to dominate our food supply, bearing little resemblance except in titular product to those we once knew as farming. At the same time, a continuing process of urbanisation has divorced the vast majority of western populations from direct experience of processes of production, and so rendered them open to innumerable influences that may shape their perspective of agri-cultural and agri-industrial functions, by which 'nature' develops a nature very different to that assumed in our historic texts. At the same time, fragmentation of traditional individual and group identities amid the general decline of fordist models of authority have opened a door for iconoclastic cultural conceptions to achieve more widespread legitimacy. And at the same time, culturally mediated intimations of environmental crisis have precipitated increasing mistrust of the entire industrial–scientific edifice that has been the cornerstone of the post-Enlightenment settlement. This leaves us searching for new identities on many levels, as is manifest in the new omnivore's paradox.

As the third millennium stares us in the face, we should not be surprised if the seeds of a new cultural compact grow into significant redefinitions of which foods are naturally good to eat, or to think with, and which less so. Diversification strategies notwithstanding, the food industry continues to assume 'business as usual' in the face of increasingly unmistakable evidence of public disenchantment with its conceptions of the nature and purpose of food provision. Even at its best, it still does not – perhaps cannot – acknowledge that people may ultimately demand something more substantial on their plates than nutrition, hygiene and promotional imagery.

The very question of our place on this planet is at stake. As nature rushes towards us as a dim and distant intimacy, our ability to appropriate it, or to tolerate others to do so on our behalf, accordingly diminishes. And the closer we approach this abstract entity which our forebears reified as 'nature', and which conditions our entire food selection paradigm, the more the historic distinction that defines it dissolves.

References   Back, K. (1977) Food, Sex and Theory. In T. K. Fitzgerald (ed) *Nutrition and Anthropology in Action*, Van Gorcum, Amsterdam. pp. 24–34.

Barthes, K. (1975) Towards a Psychosociology of Contemporary Food Consumption. In E. Furster and F. Forester (eds) *European Diet, from Pre-Industrial to Modern Times*, Harper, New York. pp. 47–59.

Béjin, A. (1976) Crise des valeurs, crise des mesures. *Communications*, **25**, 39–72.

Beck, U. (1992) *Risk Society: Towards a New Modernity*, Sage, London.

Clarke, J. J. (1993) *Nature in Question: an anthology of ideas and arguments*, Earthscan, London.

Cowart, B. J. (1981) Development of Taste Perception in Humans: Sensitivity and Preference Throughout the Life Span. *Psychological Bulletin*, **90**, 1, 43–73.

Darlington, C. (1969) *The Evolution of Man and Society*. Simon and Schuster, New York.

Douglas, M. (1975) Deciphering a Meal. In *Implicit Meanings*. Routledge & Kegan Paul, London.

Farb, P. and Armelogas, (1980) Consuming Passions. *The Anthropology of Eating*, Houghton Mifflin, Boston.

Fiddes, N. (1990) Meat in Change: rural and urban cases. In T. Marsden and J. Little (eds), *Political, Social and Economic Perspectives on the International Food System*, Gower, Avebury.

Fiddes, N. (1991) *Meat: a natural symbol*, Routledge, London.

Fischler, C. (1980) Food habits, social change and the nature/culture dilemma. *Social Science Information*, **19**, 6, 937–953.

Fischler, C. (1986) Learned versus 'spontaneous' dietetics: French mothers' views of what children should eat. *Social Science Information*, **25**, 4, 945–965.

Gofton, L. (1989) Sociology and food consumption. *British Food Journal*, **9**, 1.

Goodman, D. and Redclift, M. (1991) *Refashioning Nature*, Routledge, London.

Goody, J. (1982) *Cooking, Cuisine and Class*, Cambridge University Press, Cambridge.

Gruen, L. and Jamieson, D. (eds) (1994) *Reflecting on Nature*, Oxford University Press, Oxford.

Hirschman, A. O. (1970) *Exit, Voice & Loyalty: responses to decline in firms, organizations and states*, Harvard University Press, Cambridge, MA.

Illich, I. (1986) *H_2O and the Waters of Forgetfulness*, Marion Boyars, London.

Ingold, T. (1986) *The Appropriation of Nature*, Manchester University Press, Manchester.

Latour, B. (1993) *We have never been modern*, Harvester Wheatsheaf, London/New York.

Lévi-Strauss, C. (1970) *The Raw and the Cooked*, Cape, London.

Lévi-Strauss, C. (1973) *From Honey to Ashes*, Cape, London.

Lévi-Strauss, C. (1978) *The Origin of Table Manners*, Cape, London.

Mathews, F. (1991) *The Ecological Self*, Routledge, London.

Okely, J. (1983) *The Traveller-Gypsies*, Cambridge University Press, Cambridge.

Orwell, G. (1937) *The Road to Wigan Pier*, 1984 edition, Penguin, Harmondsworth.

Paul, E. (1994) The Representation of Animals on Children's Television. Unpublished research paper, Department of Psychology, The University of Edinburgh.

Radcliffe-Brown, A. R. (1922) *The Andaman Islanders 1964 edn.*, Free Press, New York.

Rose, D. (1994) The Evolution of Intervention. *Anthropology and Humanism*, **19**, 1, 88–103.

Rozin, P. (1976) The selection of foods by rats, humans and other animals. In J. S. Rosenblatt, R. A. Hinde, E. Shaw and C. Beer (eds), *Advances in the study of behavior*, Volume 6, Academic Press, London/New York.

Rozin, P. (1978) The use of characteristic flavourings in human culinary practise. In C. M. Apt (ed.), *Flavour: Its Chemical, Behavioural and Commercial Aspects*, Westview, Boulder, Colorado.

Rozin, P. (1982) Human Food Selection: The Interaction of Biology, Culture and Individual Experiences. In *Psychobiology of Human Food Selection*, L. M. Barher (ed) AVI, Westport, CT. pp. 225–254.

Stoller, P. (1989) *The Taste of Ethnographic Things*, University of Pennsylvania Press, Philadelphia.

Symons, M. (1993) *The Shared Table: Ideas for Australian Cuisine*, Australian Government Publishing Service, Canberra.

Tannahill, R. (1988) *Food in History*, Penguin, London.

Vialles, N. (1994) *Animal To Edible*. Cambridge University Press, Cambridge.

Visser, M. (1986) *Much Depends on Dinner*, Grove, New York.

Woodham-Smith, C. (1987) *The Great Hunger*, Hamish Hamilton, London.

7 Convenience and the moral status of consumer practices

Les Gofton

'We are confident Thailand will be a good place to conduct business. We are proud to be here – soon to be the 32nd McDonaldland country... (Thomas A. Gruber, McDonald's Vice President, Bangkok, quoted in *Sunday Nation*, 1994)

Marketing texts of twenty years ago often used to begin with the following tale:

Once upon a time, there was a perfect housewife who wanted to bake a cake for her hard-working husband and perfect, dependent children. When she went to her cupboard, although it wasn't bare, several of the ingredients which she would need were missing. In response to her anguished sobs, an inquisitive but friendly neighbour appeared as if by magic. "Here you are, my dear," she smiled. "Try this new Mary Baker© instant cake mix. Perfect, feather-light beauties at every baking, and all you need to do is add water! Isn't science marvellous...?"

"Get out of here!!!" (...recoiling in horror) "Are you trying to destroy my raison d'etre? Adding water is not cooking!! What would my hardworking husband and perfect, totally dependent children think of me? Now, if you'll excuse me, I must ring my analyst ... I feel an attack of role strain coming on..."

Luckily, a passing Mary Baker representative happened to overhear. He was admiring the apple pie cooling by her open window, having chased away a hobo who was about to steal it.

"Have no fear," he smiled. "Allow me to present – new improved Mary Baker Cake Mix. Perfect cakes every time; you have to add water and then – add an egg!!" "Now you're cookin' on the front burner, Jackson," our heroine responded, slipping into anachronistic, but apposite, jive talk from the late 1930s. "Or rather – I am, thank goodness!!"

And they all lived happily – until the middle sixties, when a great social upheaval occurred, leading to a new status and role for women which completely changed attitudes such as these, and made convenience food universally acceptable. Almost.

7.1 The rise of convenience

This tale was received wisdom among food marketers only a short time ago. Convenience was a dirty word, and woe betide the housewife whose family was fed on 'fast foods' – for she was letting them down morally, nutritionally and socially. Convenience was lazy, immoral, and, frankly, lower class. In the UK this seems to have been long established. Until fairly recently the convenience food par excellence was fish and chips, and this had long been regarded as suspect on all counts. While experts could thoroughly approve of fish and potatoes as a wholesome traditional diet for crofters, members of the urban working class, for whom it was a regular meal, were seen as unfortunates, or inadequates, who did not feed themselves, or their children, properly.

As Walton says:

> What is most remarkable, in some ways, is the persisting prejudice against fish and chips in 'respectable' and official circles, which still affects many peoples' perceptions even today … the roots of these attitudes can be found in the powerfully promoted Victorian ideal of female domesticity, the identification of the women's sphere with hearth and home; cooking, cleaning and needlework. One aspect of this was the stigmatisation of women's work outside the home; another was the condemnation of family budgets that included 'lazy' and 'expensive' foods. (Walton, 1992: p. 164; see also Dyhouse, 1981; chapter 3; Gofton, 1992)

Walton ties this moral panic over fish and chips in the late nineteenth century to the campaign to promote this feminine role through the teaching of domestic science in the (recently compulsory) schools, seeking to train girls as wives, mothers and parents. This was given added impetus by the obsession, at that time, with eugenics and the future of the race. The emergence of the first 'fast food' was seen as a threat to these 'improvements' and to the progress of the nation. The attitude is long-lived. Concerns over the decline of the family expressed in 'latchkey kids', the Bulger case, car crime and so on, or the ill-fated 'back to basics' campaign - time and again, we disinter this ideology of feminine domesticity and full-time motherhood as the 'real' foundation of order and authority. According to *This England*:

> Nowadays, children mostly arrive home from school to an empty house … The television is provided to keep (the child) quiet until (mother) arrives home from work in an office or shop. Poor mum is too tired then to listen to childish chatter. She has no time to bake a cake or prepare a nourishing meal of stew and dumplings, so she feeds her family on deep frozen microwaved convenience foods … Soaring crime rates and a breakdown of morals are the ultimate effects of evil. But the cause lies in the home … Whatever changes the family unit will eventually destroy the nation. (*This England*, 1986)

This has proved a slippery slope. In the UK, the National Food Survey (MAFF, 1994) revealed that British consumers are buying more ready-made meals than ever. Total sales rose by 10% during the thirteen years from 1980–1993, and by 1993 accounted for 35% of total food purchases. Convenience foods now provide 29% of our total energy. While sales of pizzas, frozen chips, crisps and other potato snacks rose, along with fast and easy to consume fruit and vegetables such as bananas, salad vegetables, cauliflower and broccoli, sales of beef and lamb, root vegetables and fresh potatoes have continued to decline. So too have sales of canned products, while sales of frozen products have increased.

While we eat more and more take-away, ready-prepared or easy-to-cook meals, and the ever more ubiquitous microwave extends its thrall outwards into ever-increasing areas of our kitchen, the nutritional wisdom and also the morality of this new way of eating are still suspect. Convenience foods have traditionally been thought to be 'unhealthy', to be 'junk' foods made from rubbish or inferior ingredients, to be 'full of additives and preservatives'. Even those who eat these foods do not believe that they are healthy. Bull (1985) surveyed the eating habits of 19- and 21-year-olds, and asked what foods they liked to eat, but knew they should not. Chips were first, followed by confectionery, cakes and buns, fatty foods and meat products. Fast foods were rated as 'good for your health' by only 20% of respondents in a London Food Commission survey (Carruthers, 1988).

Yet we eat more and more of such foods because, in these affluent times, although we are rich and overfed, so conventional wisdom has it, we are hungry for the extra minutes and hours which consuming such foods 'gives' to us. Health Minister Angela Browning, hard on the heels of new dietary guidelines calling for people to cut down on fat and eat more fresh fruit and vegetables, has come out in support of the trend towards convenience eating, declaring:

> Fast food is frequently criticised as unhealthy, but there is no reason why it should not form part of a healthy diet.

This spirit of realism may not be unrelated to the differential expenditure on high-cost ready meals which involves social group A spending three times as much, per person, than groups E1 and E2. Even the manufacturers are less sanguine than this. At the famous 'McLibel' trial, when environmental activists were sued for misrepresenting McDonald's products, one of the company's own experts accepted that it was 'very reasonable' to say that a diet high in fat, sugar, animal products and salt and low in fibre, vitamins and minerals is linked with cancer of the breast and bowel, and heart disease (Ehrlichman, 1994).

The reason for this ubiquity seems easy to find. 'Time will be the currency of the nineties', according to the report of the US Institute of Economists, and our everyday experience bears this out (see also Henley Research Centre, 1990; Blyton, 1985; Boden, 1987; and Carlstein *et al.*, 1978 for a broader perspective). With the increasing importance and value of time resources comes a greater demand for convenience in food, alongside other goods and services, although this creates conflict with the other major trend in food habits, that of healthy eating. Balancing both of these will continue to structure our food habits into the foreseeable future, according to many commentators (Gofton and Ness, 1991). Already, there has been considerable impact upon some of the kinds of food we eat. In the USA, for example, poultry consumption (of which 80% is chicken) rose from 19% of total meat poultry and fish consumption in 1966 to 32% in 1989. Between 40% and 45% of all chicken is consumed in fast-food stores. At the same time red meat is declining in popularity while entrees in restaurants fell from 35% in 1982 to 22% in 1988. Consumers also value time in itself; US surveys reveal that the most prized recent inventions are those such as the microwave, remote control electronic devices and automatic coffee makers, which empowered consumers and freed up time that they could subsequently control and use (Senauer *et al.*, 1991). Complexity in itself is no longer valued; goods must be simple and quick to operate, and low on maintenance/repair time, so that reliability is at a premium. Convenience foods, it is said, have grown in use within this climate. Some argue that this is because of rising demands on time for all the population, while others give more emphasis to the fact that women are still expected to carry out the majority of the household chores as well, these days, as doing a paid job within the labour force, so that they are obliged to use such foods.

According to one American commentator:

> Life seems to become steadily more hectic for many Americans. Many feel dollar rich and time poor as real incomes rise, yet the demands on their limited time availability escalate. Scarcity is relative, as we become materially more wealthy and the demands on our finite time increase, time becomes relatively more scarce and hence of greater value ... the labour force participation rate for women now is well over 50% and over 70% for married women 35–41 years old. The time pressures on these women are enormous because they not only work in the labour force but continue to do most of the work inside the home (household productive activity) particularly the cooking and food shopping. Given this environment, it is easy to understand why three-fourths or more of the households in the United States own a microwave oven and why convenience has become an increasingly important attribute for food products. (Senauer *et al.*, 1991: p. 158)

This seems like a fair description of the kinds of forces at work in the UK and most other developed nations. Most would recognise the sense of time pressure which is described here – what the Henley Centre has described as the developing 'time famines' within modern societies. Yet much of this rests on assumption and hearsay, rather than careful analysis. Is this an adequate account of the changes that are taking place? Are we 'time hungry and dollar rich'? If so, where has the time gone? If time is scarce, convenience foods seem a logical choice, but given that we seem to spend large amounts of the time available to us in non-essential – and seemingly wasteful and/or irrational – activities, that we should feel ourselves to be so short of time these days seems somewhat strange. Why should time be such an important issue now, and convenience products widely used, when the same products were regarded askance when first introduced?

The other 'conventional wisdom' of recent years has been the much vaunted move towards a 'leisure society' as people are increasingly freed from the burden of hard labour by machines. Yet the 'post-industrial' society does not seem to be a place of leisure, as such. The divisions which have grown up relate to the distribution of work and the disadvantaged are those who are either sentenced to involuntary leisure through unemployment, or struggle to maintain a reasonable income through multi-occupations in households composed of casual, part time, or black/grey economy workers who are also claiming state benefits

(Bagguley and Mann, 1992). Even for the work-endowed, changes in the economy have made virtually all occupations less secure, so that changes in occupation and periods of uncertainty or unemployment are more likely. At the same time, many of the older high-skill and labour-ntensive production industries have been replaced by 'high-tech' assembly lines and low-skill service work which have sucked large numbers of females into the workforce. Rather than a choice made in order to free time for leisure, then, convenience foods may well be a response to economic and cultural pressures. The issue then becomes – convenient for whom?

Convenience is a moral issue. Society rewards what it finds useful, and punishes what does not fit its purposes. Marketing texts of twenty years ago used to illustrate the Mary Baker story (apparently apochryphal) that consumers were far less rational than was usually assumed and were activated by guilt and morality rather than logic. The story told how a new product, which allowed housewives to achieve a perfect cake every time, simply by adding water, failed absolutely. The relaunched product became a success, according to the fable, when the user was obliged to add an egg. They were now cooking, rather than escaping from their role and moral duty by using convenience products. To be a lazy housewife was to fail, or to refuse the conventions governing the female role. Food habits here are buttressing specific values, and a particular social order. Convenience is a moral as well as a practical issue.

7.2
Convenience products

Clearly, there are great differences among the types of foods which might be described as convenient. This might be argued to embrace everything from junk foods, such as crisps, sweets and ice cream, through burgers, fries and the take-away snacks and meals, to semi or fully prepared dishes. It might be argued that any food which is bought dead, or even taken from the ground or the tree is 'convenient' in the sense that part of the acquisition, cleaning, preparation, etc. has been done before the consumer acquires it, and costs for the work will be part of the price. Time has thus been 'purchased' along with the product – paying a premium for someone else to carry out part of the provisioning work needed. As with other common-sense categories, we know what we mean, but when asked for a clear definition, the vagueness of the concept actually becomes apparent. There are no hard dividing lines between these types of foods and other common-sense categories which we employ, for example healthy foods, or 'proper' foods (see Murcott, 1982; see also Mennell *et al.*, 1992). It is possible to argue and, indeed, it was the conventional view not so long ago that all dried, tinned, or even packaged foods in general should be described as convenience foods. In short, any food which has been processed in any way could arguably be described as convenient.

When we studied the food-using behaviour of UK households in the 1980s, we found a certain ambivalence in attitudes. 'Convenience' was a very desirable attribute for certain kinds of food, while it would be used as a term of criticism when applied to other types of food. But we also found that it involved far more than the simple quality of 'time saving' in preparation and eating. Certainly, there were many types of meal occasion where ready-made meals, or foods which required no preparation time at all and simply involved heating up, were used. But 'convenience' referred to a much wider range of attributes, relating not just to time spent on preparing or cooking or the method or circumstances involved in eating the food, but also to shopping and food acquisition (where was food available, in what form), how it could be stored, and thus to what extent it was ready for use (see Figure 7.1) (Gofton and Marshall, 1988; Marshall, 1988).

Within this context, foods would be regarded as inconvenient if they did not fit readily into the overall pattern of provisioning practices. Thus, if foods could not be fitted into the typical range of food combinations which make up a plate, or course, then they would be regarded as inconvenient. Similarly, if foods could not be easily stored (frozen, in dried packs or in tins) and then brought out when needed, they would be regarded as inconvenient (see, for example, Douglas, 1982, 1984; Douglas and

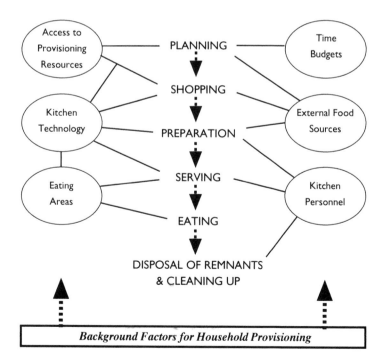

Figure 7.1 Stages in the provisioning process. (*Source:* Gofton and Marshall, 1992)

Nicod, 1974). Convenience here is being used to refer to foods which fit with a range of related practices and institutions. The typical or normal pattern which is being invoked, implicitly or explicitly, involves a particular pattern of shopping, roles for different members of the family in choosing, preparing and consuming foods on different kinds of meal occasion, particular domestic technologies, arrangements for the use of different living areas and so on. Rather than simply relating to preparation and cooking, then, convenience relates to a wide-ranging set of social institutions and living arrangements. Convenience for the households studied was a way of referring to how foods fitted into provisioning practices, which were themselves part of a set of household arrangements to provide various sorts of services to household members, and also to bring into the household itself the financial, informational, recreational, cultural, etc. resources necessary to maintain its existence.

Convenience is relative. When does food become more convenient? Foods and provisioning practices are highly class-specific. Convenience foods, until recently, were almost exclusively associated with the lower classes. Middle and upper class foods and mealtimes were composed of large amounts of relatively expensive and labour-intensive items. Indeed, for many meals, display and conformance to social norms in the form and manner of serving were extremely important. The degree of conformity, and the elaborateness of the rituals involved, varied according to the symbolic significance of the meal.

As Elias (1978) notes, in Medieval Europe, the lower classes did not eat meat not simply because of price or availability, but because it was inappropriate to their station in life. The symbolic meaning of food, rather than simply supply and demand, was the deciding factor (see also Detienne *et al.*, 1989). Food provisioning is strongly linked to the social order, power and status arrangements. The dominant groups within society exercise the greatest command over resources, and this is expressed through their food habits. As Elias argues, the growth of the centralised state and the development of mass media make it possible to promulgate a model for conduct which establishes, in the first instance, the monarch and his court as the arbiters of good taste and correct conduct, and, at the same time, a mode of regulation which involves the internalisation by the individual of mass standards, disseminated by means of the printed word, rather than face-to-face contact. 'Civilisation', then, rests on the values and ideas expressed by the dominant group within society, from whom the rest of society are to derive their aspirations, motivation and 'role models' (for a good critique see Falk, 1994). Food is particularly important here. Anthropological and historical data (Detienne and Vernant, 1989; Douglas, 1987b; Lévi Strauss, 1970) show that arrangements and conventions relating to the acquisition, division, preparation and eating of food are the basis for the most important social distinctions, divisions within society, and conceptions of social roles and personal identity. Food symbolically then is a *repository* for very important social values, and embodies ideas about how society is ordered, from the boundaries (marked by food taboos – outsiders eat what 'we' regard as disgusting, whether it is haggis, tripe or frogs' legs), while who we are, and what we are to whom, is embodied in the foods we eat, who prepares them, whom they are shared with, and so on (Cheal, 1989; Fischler, 1980, 1983). Convenience, as a principle in food provisioning, threatens this social order in the same way that dialect, jargon and slang 'threaten' the English language and its grammar. At the heart of this conception is the division of labour and, bluntly speaking, the subordination of women.

The simple idea that the division of labour arises from genetic determinants – stronger, more aggressive males providing for mate and offspring, while weaker, more emotional females filled their allotted roles as child rearers/homemakers – has been undermined by many different studies of what males and females actually did and continue to do in many different societies throughout the world. Work turns out to be much more equally divided between the gender and age groups, or even, as in the case of tribes in New Guinea, heavily skewed towards females and children, leaving the men to pursue painting (their bodies) and politics (accumulating pigs for 'Big Man' feasts), or skirmishing with other groups in pursuit of honour or revenge (Farb and Armelagos, 1980).

What is noticeable is that foods, and different ways of preparing and serving them, are given different degrees of importance or status, and that those with the highest status, or the greatest social power, have the greatest command over these resources, and also tend to display their power by 'conspicuously consuming' these resources in ritualistic feasts. In Ancient Greece, the fellowship of men and gods was celebrated in the ritual of sacrifice, where the Gods imbibed of the ritual meat through the smoke which ascended to the heavens, while the different orders of the human estate were given the appropriate cuts from the sacrificial beast, according to principles of division and allocation later to form the basis for modern mathematics (Detienne and Vernant, 1989; Svenbro, 1984).

In modern Europe, this has transmuted into the social order of the formal meal table, with the (male) head of household carving the meat (which is the centre of the most important meals) and dispensing to members of this micro society according to the age and importance. Women have always occupied a subordinate and inferior role in this system (Bourdieu, 1985). Male status, and the traditional conception of the masculine role then, rests on the work of females who have the social obligation to carry out that work. This was legitimised in a number of different ways. For instance, it derives strongly from the 'commercialisation' of time accompanying the development of industrialism, and the concomitant division between work and home. Thompson (1967: 57) describes the rise of 'time consciousness' under industrial capitalism as follows:

Mature industrial societies of all varieties are marked by a clear demarcation between 'work' and 'life' and that attitudes to time, and how it is spent, underlie this work ethic. Religion, official pronouncements and policies, and the actions of employers all concentrated on the idea of time as capital, as a resource which could be 'spent' wisely or ill.

Thompson insists that this process of disciplining is directly related to the industrialising process, and that the English working class were bombarded throughout the nineteenth century with:

... the propaganda of time thrift ... the rhetoric becoming more debased, the apostrophes to eternity more shop soiled, the homilies more mean and banal.

Generally, women became excluded from most workplaces in the course of the nineteenth century in order to provide the essential 'maintenance and support' necessary for the workforce, and also to enable them to reproduce and take care of the next generation of workers. While women are effectively excluded from the paid workforce, and assigned an essential, but highly restricted and subordinated relocate at the same time, it was argued that the role of housewife became progressively easier as social progress, and technological advance, made all the tasks associated with food provisioning and family care simpler and quicker – for example canned and dried foods, and stores which provided ready-made bread and pastries. Women who were 'at leisure' while their husbands worked in the factories, mines and offices of the new 'Coketowns' were morally obliged to spend their time caring, and were held to be bound by ethics which were rooted in society ('duty'), religion ('virtue') and, ultimately, psychobiology ('the maternal instinct' and 'romantic love'). Is this age of leisure real, however?

7.3
Whose time is being saved?

The last age of convenience for women with inaugurated at the beginning of the industrial revolution, according to some accounts. With the advent of mass production, households, according to Schwarz Cowan:

... ceased to manufacture cloth and began to buy it; ... ceased to manufacture candles and instead, bought kerosene ... ceased to butcher their own meat, but instead bought the products of the meat packers. (Schwartz Cowan, 1989: p. 53)

Although cooking and most other tasks to do with care of, and caring, within the household were thought the responsibility of women long before the modern age, like many other facets of modern life, the idea of housework was actually born with industrialism. Modern ideas of convenience, in the minds of many commentators, are also born here. It is, they argue, easier to buy these products than to have to produce them; that the move from the household as a unit of production to a unit of consumption marks a decisive improvement in living standards, and an improvement of the lot of women in particular. At the same time fertility rates were falling, and living conditions for many were steadily improving.

Yet, as Schwarz Cowan argues, the acquisition of modern technologies such as cooking stoves, cleaning machines and preserved foods did not lead to a life of leisure for the women concerned. The literature of the time is filled with accounts of women whose 'work is never done'. According to Catherine Beecher, American women at this time were certainly not enjoying a life of leisure:

> The anxieties, vexations, perplexities and even hard labour that come upon American women are endless and many a woman has, in consequence, been discouraged, disheartened and ruined in health. (Beecher, 1841: p. 18)

Schwarz Cowan argues that although labour-saving technologies are distributed throughout households in the nineteenth century, they do not save work for the housewives involved, but only reorganise the work involved. Looking at the work which housewives did with the new products and the new technologies, she finds that these typically result in a reallocation of responsibilities and the tasks associated with, for example, making bread, which shifts work from men to women, so that women's work typically increases as a result. The replacement of hearths by iron stoves, for example, typically led to the production of a more varied range of foodstuffs, as well as requiring cleaning, so that women found themselves not with free time but with more to do as a consequence of this increased capacity. The time that was saved was men's time, which was freed up to be spent in the factories and mines of the new industrial state, while women became unpaid domestic ancillaries. 'Convenience' here was economically derived. Pahl (1984) argues that what we see as a logical and 'natural' division of labour in fact represents only a partial and temporary departure from what were much more diverse and integrated role relations before industrialism. He also argues that the post-industrial lifestyles of groups most strongly affected by the disappearance of stable, long-term male employment in single occupations have returned to a more integrated pattern of household provisioning, and social roles in which male and female take part in providing a range of household services (Gershuny, 1983; Gershuny and Jones, 1987). If we look at convenience in the post-industrial age – the age of the working wife and female emancipation – then do we find a different story?

**7.4
Time famines:
explanations are not
self-evident**

Jonathan Gershuny and his colleagues have been studying time use for many years, using data gathered from a wide range of studies. Data is gathered from diaries in which subjects record details of the activities which they undertook daily, along with details of the times at which they occurred (Gershuny *et al.*, 1986). As Gershuny points out, given a longish history, and a fairly large number of actual studies yielding fundamental and comprehensive data, time budget analysis '...ought to have had a substantial effect on social science; in fact its impact has been less than impressive' (Gershuny and Jones, 1987: p. 11). Partly this is because of the nature of the data, and partly because of the lack of an adequate theoretical framework about the determinants of time use allocation.

Studies have been able to use data on time use gathered amongst the UK population on a number of occasions in the past thirty or forty years, alongside regular studies which have been conducted recently. One study based on data collected in 1961, 1975 and 1984 looked at the changes which have taken place in the amounts of time devoted to paid labour, domestic work and leisure, broken down according to socio-economic characteristics such as gender, age, employment status and so on. A number of changes are indeed apparent, but they are not entirely as we would expect, and, further, they may not be entirely self-explanatory.

The main changes for our purposes are as follows:

- Although the amount of time spent on domestic work decreases for women, and the amount of time on paid work increases, this is not a simple substitution.
- 'For equivalent amounts of paid work, women in the mid-1970s did 50–100 minutes per day less routine housework than women in the early 1960s. The change between 74–5 and 83–4 is negligible by contrast' (Gerschuny and Jones, 1987: p. 14). The earlier period saw the introduction of the majority of 'timesaving' technologies into the home, while the later period did not.
- Women do between two and three times as much domestic work as men, but there is a clear rise in the amount of domestic work done by men between the mid-1970s and mid-1980s.

First, Gershuny cautions against assuming that the descriptions of time use will provide a good basis on which to construct explanations. For example, noting the time spent on paid work by women and the time spent on child care or food provisioning will not *in itself* provide a useful basis for making inferences about what that means for the people in question:

> at a given historical juncture we have a set of cross cutting explanatory classifications, some of them determined wholly externally (exogenously) to the individual (sex, age) some of them ('endogenously') determined in part at least by the individual (employment status, stages in the family life cycle). And associated with each of the categories defined by these cross cutting classifications is a particular pattern of behaviour. (Gershuny and Jones, 1987: p. 16)

Gershuny makes the point that, for instance, changes in *proportions* within the population which fall into a particular category may create the *appearance* of change, while the pattern of behaviour associated with that category actually remains unchanged. For example, as the number of single people rises and the number of nuclear families falls, the time allocated to activities in which the enlarging group are involved will increase, even though the behaviour of people within population categories is exactly the same. Simply looking at the time spent on such activity by the population *as a whole* would not give us this information. The aggregated figures do not provide an adequate basis for drawing inferences. Women are in different circumstances if they are single parents, or living alone in old age, or coping with a large family and a husband and a job. If, for example, the number of young single people increases, this will have implications for eating patterns, how people shop, what they buy and how they provide food for themselves. 'Convenience' and eating out, for example, for this group will mean something quite different than for a working mother, or a pensioner. Even when the data is disaggregated, the descriptive social categories involved may only offer a tautologous explanation: '...we do child care "because" we have children, we do paid work "because" we fall into the category of the full time employed' (Gershuny and Jones, 1987): p. 16), while other categories function as 'explanatory' variables on the assumption of normative implications (e.g. female implies housework on the basis of a gender norm). Changes in time use, according to Gershuny, may be accounted for in a number of different ways. Descriptions do not logically precede explanations here – the descriptions are not in themselves self-evident, but require careful exegesis. Apparent changes in the ways in which time is used may be a result of:

- changes in the composition of the population (e.g. more fast food because more single-person households – less cooking);
- changes in patterns of behaviour specific to particular categories (young people eat less meat than in the past; meat is the important aspect of main/formal meals);
- changes in the 'sorts of people' recruited into a category (e.g. more single parents amongst employed women – increased use of children's convenience food/snacks/separate meals, etc.).

Depending on the sorts of initial judgements ('theories') which inform the initial choice of classificatory dimensions (and which in turn result from the choice of boundary of what is to be explained) the same decrease in domestic work over a historical period could be viewed in two quite different ways; as a result of a change in the composition of the population (fewer housewives, more part-time employed women with children) with no change in behaviour of either category; or as a change in women's behaviour (women deciding to have jobs and children). Neither explanation is self evidently better than the other; which we choose depends on whether we take the choice of family and employment status as externally determined givens – or as a lifestyle requiring explanation. (Gershuny and Jones, 1987: p. 17–18)

In the same way, clearly, if we accept that there is less time available now, this may be attributable to there being a larger number of households in which, for example, one of the parents is missing so that the remaining parent may be taking on virtually a dual role, or there is only one person to carry out all the tasks shared between the family. There may not have been a significant change of behaviour in different segments of the population (for instance, older people) but the balance of population has shifted. The time pressures on which explanations of the use of convenience food rests, then, are certainly not self-evident. Such changes in time use as have taken place require explanation, and the data suggest that such explanations are unlikely to be simple and transparent.

7.5
Why have our attitudes towards convenience food changed?

The old ideas concerning the importance of certain kinds of foods, and the role of females in the production of such foods, has certainly not disappeared. As Anne Murcott (1982) found in her study of food provisioning arrangements within violent, wife abusing households, the idea of a 'proper' meal, and the insistence on the wife's duty in preparing it, were often the source of conflict and violence in households where, in many instances, the husband was no longer the financial breadwinner, and where wives found themselves having to carry out a new working role in addition held to the responsibilities of the old (see also Glyptis, 1989; Hinrichs and Sirianni, 1993).

A range of research into time budgets by Gershuny and his colleagues has borne this out. Although there have been changes, and the majority of women now spend most of their adult life in paid employment, many 'domains' of domestic life remain predominantly female. Most married men still do not know how to operate the washing machine, few ever clean the bathroom, and women now spend more time with their children than they did in the early 1950s. As Thrall has commented:

> Rather than break down the traditional role assignments, modern household equipment seems to reinforce them by making it easier for those who are stereotyped as doing particular tasks to do them without help from others. (Thrall, 1982: 192)

Yet food habits have changed. In their analysis of National Food Survey (NFS) data over a thirty-year period, Ritson and Hutchins (1990) point to a clear shift in demand towards foods which seem 'healthier' and also 'more convenient'. Demand for foods which were previously associated with status and affluence, such as meat and dairy products, has remained static or has declined, while foods requiring little preparation, snacking and grazing foods, junk foods and ready meals have all increased (see Ritson and Hutchins, chapter 2 of this volume). The usual explanation for this revolves around women's emancipation, on the one hand, and increasing affluence on the other. It is argued that affluence and the ownership of labour-saving technologies have released females from domestic labour, enabling them to earn wages which can then become part of household resources. Increasing numbers of 'working wives' has had a 'knock-on' effect on the demand for all kinds of goods and services which are more time economical. McKenzie's (1986) diagram illustrates this very well (see Figure 7.2).

Yet, of course, other sorts of motives are at work, as they have always been. To reduce the 'utility' of food to the time which it takes to prepare

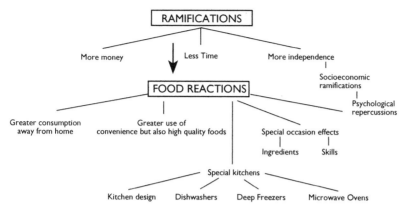

Figure 7.2 Ramifications of women at work. (*Source*: McKenzie, 1986; p. 164)

it seems to make consumers, in the phrase of Harold Garfinkel (1967), into 'judgemental dopes'. Our reasons for food choice are seldom divorced from the relationship of provisioning practices within a nexus of social institutions which all influence how choices are made, what they mean to those involved and so on. In fact, a strong argument in support of this view has been produced by an economist, dealing with the difficulties involved in the traditional economic theory of consumer behaviour.

Under the traditional economic theory of choice, any observed consumer behaviour which is not explained by income and prices is to be explained by 'variation in tastes, since they are the portmanteau in the demand curve' involved in the function:

$$X_i = {}^{D_1}\left[\frac{I}{P}, \frac{P_I}{P}, \frac{P_S}{P}, {}^{T} \right]$$

where

X_i = the demand for goods
I = money income
P_I = the money price of goods $X1$
P_S = vector of prices of substitute goods
P = a price index
T = time trend

Yet, as Becker (1965, 1973, 1976) points out, economists have no real theory about how tastes develop, nor any way to predict their effects. When grouped data are compared in terms of the influence of family size, family age structure, education, housing tenure, occupation, race or socioeconomic status, systematic effects are often observed – but, of course, these factors are standing as proxies for the effects of 'tastes'. Thus, for example, demand for fish fingers, or hamburgers, stands as a proxy for a changed taste for 'convenience', while increased demand for fresh produce and wholemeal bread indicates a changed taste for 'health' products. This is the position taken by, for example, Ritson and Hutchins (1990) when commenting on the failure of prices and income to account for changes in the demand for various products such as red meat.

Becker himself argues for the attempt to develop an economic theory which incorporates a broader set of factors than simply those which can be easily quantified by '...the measuring rod of money'. Behavioural decisions involving choices made with limited resources among competing ends might include issues such as the allocation of the consumer's non-market time, choice of religion, a marriage mate, a family size, a divorce, a political party or a lifestyle, all of which involve the allocation of scarce resources amongst competing ends.

The basic allocation of time, for instance, involves work in the labour force, household production activity and leisure, although other categories might be added (see also Blyton, 1985; Gronau, 1977; Hewitt, 1993). Typically, the value of time, or the 'opportunity cost (what must be foregone by making one choice rather than an alternative)' is based on the individual's wage rate. Those whose time is not spent in the workforce can be said to have foregone opportunities to earn money, according to the wage rates they command, or their potential, based on qualifications, experience, etc.

Becker combines budgetary and time constraints together as a single 'full income' constraint. This involves the sum of all non-labour income (e.g. rent or interest) plus the total time allotment of each household member valued at his or her opportunity cost of time. In this sense, it is a measure of *all* the household's resources, which can then be allocated to leisure, household production activities and work within the labour market. According to neo-classical theory, maximum efficiency will be achieved when time is allocated so that its value in different activities is equal, although this is seldom achieved within the constraints of the real world, in which individuals must meet conflicting requirements for the way in which they disburse their time. According to this approach, foods may be characterised in terms of 'time intensity' (ratio of time costs to full price) or 'expenditure intensity' (ratio of purchase price to full price).

In this model, then, the demand for a particular good is affected by not only its price, the price of other goods and the household's full income, but by the value of time. The full price of something comprises the cash expenditures and also the time costs to make the item consumable. In the case of food, the 'full price' will be determined by the cost to buy and prepare, but also by the value of the time which is put into the total process of acquiring and consuming the food involved.

It is quite clear too that other issues affect the value of time – for example, competitive demands upon the time of the 'kitchen person' will be highly relevant to the 'price' of particular types of foods. Thus, for example, the perceived responsibilities of various household members will strongly influence the way in which time is allocated, so that, for example, a general perception of the need for parents to 'spend time with their children' or to 'improve their qualifications' or 'quality of life' will have implications for the value that is placed upon time resources. Becker argues that we cannot begin to understand how choices are made in relation to food unless we see time, and other non-money factors, as resources which are implicated in the provisioning process.

While traditional economics sees the consumer receiving utility directly from products and services as they are consumed, Becker proposes that actual consumables which satisfy human wants and needs are produced within the household by the combination of purchased goods, members time, household and human capital. Thus, for example, *the utility of food is composed of the satisfactions that it can provide which can only be brought out by using a whole range of household resources* (an automobile for shopping, equipment for storage and cooking, condiments, sauces and complementary foods), as well as the time, effort and skills which have to be acquired by the people involved in the provisioning process. Convenience, then, rather than simply relating to 'time saving' may well involve different characteristics for different products. Convenience in a

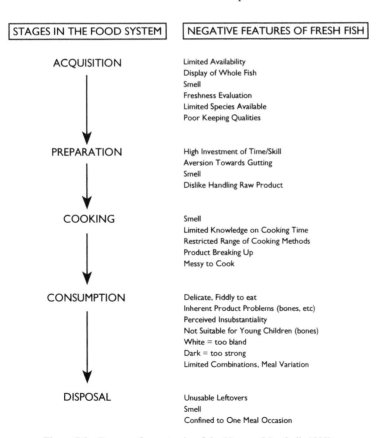

Figure 7.3 Reasons for not using fish. (*Source*: Marshall, 1988)

food product may well reside in the fact that it does not combine with standard vegetables, or is a minority taste.

When we investigated consumer food habits in the UK (Gofton and Marshall, 1988, 1992) we found (Figure 7.3) that fish was regarded as very inconvenient, because of the need to invest large amounts of time and effort, and devote special resources to the various stages of the provisioning process, but also because it necessitated non-usual complementary vegetables.

In large part, of course, aspects of the 'inconvenience' of fish were cultural specifics, related to the conventions of taste and custom – for example, the non-substitutability with meat, mutually exclusive patterns of combination with sauces and vegetables, etc. Yet neither potatoes nor meat were regarded as inconvenient products, despite the fact that they also require significant amounts of preparation time and 'trouble' in cooking and disposal of remains, for example. According to Mary Douglas, this seems likely to be because of the role that such foods play in our food system:

> Cheap labour-saving substitutes for or additions to a diet can be introduced in the unstructured parts of the system ... (the most structured parts) ... combine formality and intimacy, extending from Christmas Day to the weekday six o'clock meal. Innovations (in the working class meal system) have involved upgrading the more modest forms to a closer copy of the best ... In the most highly structured areas of the dietary system taste discrimination is most developed and standards of presentation strictest. This is why powdered potato never threatens the full place of the cooked potato in the working class diet in spite of the heavy demands that the latter makes (in cleaning, peeling, cooking and pan washing) upon the housewife's limited time. (Douglas, 1982: p. 103)

Recent changes here may well reflect changes in the social relations within the family – for example, more joint decision making, and/or more consultation, or even more autonomous food choices by family members. Becker's account, then, would see convenience food use as a shift in the significance of the elements involved in the full cost in terms of household resources. As the value of women's time increases, given the increased pressure on household budgets, and various social and economic pressures, the use of convenience food more than offsets the increased money expenditure.

Research has shown that as the value of (women's) time rises, consumption of fast food and food eaten away from home rises, and more convenience food is bought (Hull et al., 1983; Robinson, 1972). One US study (Hull et al., 1983) looked at the relation between consumption of convenience food and different kinds of households: where married women were full-time homemakers, a calculation of the value of their time based on their potential earnings in the labour force revealed that the higher that value, the more the household spent on convenience food; if the primary meal preparer was in paid employment, her rate of pay was directly and positively related to the rate of consumption of complex and manufactured convenience foods. Becker is essentially arguing that convenience is not an attribute of products in themselves, but is an outcome of the ways in which these are used in household production processes. Thus higher levels of convenience in some foods are achieved, not simply by using certain kinds of foods, but by bringing to bear appropriate household resources – which may involve specific technologies (for storage, cooking, preparation, etc.), special skills (in preparation, serving, cooking etc.), combination with other ingredients or special satisfactions related to the form of the food, the manner of use, the setting/occasion, or the user (see Marshall, chapter 11, and Bell and Meiselman, chapter 12 of this volume).

Conveniences relates, then, to the capacity of consumers to employ particular resources, as well as simple time availability. Thus, for example, Marshall and Gofton (1992) found that fish fingers provided a range of benefits related to their use by children – their ease of storage and availability, their healthiness, safety and predictability, their suitability for lap meals, ease of handling on the plate, and so on – apart from their speed and ease of preparation.

Becker, however, prepares an advocacy of the full cost model which is analytic and formal, but also recognises the limits to its empirical application. Time costs involve elements which would be analytically important, but almost impossible to quantify. It is not, he says, simply a question of time available, but the quality of that time.

Human capital, the knowledge, skills and abilities of members of the household, clearly affects the quality of time input, and will increase production of household goods and services. Historic factors, then, such as time invested in training and education, or learning new skills, is also directly relevant to the calculation of full cost. Rather than simply deriving from the value of time available, then, full cost will be affected by the time and resources devoted to training or not training members of society in the necessary provisioning skills required to plan, choose, prepare, consume, etc. food and drink. Our research (Gofton and Marshall, 1988) showed very clearly that in the use of (healthy) fresh fish as a food resource, the food was regarded as highly risky by younger housewives because most did not possess the necessary skills to discriminate fish which was really fresh, had no knowledge or experience of any but a small number of fish species, and thought fish very difficult and inconvenient to prepare and cook. What households can produce, then, is affected not simply by monetary resources, and time availability of household members, but by the *quality* of their input, itself a product of the investment made by individuals, and by society at large, in training and providing experience of the necessary skills and knowledge.

Convenience culture, then, must also be seen as the outcome of an historic process in which consumers are trained and educated, and, hence, to the values and ideologies which underlie these arrangements. Time is, then, an important resource, and it is quantifiable, but it is also qualitative. Different types of time availability will have different implications for the provisioning arrangements of the household (Morris, 1989). In other words, convenience will have different implications, according to the characteristics of the household itself. To assume a simple notion of 'time saving' is to assume a uniformity in household structures and roles which seems dubious at the outset. Social roles are changing, but how?

7.6
The ways in which households and social roles are changing is complex

According to Ray Pahl, it is excessively simple-minded to assume that there can be a modal form for the changes in social roles, and the arrangements made within individual households as a consequence for the profound range of social changes which have taken place. Researchers are focusing on:

> ...how the social relations of employment intermesh with gender relations, with the workings of the household and with the distinctive patterns of lifestyles that men and women wish to create for themselves in a society where a belief that such choices are possible is now more widely held by women and by men. (Pahl, 1993: 630)

In fact, such relations (between work, employment, kin and household arrangements) are highly complex, and arrangements are likely to be highly dynamic as situations change and develop:

> Children, elderly parents, new partners, siblings and au pairs come and go. More serial monogamy, more stepchildren, more 'care in the community', all have to be meshed with just-in-time production systems, flexitime, job sharing, round the clock provision of services and greater demands for travel and mobility consequent upon enlarged markets and more personal negotiation based on increased flexibility. (Pahl, 1993: 630)

Most of the research suggests that heralds of gender equality are responding to a false dawn. For various reasons, work relations and domestic division of labour are still heavily gendered. Indeed, Charles (1993) claims, contrary to the prevailing view, that the general 'improvement' in the condition of women, rather than coming from the political movements, has actually come from the economy, which has consistently increased its demand for labour in the 20th century. In fact, she argues, the rhetoric of liberal 'improvement' (individualistic, equal rights, feminism) throughout the 1980s fitted, in the Reaganite USA and Thatcher's Britain, perfectly alongside the individualist, 'enterpreneurialism' which was also calling for the dismantling of the old masculine, union-dominated work system. The women's movement apparently:

> ...had the effect of accelerating changes that would probably, sooner or later, have occurred anyway. Thus, it could be argued that the women's movements have played a part in changing legal and social policies such that they are more in line with economic developments. (Charles, 1993: p. 249)

It follows that the domestic division of labour has hardly been changed; it could, indeed, be seen as reinforced, as women's responsibility for home and children has been reasserted with ideas such as 'family values' and 'community care'.

That the subordination of women has not been materially affected by the changes of the past few years may be contentious, but most of the studies which have been done show that statistics on women's working paint far too rosy a picture, since the most disadvantaged, notably the low paid workers, homeworkers and the unemployed are not always included and those with responsibilities for caring for the young, the sick, the elderly and the handicapped will certainly never be integrated into the world of work in the way that men without such responsibilities will. Such men are integrated into the economy precisely on the assumption of unwaged servicing by adult women. Cultural conventions and real differences in patterns of training, ideologies of work and gender, and channels of opportunity are profoundly important.

It seems likely, then, that the use of convenience foods is related more to economic pressures than material improvements in the condition of women and that the changed moral status of convenience foods reflects an ideological shift which recognises the productive benefits of that pattern of usage. At the same time, food is a powerful social symbol; and changes here, as we saw earlier, strike at our deepest values related to social order and social relationships.

Food symbolises many things – the care of a mother for the different members of her family, friendship or kinship, respect or duty. These usages become established through convention, through the association of a particular kind of foodstuff or meal with a particular kind of context, ritual or exchange. These conventions can be very powerful. Lévi-Strauss describes how reluctant are the French to eat certain kinds of foods – turkey, large joints of meat, ham, and so on – without sharing them with other people, almost as though the form of the food implied how many people must consume it (Lévi-Strauss, 1970). Clearly, particular occasions imply certain kinds of dishes: business lunches, or inviting the boss to dinner, require a degree of formality, fuss and ceremonial display, while friends or family may be treated more intimately and casually. As many social analysts have pointed out (Douglas, 1984; Douglas and Nicod, 1974; Goffman, 1959), the greater the degree of intimacy in a social relationship, the less important it is to retain propriety and maintain social distance through putting on a more or less formal marking of social boundaries. Within the family, or with close friends, food occasions in the home can involve sharing full meals, including items such as potatoes, meats, vegetables and so on. Visitors or service personnel will be offered a more limited and restricted form of hospitality. Food here is expressing something about the quality of a social relationship in a material form. Change is signalled by changes in what is eaten, rather than in any explicit statement or formal declaration.

According to Mauss's (1954) famous essay on the nature of gifts and their social significance, in general terms, the less practical an object is in terms of mere subsistence, the more power it has to convey symbolic meaning. Those things which are sold in 'gift' shops have almost no practical use whatsoever, but serve entirely symbolic functions. When we buy message balloons, huge plastic keys or heart-shaped cushions, we are buying the means to express our feelings, to acknowledge propriety, to fulfil an obligation, to reciprocate a similar gesture or whatever. This may also be accomplished by offering or consuming food or drink in forms or amounts which far exceed our appetites or needs. We recognise that foodstuffs in many settings are expressions of caring, respect or duty, and these must be consumed, to avoid giving offence by denying or refusing to recognise the relationship that is implied, or the moral authority of the conventions governing behaviour within the social setting. Often the symbolic association may come to outweigh the utilitarian aspects.

In the case of food, the encroachment of 'trash' and 'junk' into the careful formality of 'family meals' has been seen as a threat to those institutions which are at the heart of our civilisation – the family, our sense of history and continuity with the past, and even our notion of how public and domestic spaces are properly used. Junk food, along with other innovations, fads and fashions within popular culture, have been seen as evidence of our moral disintegration – the family which does not eat together is the family which causes many of our most alarming modern social problems. The 'dysfunctional family' has become a modern bugbear (see Dennis and Erdos, 1993; see also James, 1990).

Falk (1994) sees modern eating cultures as reflecting even more profound changes. The meal as a social ritual, the utopian idea of the 'eating community' which is resuscitated in the 19th century, promoting ideas of sharing and reciprocity (for example, the invention of Christmas feasting in Dickens; see Hobsbawm and Ranger, 1979) collapses as a structuring principle of social life. Falk argues that the marginalisation of the meal as a collective social event, accompanied by the rise of different forms of non-ritual eating involving: '. . . oral ingestive activity which concerns substances that are not considered to be food – sweets, titbits, soft and alcoholic drinks – or which actually fall outside the category of nutrition (tobacco, chewing gum)', which he characterises, following Goffman (1967; p. 146) as 'oral side involvements', can be traced to the impact of mass production and consumption of these items in the USA at the end of the 19th century. He uses George Simmel's analysis to argue that the grounds of sociality, originally formed around the rituals of food sharing and partaking, have moved through the stylised exchange of words in the modern meal occasion, underpinned by etiquette, and towards the marginalised modern meal, bounded by acts of individual, solitary eating. The replacement of food from a central role (in the Kwakiutl culture, talking about food while eating is rigorously forbidden) in the modern meal, where 'Every meal, every food is experienced by individual persons as a "taste event"' (Rath, 1984: p. 321) every food evokes and sustains conversations, sharing of words, and on towards eating as a peripheral, side activity, no longer the cement which binds the group together. This is, he argues, a model for the role of all consumption in our times – from the incorporative passivity of 'eating the world', taking it in, using it, to the active, expressive role of consumption in the (reflexive,) practices of 'symbolic creation':

> ...The great civilizing move from communion to communication characterises not only meal rituals and foodways but the whole modern culture and the ways people relate to others, to themselves and ... to the objects of consumption. (Falk, 1994: p. 36)

The modern self, then, rather than being bounded by the rituals of the meal/social group, and silenced, consumes to mark off the boundaries of the self, is creative and 'external', outgoing. The individuality of snacking, and non-food oral stimulation, represents a tendency in the development of eating culture towards an active, creative, expressive form of consumption, rather than the rituals of what Durkheim referred to as 'mechanical solidarity' (see also Warde, 1994). This theme has also been celebrated in the work of Paul Willis (1990).

For many years, popular culture was seen as an oppressive and dehumanising product of capitalism. Fierce critics such as Theodore Adorno and Walter Benjamin saw consumers as manipulated and passive, and cultural forms such as popular music as mindless pap, which, quite literally, anaesthetised popular sensibilities. Post-Marxist theories have, however, rethought their views; previous critics are now being seen as élitist, and the freedom of consumers to make choices is being emphasised. The practices of everyday life are certainly influenced by structural constraints – for example, factors such as income and educational opportunities and by what the culture industry in all its myriad forms offers, such as the demand for soft drinks, or fast foods. Nevertheless, people can always actively choose and create their own 'cultural home'. John Fiske (1990) argues that, within the post-modern world, rather than serving simply as a means to mould and manipulate, consumption offers the possibility for:

- collective resistance;
- personal pleasure;
- individual freedom within everyday life.

> Consumers of media messages or cultural commodities are not just passive dupes of advertising, soap operas or fads generated by 'keen capitalists'. (Fiske, 1990: p. 23)

But rather there is an emphasis on freedom and choice. Paul Willis uses the expression 'practical existentialists' to describe the workings of the 'common culture' inhabited by the young. Willis uses research with young people in Wolverhampton and how they deal with television, magazines and videos, pop music, clothes and fashion in order to develop a general theory of consumer practices. He is firmly anti-élitist:

> People are not manipulated consumers whose everyday lives disappear in endless streams of loud sounds, vibrant images and cheap but fashionable commodities. (Willis, 1990: p. 68)

Cultural commodities are actively used as symbolic resources within specific individual and/or social contexts. These construct or sustain a particular personal or social identity. Images, sounds, words, etc. are invested with specific meanings and 'grounded' in everyday life. Consumer culture is, then, the medium through which everyday life is given meaning and order, and understandings of its artefacts and their usage is the medium through which individuals within their social settings interpret what is going on, and also transform its meaning in the creative process. Commodities are used as a resource in 'practical creativity' and function as raw materials which are constantly encoded, decoded and recoded. They never have a fixed, final meaning, but are always being used in novel ways which undermine and subvert pre-existing conventions and norms. In many ways, Willis argues, common culture is being used in place of the central institutions of society – religion, art and literature, political beliefs, occupations – which formerly provided guidance and reference points in the constitution of the individual identity and the group's consciousness of itself. Large numbers of people find that they can no longer rely on organised religion, high culture, general ideologies or politically articulated forms of class consciousness to define themselves. Consequently, everyday symbolic creativity has become more important, as a result of the '...vast interplay between secularisation, consumerism and value systems'. Consumers construct their individual identity out of many 'commercial resources' or cultural commodities:

> This bricolage is reinforced within the lower strata by the ongoing mechanization of work and industry. (Willis, 1990: p. 72)

Realisation of autonomy, or self-actualisation, which was previously part of occupational life, is now closed off to many. As a consequence, Willis argues, deskilled people use the realm of leisure to demonstrate their personal abilities (see also Hirschman, 1992). In the realm of consumption, affiliations are also established – consuming a product means being included, being 'one of the in crowd' as many advertisements promise. Some advertising sets out to appeal to this need to belong by attempting to generate anxiety (Robins, 1994). In a sense, this represents a new kind of association. Willis also sees earlier forms of sociability ('fixed sociability' related to objective factors such as neighbourhood or work) being replaced by 'proto communities' or floating communities, revolving around these kinds of shared leisure and consumer interests, fashions, musical tastes.

Our research into markets for alcoholic drinks and drugs (Coffield and Gofton, 1994; Douglas, 1987a; Gofton, 1990, 1993) illustrates this sense of community born of consuming. Young people at a Rave or in town centre pubs on a Saturday night are united precisely by their consumption, where previously they would have been formed around neighbourhood, peer group and occupational groups. 'Being together', being with people like myself' were common reasons for attendance amongst both groups. Willis argues that leisure interests are now much more important to young people than class interests or these kinds of objective elements. At the same time he criticises the 'symbolic violence' which imposes, through the educational system and official cultural policy, a very narrow definition of symbolic creativity and of a very disputable and even refutable hierarchy of high art and low cultural forms. Although practical freedom is subject to constraints, these do not determine actions within leisure, but only structure leisure or consumer opportunities. Through this creativity, young people are taking the opportunity to form their own life histories. Consumer culture, he argues, is a form of 'grounded aesthetics', creatively transforming everyday life. In this case the 'added value' attached to convenience foods must be seen as a symbolic *resource* in self-constitution, an option by means of which the consumer can *choose* to constitute a self rather than a 'gull' through which passive dupes are hoodwinked (see also Chaney, 1990; Jameson, 1983; Lane, 1993).

7.7 Fast food as popular culture and practical aesthetics

A move towards much less structured, more individualistic eating – in Durkheimian terms, a more 'anomic' form of eating – may well indicate changes in the ways in which household relations are articulated. For example, sharing entertainment, or participating in the consumption of information, is surely replacing food consumption as a form of sociality. Food has moved from being the focus of household ritual to being an adjunct, or an embellishment, to other kinds of household activities. Symbolically, food is now playing the kind of part which Mauss assigned to other 'non-functional' goods in exchange processes. As Falk suggests, food is no longer the primary means by which the group is bound together, but plays a different role (Figure 7.4).

As the micro culture of the modern household changes, food occasions must take their place alongside other home-based activities. Reducing the motivations involved in the consumption of modern food items to 'timesaving' ignores the range of other meanings and satisfactions provided. Fast foods are, as critics are quick to point out, heavily marketed. Such marketing typically strives to 'add value'; to create a 'brand personality'; to present products as 'solutions to consumer needs'. When we investigated the market for alcoholic drinks (Gofton, 1986, 1988, 1990), we found that amongst younger consumers, the appearance of the product was extremely important, and the container was a kind of fashion accessory. Drinks went in and out of fashion, and consumers were prepared to pay premium prices for 'designer label' products or 'liquid confectionery'. In the trade, this is referred to as 'badge drinking'.

Foods are clearly regarded in very similar ways. Eating 'home delivery pizza' while wearing a Malcolm X baseball cap and watching American

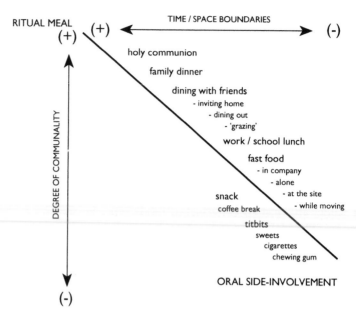

RITUAL MEAL (+)

TIME / SPACE BOUNDARIES (+) ←—————————→ (-)

holy communion

family dinner

dining with friends
- inviting home
- dining out
- 'grazing'

work / school lunch

fast food
- in company
- alone
- at the site
- while moving

snack
coffee break

titbits
sweets
cigarettes
chewing gum

DEGREE OF COMMUNALITY

(-)

ORAL SIDE-INVOLVEMENT

Figure 7.4 Transformation of the eating-culture. (*Source*: Falk, 1994: p. 31)

football on television are, of course, to do with the ways in which consumer choices constitute the basis for personal identity in the world after work has disappeared (Lash, 1993; Laermans, 1993). Making ready meals available to single-person households for whom the structures, routines and focus provided by partners and families, and others in general, is largely absent is as much about removing the need to 'bother about' food provisioning (and recall that those activities involved in 'bothering' have in the past been the means by which the forms of relationships within the household were daily acknowledged) as it is timesaving. Singles may well have plenty of time, but little inclination or incentive to spend it shopping, preparing food and cooking simply for themselves. In addition, we could hypothesise that the social dimension of meals within households is being replaced by the sociality of McDonaldland, of the fast food bars, the 'Happy Eater', the drinking place, the Rave, the nightclub and so on, where food and drink are complementary to the exchange of words and other tastes and symbols – where sociability is consumed above all else. The de-skilling and the functionality which is involved has an important dimension too; for men and women in the post-feminist age, food provisioning has a whole set of resonances which relate to traditional gender roles and patriarchal family structures. Post-modern provisioning defuses these issues; formal eating shrinks to the most important ritual feasts – weekly, yearly and at rites of passage, rather than daily manifestations of subordination and dominance.

In the same way, young people's increasing rejection of meat has very clear moral and aesthetic roots. In a culture which anthropomorphises machines, let alone animals, and which is increasingly seeing the natural world as a leisure opportunity and theme park, rather than a source of protein, the added value imparted by convenience food involves selling the sizzle, rather than the burger. The sociality of the fast food bar is the main theme of advertising and promotion; McDonald's have long been involved in selling the service and friendliness of their establishments, rather than the food as such.

7.8
What is 'convenience'?

Convenience, then, does not simply mean 'labour saving'. At the very least, it raises the question of whose labour is actually being saved. Schwartz Cowan's work shows us that the real issue is, in the words of Lewis Carroll's Humpty Dumpty, not what convenience really means, but who is to be master. Changes in household technology created the possibility of there being 'separate spheres' for men and women (Schwartz Cowan, 1989). Male involvement in provisioning and other domestic areas was reduced while chores for women increased. Conventional wisdom, which typically represents the process of industrialisation as the period when the family moved from being a unit of production to becoming a unit of consumption, actually misrepresents the process, since as Schwartz Cowan argues, with the Marxists, what is actually going on is the industrialisation of domestic production. Goods and services consumed by the family, for example cooked meals, child care, rest, psychological support, clothes, shelter and so on, are still produced within the home, and the home can be seen as producing healthy workers and future citizens by provisioning, caring for their health and wellbeing, and providing them with rest and psychic support. All of these tasks, under industrialism, become the province of the women of the household, however; the first phase of industrialism removes responsibility from the males.

Time availability, time as a resource, reflects social values. It becomes a moral issue when it functions as a discipline for the family unit under industrialism. Falk's notion of this 'utopian' vision of the sociality of the meal event, yet another moral code invented and disseminated to legitimate the social order (with Dickens and Thackeray in the place of Erasmus, see Elias, 1978) has survived almost till today. The lazy housewife who served her family junk food and failed to do her duty in the kitchen was a moral fable underpinning what was actually the transfer of work and responsibility for domestic work from men to women. The corollary of the moral responsibility of women for unpaid domestic work was the commonly expressed view that labour-saving technologies left women with nothing to do, compared with the past, and denied them legitimate leisure, or command over household income. Industrial technology in the kitchen was really there to increase the amount of work women could do, and could thus be made responsible for, while their men were in the factories and mines (see Mintz, 1980, for a parallel case; see also Reilly, 1982; Rees, 1993; Pahl, 1989).

In the post-industrial world, women's roles have been extended even further. The general perception that dual income families, or families where the wife works while the husband is unemployed, have led to more emancipation for women, the dismantling of gender divisions in domestic responsibilities, and more free time, is, generally speaking, wrong. Gershuny's time budget studies show that men have more free time but women generally less: that many areas of domestic responsibility remain exclusively female; that women are spending more time looking after their children now than they did in 1950, not less; and that most women work and are also still responsible for domestic care systems of all kinds. It could be argued that women have been, once again, used to buttress the social dislocation resulting from profound social and economic changes. 'Time famines' hinge, it appears, on women's lack of entitlement to use time and not simply its general availability or scarcity. As Sen (1981) has pointed out in relation to food famines, these often arise as a result of groups being deprived of entitlements, or because of society failing to distribute food resources equitably rather than because of crop failures or real shortages in total resources. Time famines are result of women being used now in order to provide flexibility in the economy, and to buttress the shift in political and economic power which that labour market ideology involves.

Changes in attitudes towards, and usage of, convenience foods reflects a shift in control over household income, and responsibility for domestic work. Cooking has been de-skilled, or transferred out into the commercial sector, either into the burgers, chips, coke and ketchup on which the younger parts of the family graze, the ready-to-eat food which it brings home, or the chilled and frozen products which make up the meals of the older people. Women who do not cook are no longer subject to Society's approbium and this represents a dual change. It represents a reordering of the economic arrangements which underlie both the macro-economy of society and the micro-economy of the household; and it also represents a cultural change, under which social life and collective membership is realised not within the passive, mechanical discipline of formal meals, but in the active, creative aestheticisation of the world in the act of consuming, in which the symbolic dimensions predominate over the materiality of eating. The preceding analysis has aimed to show that 'convenience' eating is a complex and contentious object of analysis. To assume that it can be dismissed as 'timesaving' is misleading, simplistic and truly a waste of time.

References Bagguley, K. and Mann, K. (1992) Idle Thieving Bastards? Scholarly Representations of the Underclass. *Work Employment and Society*, **6**, 1, 113–126.
Becker, G. (1965) A Theory of the Allocation of Time. *The Economic Journal*, **LXX**, 299.

Becker, G. (1973) On the New Theory of Consumer Behaviour. *Swedish Journal of Economics*, **75**, 378–395.

Becker, G. S. (1976) *The Economic Approach to Human Behaviour*, University of Chicago Press, Chicago.

Beecher, C. (1841) *A Treatise on Domestic Economy*, Boston.

Blyton, P. (1985) *Changes in Working Time; An International Review*, Croom Helm, London.

Boden, D. (1987) *Temporal Frames; Time, Talk and Organisations*, Dept of Sociology, Washington University, St Louis.

Bourdieu, P. (1985) *Distinction*, Routledge, London.

Bull, N. (1985) Dietary Methods of 15 to 25 Year Olds, Supplement 1 to *Human Nutrition, Applied Nutrition*, **39a**, 1–68.

Carlstein, T., Parkes, D. and Thrift, N. (eds) (1978) *Making Sense of Time*, Wiley, New York.

Carruthers, F. (1988) *Grazing in Peckham*, London Food Commission Report, London.

Chaney, D. (1990) Subtopia in Gateshead; The MetroCentre as a Cultural Form. *Theory Culture and Society*, **7**, 49–68.

Charles, N. (1993) *Gender Divisions and Social Change*, Harvester Wheatsheaf, Hemel Hempstead.

Cheal, D. (1989) The Post-modern Origin of Ritual. *Journal for the Theory of Social Behaviour*, **18**, 3, 269–290.

Coffield, F. and Gofton, L. (1994) *Young People and Drugs*, Institute of Public Policy Research, London.

Dennis, N. and Erdos, G. (1993) *Families Without Fatherhood*, Institute for Economic Affairs. London.

Detienne, M. and Vernant, J. L. (eds) (1989) *The Cuisine of Sacrifice among the Greeks*, University of Chicago Press, Chicago.

Douglas, M. (1982) Culture and Food. In M. Douglas, *The Active Voice*, Allen Lane, London.

Douglas, M. (1984) *Food in the Social Order*, Basic Books, New York.

Douglas, M. (1987a) *How Institutions Think*, Routledge and Kegan Paul, London.

Douglas, M. (1987b) *Constructive Drinking; Perspectives on Drinking from Anthropology*, Cambridge University Press, Cambridge.

Douglas, M. and Nicod, M. (1974) Taking the Biscuit: The Structure of British Meals. *New Society*, **19** (30 December), 774.

Dyhouse, C. (1981) *Girls Growing Up in Late Victorian and Edwardian England*, Routledge and Kegan Paul, London.

Elias, N. (1978) *The History of Manners*, Basil Blackwell, Oxford.

Erlichman, J. (1994) How Mac Grew Big Fats after Fast Food's Slow Start. *Guardian*, October 26, 6.

Falk, P. (1994) *The Consuming Body*, Sage, London.

Farb, P. and Armelagos, G. (1980) *Consuming Passions: The Anthropology of Eating*, McGraw Hill, New York.

Fischler, C. (1980) Food Habits and Social Change: The Nature Culture Dilemma, *Social Science Information*, **19**, 6, 937–953.

Fischler, C. (1983) Le Ketchup et la Pillule. *Perspective et Sant*, **25**, 110–119.

Fiske, J. (1990) *Understanding Popular Culture*, Unwin Hyman, London.

Garfinkel, H. (1967) *Studies in Ethnomethodology*, Prentice-Hall, Englewood Cliffs.

Gershuny, J. (1983) *Social Innovation and the Division of Labour*, Oxford University Press, Oxford.

Gershuny, J. and Jones, S. (1987) The Changing Work/Leisure Balance. In J. Horne and T. Horne (eds), *Sport, Leisure and Social Relations*, Routledge and Kegan Paul, London. pp. 9–50.

Gershuny, J., Miles, I., Joes, S., Mullings, C., Thomas, G. and Wyatt, S. (1986) Time Budgets; Preliminary Analysis of a National Survey. *Quarterly Journal of Social Affairs*, **2**, 1, 13–39.

Glyptis, S. (1989) *Leisure and Unemployment*, Open University Press, Milton Keynes.

Goffman, E. (1959) *The Presentation of Self in Everyday Life*, Allen Lane, London.

Goffman, E. (1967) *Frame Analysis*, Allen Lane, London.

Gofton, L. R. (1986) Social Change, Market Change. *Food and Foodways*, **1**, 3, 253–277.

Gofton, L. R. (1988) Folk Devils and the Demon Drink; Drinking Rituals and Social Integration in North East England. *Drogalkohol; Alkohol und Drogen*, **12**, 181–196.

Gofton, L. R. (1990) On the Town; Drink and the New Lawlessness. *Youth and Policy*, **29**, 33–39.

Gofton, L. (1992) The Food of a Battered Class. *Times Higher Educational Supplement*, April 10th.

Gofton, L. (1993) *Drug Use Amongst Young People in Newcastle*, Home Office Drugs Prevention Initiative.

Gofton, L. and Marshall, D. W. (1988) *A Comprehensive Scientific Study of the Behavioural Variables Affecting the Acceptability of Fish Products as a Basis for Determining Options in Fish Utilisation Research and Development at Torry Research Station*, University of Newcastle Upon Tyne.

Gofton, L. and Marshall, D. W. (1992) *Fish: A Marketing Problem*, Horton Publishing, Bradford.

Gofton, L. R. and Ness, M. R. (1991) Twin Trends; Health and Convenience in Food Change or Who Killed the Lazy Housewife. *British Food Journal*, **93**, 7, 17–23.

Gronau, R. (1977) Leisure, Home Production and Work; the Theory of the Allocation of Time Revisited. *Journal of Political Economy*, **85**, 117–129.

Henley Centre For Forecasting (1990) *Time Famines*.

Hewitt, P. (1993) *About Time: The Revolution in Work and Family Life*, IPPR/Rivers Oram Press, London.

Hinrichs, R. and Sirianni, J. (eds) (1993) *Working Time in Transition*, Temple University Press, Philadelphia.

Hirschman, E. (1992) *Post-Modern Consumer Research*, Sage, London.

Hobsbawm, E. and Ranger, T. (ed.) (1979) *The Invention of Tradition*, Mentor, New York.

Hull, D. B., Capps, O. and Havlicek, J. (1983) Demand for Convenience Foods in the United States. In K. P. Goebel (ed.), *Proceedings of 29th Annual Conference of the American Council on Consumer interest in Kansas City*. pp. 44–50.

James, A. (1990) The Good, the Bad and the Delicious: The Role of Confectionery in British Society. *The Sociological Review*, **33**, 4, November, 666–688.

Jameson, F. (1983) Post Modernism and Consumer Society. In F. Jameson (ed.), *The Anti-Aesthetic*, Port Townsend, Bay Press.

Laermans, R. (1993) Bringing the Consumer Back. In *Theory Culture and Society*, **10**, 153–161.

Lane, R. E. (1993) *The Market Experience*, Cambridge University Press, Cambridge.

Lash, S. (1993) Reflexive Modernisation; the Aesthetic Dimension. *Theory Culture and Society*, **10**, 1–23.

Lévi-Strauss, C. (1970) *The Raw and the Cooked*, Jonathan Cape, London.

McKenzie, J. C. (1986) An Integrated Approach – With Special Reference to Changing Food Habits in the UK. In C. Ritson, L. Gofton and J. McKenzie (eds), *The Food Consumer*, John Wiley and Sons, London.

Marshall, D. W. (1988) Behavioural Variables Influencing the Consumption of Fish and Fish Products. In D. M. Thomson (ed.), *Food Acceptability*, Elsevier, London. pp. 219–233.

Mauss, M. (1954) *The Gift*, Routledge, London.

Mennell, S., Murcott, A. and Van Otterloo, A. H. (1992) *The Sociology of Food: Eating Diet and Culture*, Sage/International Sociological Association, London,

Ministry of Agriculture Fisheries and Food (MAFF) (1994) *Household Food Consumption and Expenditure: Annual Report of the National Food Survey*, HMSO, London.

Mintz, S. (1980) Time, Sugar and Sweetness. *Marxist Perspectives*, 56–73.

Morris, L. (1989) *The Workings of the Household*, Polity, Oxford.

Murcott, A. (1982) On the Social Significance of the 'Cooked Dinner' in South Wales. *Social Science Information*, **21**, 4/5, 677–696.

Pahl, R. (1984) *Divisions of Labour*, Basil Blackwell, Oxford.

Pahl, R. (1989) From 'Informal Economy' to 'Forms of Work': Cross National Patterns and Trends. *Industrial Societies*, Blackwell, London.

Pahl, R. (1993) Rigid Flexibilities? – Work Between Men and Women. *Work, Employment and Society*, **7**, 4, 629–642.

Rath, Claus Dieter. (1984) *Reste der Tafelrund. Das Abenteuer der Esskultur*, Rowohlt, Hamburg.

Rees, T. (1993) *Women and the Labour Market*, Routledge, London.

Reilly, M. (1982) Working Wives And Convenience Consumption. *Journal of Consumer Research*, **8**, 407–418.

Ritson, C. A. and Hutchins, R. (1990) The Consumption Revolution, Paper presented to The Royal Society, December 14th. London.

Robins, K. (1994) Forces of Consumption; from the Symbolic to the Psychotic. *Media Culture and Society*, **16**, 449–468.

Robinson, T. (1972) *Agricultural Product Prices*, Cornell University Press, Ithaca.

Schwartz Cowan, R. (1989) More Work for Mother: *The Ironies of Household technology from the open hearth to the Microwave*, Free Association Books, London.

Sen, A. (1981) Ingredients of Famine Analysis: Availability and Entitlement. *Quarterly Journal of Economics*, August, **XCVI** (3), 433–464.

Senauer, A., Asp, E. and Kinsey, J. (1991) *Food Trends and the Changing Consumer*, Egan Press, Minnesota.

Svenbro, J. (1984) The Division of Meat in Classical Antiquity. In I. de Garine (ed.), *Food Sharing*, Weiner Raumer Stiftung, Bad Homburg.

This England (1986) Cited in Walton (1992).

Thompson, E. P. (1967) Time, Work, Discipline and Industrial Capitalism. *Present and Past*, **38**.

Thrall, C. A. (1982) The Conservative Use of Modern Household Technology. *Technology and Culture*, **23**, 175–194.

Walton, J. (1992) *Fish and Chips and British Working Class*, Wiley, London.

Warde, A. (1994) Consumption, Identity Formation and Uncertainty. *Sociology*, **28**, 4, 887–898.

Willis, P. (1990) *Common Culture*, Open University, Milton Keynes.

8 Developing new products for the consumer

David Buisson

David Buisson

8.1 Introduction: overview of the food market

The developing and marketing of successful new food products is recognised as an important competitive strategy, albeit one of the riskiest, of a modern food company. In consumer food markets, new and improved products have become important symbols of economic growth, improved living standards and technological progress. Manufacturers do not expect to be able to sustain, let alone improve, their market positions without product initiatives. The markets themselves are seen as vulnerable if they cannot be revitalised with new products.

What may look as if it is a homogeneous world market for food products is not. The European Union (EU), US and Asian markets are a series of diverse markets each with their own cultural and business characteristics with consumers worldwide increasingly eating cross-culturally (Buisson and Garrett, 1992, 1994). The food industry is said to be undergoing a 'paradigm of shift' (McPhee, 1993) in consumer mood and market realities. The public's diversity of ethnic origin, age, and individual needs is growing. The mass markets of the 1950s and 1960s gave way to the segment markets of the 1970s, the niche markets of the 1980s, and the particle markets of the 1990s (Anon, 1991).

Four factors stand out in recognising the need for new product development:

1. Market places are in turmoil. Consumers are ever-changing, being influenced by global currents, lifestyle changes, economic conditions, shifting social values and curiosity. Consumers have come to expect new products.
2. Technological progress never ceases making products possible that were not considered feasible a few years ago, leading to a conflict of what the consumer wants and needs in a food product.
3. Competitive pressure, now worldwide, is relentless with enormous international market opportunities. Increasing internationalisation of food markets is compounding this.

4. Shortening product life cycles has necessitated a speeding up of product innovation to maintain a competitive advantage.

New technologies are enhancing food and quickening its distribution, while different cultures are continuing to bring their own unique tastes into the mainstream. In the future a recognised pattern of five meals per day is conceivable. Other concepts coming into the forefront are the purchasing of food at alternative locations and cross-cultural cooking. Many changes are occurring in the world of food and many more are to come (Wolf, 1994).

In the USA in 1992 there were 12 312 new food product introductions – most doomed to failure (Dornblaser and Friedman, 1993). These new introductions were described as being of 'weak quality' and there were few entries that could be classified as being truly new – i.e. never offered before. Companies avoided risky, breakthrough concepts and stuck to tried and true line extensions, especially new flavours of existing brands. The same comments can be made about other food markets of the world. There are some notable exceptions, with Japanese producers being innovative in their approach to product development – albeit by flooding the market with new products to ascertain which will succeed (Lehr, 1991). In the UK, major retailers have been the driving force behind the growth in new product introductions through the 1980s and 1990s, whereas in the USA food manufacturers have been the major stimulus (Hughes, 1994).

Numerous authors have defined what a new product is (e.g. Cooper, 1993; Gruenwald, 1988). Booze, Allen and Hamilton (1982) identify six different types or classes of new products:

1. *New-to-the-world products* These products are the first of their kind and create an entirely new market.
2. *New product lines* These products, although not new to the marketplace, are new to a particular firm, allowing a company to enter an established market for the first time.
3. *Additions to existing product lines* These are new items to the firm, but fit within the existing product line the firm makes. They also represent a fairly new product line to the marketplace.
4. *Improvements and revisions to existing products* These are not so new products and are essentially replacements of existing products in a firm's product line. They offer improved performance or greater perceived value over the 'old' product.
5. *Repositioning* These are essentially new applications for existing products, and are often the retargeting of an old product to a new market segment.
6. *Cost reductions* These are the least new of all the product categories. They are new products designed to replace an existing product in a line, but yield similar benefits and performance at a lower cost. From a marketing viewpoint they are not new products, but from a design

and production viewpoint they could represent a significant change for the firm.

In the food industry, current developments would indicate that the vast majority of new product introductions are additions or improvements to existing products with a significant contribution from cost reductions in processing that usually offer products with few benefits to the consumer. As the food industry becomes increasingly more competitive and own brands proliferate, producers will seek a greater contribution to new product developments from cost reductions. In the USA, Gorman's *New Product News* states that 'while defining new products is difficult they estimate that, on average, only 10 to 15% of new products reported are available to the consumer for the first time' (Friedman, 1990).

Internationalisation of the food industry as global take-overs occur, or companies enter export markets, have added a dimension of complexity to product development. Levitt (1983) and others have contended that products and brands are transferable around the world with little product modification. This is not necessarily the case with food products, owing to the very emotional nature of the product; food usage being based on centuries of cultural and religious habit. Product change for different markets is required as food preferences and habits are not easily changed.

The lack of genuine innovations and new food products in the food industry has been attributed to factors such as the segmenting food market, the rapidly changing nature of consumers, the lack of understanding of consumers' needs and increases in own-label brands. Few retailers can afford to pay for the research and development to come up with new products. The Boston Consulting Group pointed out that food categories that have seen the most new product launches in recent years are those where the own-label share is low (Anon, 1993a), such as cereals, soups, soft drinks and coffee as compared with juices, natural cheeses and jams. Many more innovations in the food industry have occurred in packaging and design.

The great majority of new products never make it to the market, and of those that do the failure rate is high. In both the USA and UK 75% to 85% of new products fail to maintain retail presence past one year, and significantly, some industry analysts identify that 25% or more of retail food sales consist of products introduced within the past five years (Hughes, 1994). According to the Leatherhead Food Research Association there were 2824 food and drink product launches in the United Kingdom in 1991, typically with two problems (Meenderink, 1992). First, the economics of bringing product to market at the pace consumers require is becoming more difficult; and secondly the process of introducing and improving products in the majority of companies is not working as well as it should, with 37% of new products being late and 46% over budget. It is not possible to conclude with existing knowledge whether the food industry differs in the application of success and failure factors from other industries. A number of the general success factor studies have included food companies as well as other industries (e.g. Cooper, 1984) without drawing particular attention to the food industry.

In view of the high failure rate of new products and frequent consumer criticism and/or rejection of new product offerings, the methodology for new product development badly needs improving. The most important contributors to the success or failure of a food product are the consumers, and frequently the development process is well down the track before they are included.

The involvement of consumers in the product development process is critical to the understanding of consumer needs and wants. Food manufacturers and marketers have strived to develop products responsive to consumers' changing tastes; however, the involvement of consumers in the product development process has met with mixed success.

Coca-Cola spent millions on market research before it launched its new formula Coke. In test after test consumers said they preferred the taste of new Coke. Notwithstanding one of the biggest promotion expenditures in history, the product bombed because of a lack of understanding of the Coke consumer – the consumer saying they liked the new Coke but not buying the product. Kirk *et al.* (1987) discussed the creation of perceived or real consumer demands and how they can be identified. They stated that it was not possible to respond just to what consumers say they are doing, because consumers frequently say one thing about what they consume but actual consumption patterns may be very different. Campbell's introduced a low sodium soup but found that consumers did not accept it owing to a lack of taste (Duniaf and Khoo, 1986). A similar study (Weller, 1992) of the bakery industry in the USA showed that whereas consumers were saying that they were willing to buy nutritionally improved bakery products their spending habits did not reflect this.

Overall, the literature suggests that the non-incorporation of critical variable consumer problems in the development process is partially responsible for the high failure rate amongst new food products. An understanding of consumer problems and an involvement of consumers at early stages of product development could lead to a better understanding of consumer behaviour as well as the development of better consumer products and marketing programmes.

The world food industry is in a state of change, with globalisation of the food industry and consumers occurring rapidly. From a producer perspective, to compete and prosper companies are taking over their domestic competition, buying foreign firms, and establishing joint ventures to gain entry into new markets (Gerrity, 1991; Swientek, 1993).

Food marketing development efforts have accelerated in many regions, notably the Asia-Pacific region with its large size, coupled with increased consumer purchasing power and changing food preferences resulting in a rapidly growing market. The growth of urban supermarkets, emergence of the middle class, increasing urbanisation and the changing Asian diet has made consumers of this region a target for many companies (Crippen and Oates, 1993).

Selwyn (1991), however, has warned that if the markets are to be penetrated, there is a narrow window of opportunity in this region's markets. Due to the short time frame for entry, the need for close interaction of consumers with producers is vital for product development in this market. This has led to Campbell's Soups setting up a research and development office in Hong Kong and Nestlé and the New Zealand Dairy Board setting up similar offices in Singapore where panels of local taste testers can be utilised. In order to be successful in markets where ethnic, religious and other cultural variables are characteristics of product preferences, 'in-market' research is required for product development.

Product modification is a strategy used by many companies in order to gain quick market access and not to miss a chance to rapidly establish an identity in a growing market. Campbell's Soups, for example, use a global regionalisation strategy for their soup line, with product modifications for a particular market.

With the advent of the European Union (EU) of 1992 the issue of pan-European products and brands with uniform standards and product acceptability has led to a number of developments. European retailer buying groups such as Association Marketing Services are developing common products, including European own-label brands. This has begun with the mutual export and import of each partner's 'own brands' developing into a common product sourcing taking advantage of the EU 'mutual recognition' directive. Production is becoming concentrated in a few companies that dominate the food market leading to significant increases in own-label products on the market, with the result that these products are increasingly dominating new product introductions.

Much debate has been going on as to whether a euroconsumer exists in the EU. While there are some clear similarities between the different nationals, they come from different cultures, geographic regions, languages and regulatory and business environments that make them still quite diverse segments within the overall European market. Vander-Menwe and L'Huillier (1989), however, maintain that greater Europe can be divided into six regional and cultural clusters with similar consumer profiles. From a product development perspective, producers worldwide will need to look beyond their national boundaries to cultural consumer groupings that transcend national boundaries. Cultural differences may only be surface deep in certain consumer markets, with other factors such as price, quality and availability becoming increasingly important (see Ritson and Hutchins, chapter 2).

All these factors are impacting on what is needed to undertake successful new product development. In the competition to win in the consumer food marketplace, companies need to respond quickly and accurately to constantly shifting consumer needs, wants and expectations. Market-driven managers are bringing consumers into their new product development process earlier and using them as a source for greater opportunities not only as input to new product ideas, but more importantly to the innovation process. Technical and marketing managers are learning that using customers as research laboratories is the future of innovation in the business. Using consumers can be the easiest way of increasing quality, decreasing cost and speeding up the cycle time for introducing new products. It can be considered a key strategic move ensuring that each new product launch is part of a continuum as well as part of the integrated closed loop strategic process. A significant impact on total consumer satisfaction is achieved, with the integration of consumers in this process.

The need today to develop products for the world means it is critical to look at product development from a global perspective. Food consumers worldwide will produce new and vital markets and offer great product development and marketing challenges into the next century. The real advantage will accrue to those producers who review, analyse and evaluate consumers worldwide and who plan strategies that transcend their national boundaries.

8.2 The new product development process

During the past three decades the new product development process has undergone significant change in thinking and approach. In the 1960s, companies largely created new products internally believing that they had the resources within the company to develop successful new products. In the 1970s the focus shifted to identifying external consumer needs in the marketplace using sophisticated market research techniques seeking high growth markets and niches that new products could be successful in. In the 1980s and the 1990s the underlying principle of new product development combines both these views, allowing companies to produce products that meet customer needs and meet strategic objectives.

Meyer (1984) outlines a systematic method of eleven sequential stages (Figure 8.1) for developing new products.

This pathway is idealistic, requiring the combined efforts of an interdisciplinary team from throughout the company. The development timetable differs depending on product type, with typical figures being 38 months for a reported or hot fill product and 33 months for fresh, refrigerated, frozen and dry mix products from conception to national roll-out in the USA. The timing for each stage may vary for a variety of reasons. This process of product development is very structured and formal, with many gates for approval of the project to pass through. There is a tendency for increased formalisation of product development the larger the company, in particular in the US food industry.

In contrast, companies in Denmark (Harmsen, 1994) and New Zealand,

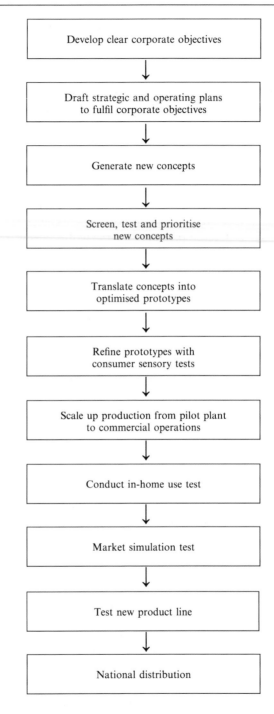

Figure 8.1 Eleven stages of the new product development process. (*Source*: Meyer, 1984)

where food companies tend to be small by international standards, the product development process is less formal and often placed within the laboratory, marketing or quality assurance function. Most companies have some kind of product development committee, usually made up of management, product development, production and sales and/or marketing personnel, for discussion and evaluation of all major projects. Generally product development is well integrated and builds on broad, informal, cross functional cooperation, unlike in the USA where it is formal. Significant amounts of the company's product knowledge is not documented but resides with a single person or is considered joint informal knowledge. Such knowledge usually will remain with the company with stable personnel, unlike in the USA where there is considerable turnover of personnel. Such a product development mechanism is typical in smaller companies and countries and often leads to more innovative food products.

The Japanese approach to product development emphasises a continual technological and marketing improvement aimed at making an already successful product even better for consumers (Czinkota and Kotabe, 1990). Due to the incrementalist product development approach, Japanese firms have been able to increase the speed of new product introductions, meet the competitive demands of a rapidly changing market place, and capture market share. Japanese firms adopt emerging technologies first in existing products to satisfy consumers' needs better than their competitors. This affords the opportunity to gain experience, solve technological problems, reduce costs and adapt products to consumer use but requires close collaboration with consumers, whose inputs help Japanese firms improve their products on an ongoing basis. The Japanese largely use the marketplace as a product development laboratory.

This continuous introduction of newer products brings greater likelihood of market success. Ideal products often require a giant leap in technology and product development, hence higher risk of consumer rejection, but the Japanese approach allows for continual improvement, a stream of new products, and quicker consumer adoption. Often a new marketing approach is developed around the incremental development. Cost reduction and quality improvement are parallel objectives that go in tandem with the incremental development in the Japanese approach rather than being a separate new product development as is often the case in the USA and Europe.

Whereas US market research and product developments are sophisticated, the advantages of the Japanese incrementalist approach are in the areas of cost, speed, learning and consumer acceptance. A continuous understanding of current and changing consumer needs is obtained with clear contextual usage, and usage conditions of products being undertaken for market research (see Bell and Meiselman, chapter 12). These three approaches of the USA, smaller countries and Japan are quite dif-

Table 8.1   Activities in a well-designed new product development game plan

Initial screening: the initial decision to spend time and money on the project

Preliminary technical assessment: an initial attempt to prove technical feasibility, assess manufacturing implications, identify technical risks and issues

Preliminary market assessment: highlighting above, this is the first-pass market study

Detailed technical assessment: detailed technical work to prove technical feasibility, and address technical risks.

Manufacturing assessment: technical work to determine manufacturing implications, capital expenditures, and probably manufacturing costs

Detailed market studies: includes the user needs and wants study, competitive analysis, and concept tests

Financial analysis: probes the expected financial consequences and risks of the project

Product definition and business case: integrates the results of the technical, manufacturing, marketing and financial analyses into a product definition, project justification and project plan

Decision on business case: a thorough project evaluation and decision to go to full development

Source: Cooper (1993)

ferent and reflect the type of new product development undertaken in each of these markets.

Cooper (1993) outlines the essential up-front activities in a well-designed game plan for new product development shown in Table 8.1.

Undertaking such an analysis rapidly before product development starts allows for greater risk-taking in product development and the opportunity for truly new food products rather than product modifications.

Successful new product development is impacted by three generally agreed upon factors. These are:

1. The degree of market orientation in the product development process.
2. The strategic role of product development.
3. The organisation of product development activities.

Harmsen (1994) has drawn together the elements in each of these areas that contribute to success from a number of studies, as shown in Table 8.2.

The development of market-oriented products requires close interaction with consumers. Rarely do products appeal to consumers because of one single attribute. More often their appeal is based upon a combination of attributes, such as taste, aroma, texture, colour, packaging appeal, convenience, status appeal, etc. all of which impact upon the buying decisions. Missing key customer attributes, when working under pressure or with poorly defined marketing objectives, can mean that customers are

Table 8.2 Elements contributing to success in the product development process

1. **Market orientation**
    – The degree of product superiority from the consumer's point of view
    – The degree of contact with the market/consumers during development
    – The degree of up-front marketing, market function representation in the
      development process, knowledge about the market/competitors, and marketing
      competence

2. **Product development strategy**
    – An explicit product development strategy defines what type of product to develop
    – The product development strategy is part of the overall strategy of the firm
    – An understanding of the need for product development is motivated by the
      management
    – Top management actively supports the development teams

3. **Organisation of product development**
    – Cooperation between product development and marketing
    – The use of temporary multifunctional groups
    – Clear goal setting
    – A participative management style
    – Loose structuring in the early stages (idea generation, screening, concept
      development, and testing)
    – Product development as a learning process
    – More formalisation in the implementation stages
    – An emphasis on up-front activities
    – Formalised product development process
    – An emphasis on accomplishing all the phases and activities in the process

Source: Harmsen (1994)

more likely to focus on what they don't like. Quality is another attribute
clearly defined by the consumer, not the product developer or market.
The consumer affects quality, e.g. children's tastes differ from adults, or
the ethnic origin of the consumer; and eating occasion may alter the perception of quality.

Markets are increasingly becoming more downsized into particle markets (Gorman, 1991). The ability to discern and describe consumer groups
is quite refined today, and marketers can detect and identify minute distinctions between demographic groupings and do their target marketing
very pointedly. The future for developing niche markets will be challenging, trying and profitable. Miller's clear beer was not successful in the
USA, but other clear products have been.

The continual evolution of consumers' eating, spending and lifestyle
habits require food producers not only to respond but to anticipate future
changes. Consumer behaviour experts identified a growing trend in the
1990s toward do-it-yourself products kits (Anon, 1993a). Close interaction with consumers revealed that they consistently asked for food that
takes less time to cook. A number of companies responded to this with
products that formed a base for meals such as Hormels Light and Lean
97 cuts of turkey, Campbell's Soups Swanson Vegetable Broth as a base

for home-made soups, and Dr August Oetker of Germany's Creme Bavorise. Consumer trends must be evaluated and anticipated if opportunities are to be realised.

This global perspective further complicates market diversity (Cannon, 1985). Generalities can be made, for example about the Asian market, but the considerable diversity of the market cannot be over-emphasised. Asian consumers, in contrast to their western counterparts, are traditionally more adventurous, opening many niche opportunities, given the Asian curiosity for variety, but such products are unlikely to achieve market dominance. This trait for trying new products is partially attributable to the influence of past cultures on a country, e.g. the Dutch in Indonesia, but more so today because of greater westernisation, exposure to new foods, youth equating western foods as having 'social value', to the increasing number of women in employment, and the perceived high and safe qualities of western foods.

Lack of commitment to a product development strategy is exhibited in many ways, for example as a lack of focus and erratic new product appearance in the marketplace. The Japanese have been able to penetrate new markets successfully through focusing on strategic objectives and a long-term commitment to achieving those objectives. The microwave industry is an excellent example of commitment to new products and technologies (Larson, 1991). Con Agra's leading 'Healthy Choice' ready-prepared meal line had a direct commitment from the Chief Executive and was rapidly expanded into products such as Swift-Eckrich Healthy Choice Luncheon Meats and Hunt Wesson Healthy Choice Pasta Sauces.

Poor organisation of product development is exemplified by delays in market launch and products that do not meet consumer expectations from a technical or market perspective. Balancing speed of development with appropriate checks in product development is critical to achieve rapid time to market with high quality products. Timing is an often neglected key component of a strategy. Other companies, it can be assumed, will be working on a similar product and a delay in development must be avoided. Other products are rushed to the marketplace before their quality attributes have been properly refined with the outcome being disappointed consumers (Antony and McKay, 1992).

A number of approaches have been proposed to speed up the process of product development. Takeuchi and Nonaka (1986) replaced the old sequential process of product development, which simply would not get the job done, with a holistic approach, as in rugby, with the ball getting passed within the team as it moves as a unit up the field. This holistic approach has six characteristics: built-in instability, self-organising project teams, overlapping development phases, 'multilearning', subtle control and organisational transfer of learning. The six pieces fit together like a jigsaw puzzle, forming a fast and flexible process for new product development. This approach also has the advantage of introducing creative market-driven ideas and processes into an old rigid organisation. This process, adopted by many successful organisations today, allows speed and flexibility in developing products.

Bingham and Quigley (1990) propose a new product implementation process designed to reduce the risk inherent in new product introductions to consumer markets. Centralised around the marketing function, methods are adopted that exert greater control over internal and external factors critical to the pace of product development and implementation. Market research and its application to all stages of a development are critical to the success of this approach.

Wheelwright and Clark (1992) propose an overall development process that starts with a broad range of inputs from consumers, the company, etc. and gradually refines and selects from among them, creating a handful of development projects that can be pushed to rapid completion and introduction. They propose the notion of a converging funnel that in its simplest form provides a structure for thinking about the generation and screening of alternative development options, and combining a subset of these into a product concept. A variety of different product and process ideas enter the funnel for investigation, but only a fraction become part of the fully fledged development project. Those that do are examined carefully before entering the narrow neck of the funnel, where significant resources are expended in transforming them into a commercial product and/or process.

The need to speed up the introduction of new products is therefore changing the approach to product development. Past are the days when a project would pass from research to engineering to manufacturing and finally to marketing. Instead we are seeing internationally the adoption of the Japanese practice of cross-functional teams, with all the functional groups working together. Fast execution of product development is a key component of a company's strategy. Betty Crocker's Dessert Bar's gestation, from start of product development to marketplace, was only nine months.

Overall, the most successful new product developers will recognise that growing consumer awareness and needs, and growing retailer influence, must be considered for the future organisation, innovation and marketing of new food products. Much of the debate in the food industry today is how to interact marketing teams closely with research and development, and above all how to establish close contact with the consumer.

The Economist magazine (Anon, 1993b) predicts that warmer relations between food retailers and manufacturers would help to achieve some of these goals. By combining grocers' wealth of data on consumer buying habits with their own knowledge in product development, the best food companies should be able to develop successful products and avoid failures. The mutual antipathy that hinders cooperation between the two could take time to disappear, though some American and Japanese food manufacturers seem to be moving in the right direction. In addition, in the creation of new products food marketers will also have to look at new ways of distributing them, perhaps bypassing the retailer altogether.

As consumer bases are widening, the best food companies will narrow their product base and a reshuffling of food assets has already begun with companies divesting parts that are not a core business. Tony O'Reilly, the Chief Executive of Heinz, predicts that there will be 'major and exciting changes in food portfolios' in the years ahead (O'Reilly, 1991). Many of these deals will cross borders in order to take advantage of pan-regional economies of scale in manufacturing made possible by trade pacts such as NAFTA or the EU. Government regulations and trade restrictions will, however, impede developments in many parts of the world.

The Asia-Pacific region will offer significant opportunities, but differences in culture, religion and taste will make food a difficult product to sell internationally and even global brands will have to be tailored to local taste in this and other developing markets such as Latin America and the Middle East (Buisson and Garrett, 1994).

8.3 Technology and product development

Two basic types of technological development are of importance to the food industry and impact on consumers in different ways. The first is technical innovation that retains or improves the quality of a product through a process innovation, ingredient substitution or a packaging innovation. Innovations such as a better freeze-drying process for coffee, or dough conditioning that gives a better-tasting bread, fall into this category. These are incremental product changes that the consumer may not even perceive as a change in the product but to a producer may result in cost reduction or greater throughput. The second is radical technological change that produces a product new to the consumer. Products such as sous vide meals, tetrabrix packaging and biotechnologically engineered food products fall into this category.

The incremental technological change must only improve the product to the consumer. Too often the change has led to changes in taste or in texture, as exemplified by the lack of taste in the early low-salt products and the lack of texture in many fat substitute products. Duniaf and Khoo (1986) found that consumers rejected low-salt soups, setting up a risk/benefit equation: if the reward is perceived to outweigh the risk, then they often accept the risk. Here the reward is perceived as taste and market research indicates the consumer believed that low-sodium, low-fat foods did not have a good taste. Today technical staff are being directed toward the mundane tasks of formulating products with reduced fat, low calorie and all natural attributes but maintaining taste and texture.

The Japanese approach of taking greater technological risk for a long-term benefit and flooding the market with new products lends itself to the introduction of 'new to the market' products. The products will be tried in Japan before taking them internationally. Examples of these are Suntory flower flavoured wines, available in rose or mimosa varieties; healthy beverages such as peach and green pepper juice, rich in vitamin C; and Red Zone, a Japanese hot spicy instant snack noodle marketed in a pack with an aluminium lid with a photograph on it – after hot water is poured into the pack, the photograph develops. The Japanese successfully transplanted their centuries old surimi technology into a very successful product range of imitation crab, prawn, scallop and lobster for western markets. They have not, however been successful in transferring their ranges of seaweed-based cereals and drinks to western markets although France's Now Vegetable launched the Aquatable range based on a seaweed-like vegetable harvested off the coast of Brittany (Lehr, 1991). The Japanese consumer market tends to lend itself to trials of new products because of its size and diversity, and like most Asians the Japanese tend to be more adventurous in trying new foods than their western counterparts. From these trials the Japanese producers use the concepts for worldwide development.

Notable technologically driven product developments in Europe have been the development of types of pharmaceutical food products that have met with reasonable success in the European market. Yoghurt and other fermented milk products that contain 'probiotic' bacteria accounted for 25% and 9% of the fermented milk market in France and Germany, respectively. In the UK they comprised 4% of the total yoghurt market but were predicted to have a market share of 15% by 1993 (Anon, 1992a).

The USA tends to lag behind other world markets in many of these developments. This is largely attributable to the need, often, for FDA approval, or to companies not being prepared to take the risk of launch because of potential product liability claims. This approach delayed the entry of tetrabrix packaging for some years in the USA. There is considered to be a quiet revolution occurring in the US food industry (Barkema et al., 1991), allowing food processors to target specific consumer niches with technologically advanced products. Note that these are ini tially niche markets containing innovative consumers which emerging technologies make it possible to hit. For example, technologies aimed at reducing fat and cholesterol using supercritical fluid extraction in eggs or red meat are being perfected and tested in trial markets, as are restructured foods in which unsaturated fats replace saturated fats, and foods produced under ultra pressure processing conditions (Anon, 1992b).

The involvement of consumers in technologically driven developments in the early stages is difficult because of the conflict that as the technology develops the product produced improves. At what point do you look at consumer acceptance of the product? The questions are: how technologically advanced can you be with the development of a new product; will the consumer want the product; what are the perceived benefits of it in consumers' eyes; and is it worth the risk of development for what may be a limited market?

Major emerging technologies are being developed that could have a significant impact on product developers. Examples of significant developments are as follows:

1 Genetically engineered foods

Biotechnology, defined loosely as the manipulation of life forms to generate new products, is already resulting in many food products and ingredients being materially different in either quality, nutritional value or cost. Some will occur beyond the consumers' awareness but many will be fraught with problems of consumer acceptance.

Some genetically developed foods have been readily accepted, such as the yoghurts mentioned above, but many have not. Most of the developments in food-related genetic engineering have been in the area of plant breeding, with some merely accelerating conventional breeding techniques under controlled conditions. Calgene Fresh Inc. in the USA (Best, 1993) produced a FLAVR SAVE tomato with delayed softening properties that allow it to be picked ripe off the vine, but to retain its quality. This would result in substantially lower year-round production and distribution costs. The direct benefit to consumers included better flavour and keeping qualities. Activist groups were quick to react to the technology and food companies were quick to assert their intention not to buy for processing. The product has well-defined consumer benefits and, if these benefits are properly communicated to consumers, should succeed. Other products such as a delayed ripening tomato expected to be introduced in 1996, a high-solid potato that offers improved processing costs to dehydrated potato processors and reduced oil absorption in fried potatoes, and soybean oils with improved nutritional properties will all offer well-defined consumer and economic benefits.

The development of biotechnologically modified products will require well defined consumer benefits and education for them to gain full acceptance. Advances in biotechnology will need to be accompanied by strong promotional and educational elements if consumers are to accept foods that are safer, more nutritious, better tasting and at a lesser cost. Food products, however will be increasingly developed to satisfy the needs of the discerning consumer, and it is likely that well before the end of the decade custom-designed fresh produce will be exclusively available in certain outlets (Hughes, 1994).

2 Packaging

Today's consumer expects a package to not only contain the product, but to meet their needs for convenience, product freshness, value, and more recently be environmentally friendly. From a marketing perspective it must be robust for handling, stack well and have eye appeal. Significant packaging breakthroughs such as the tetrabrix, controlled atmosphere storage, and retortable packs have been successful developments for producers and consumers alike.

The most profound developments occurring in packaging are in the areas of microwave and flexible packaging. Microwave packaging developments are essential since consumers increasingly have questioned that microwave foods deliver good taste and good value. The new generations of microwave packaging show a much more carefully designed relationship between the food product and the packaging material. New susceptor packaging and the ability to selectively cook different foods within one tray to different levels open many opportunities for development but only if the consumers are informed as to how to use these technological developments and are involved in the developing of the products. Flexible packaging today is characterised by dramatic innovation, which has been teamed up to meet consumer demands for convenience and environmentalism giving consumers many advantages over existing packaging (Spaulding, 1992).

3   Foodaceuticals/nutriceuticals

Functional foods, known as nutriceuticals, pharmafoods or foodaceuticals, represent the next big development in food product development. Consumers have trouble grasping the concept of what these foods actually are.

Functional foods include a whole range of foods purported to have enhanced health benefits. The role they play is likely to increase as their profile builds amongst consumers and they could significantly alter what we eat. Consumer concerns about their health and wellbeing mean, for food companies, that these foods are too great an opportunity to miss. Sales of foods with health claims are forecast as a major growth area of the food industry. Forty per cent of new food products in the USA carry some kind of general or specific health claim (Shakla, 1992). From reduced calorie foods to foods purported to prevent cancer and cardiovascular diseases the current markets for functional foods in the USA are estimated at $2.5 billion in 1988, with a projected growth of 17% to 20% (Shakla, 1992).

'Health' products have appeared, such as a whole range of drinks with extra food fibre for those concerned with the health risk of colon cancer and biscuits or cereals with added fibre food that aids digestion targeted at the ageing population. Japan and Singapore are particularly receptive to functional foods and in Japan the range of these types of foods is immense and continues to increase (Anon, 1990).

Regulatory authorities are in some dilemma as to how to handle these products that cross over between a food and a pharmaceutical (Hunt, 1994). The Japanese have had more time than the rest of the world to get used to these foods. The Japanese Ministry of Education, Science and Culture in 1984 realised that functional foods (as they later called them) could be the most important new market of the late 1990s and early into the next millennium and encouraged their development. The Japanese government has gone further and is said to be steering young people toward functional foods in a long-term bid to cut the pharmaceutical bill. The real functional food explosion will come early in the next century.

Consumers do not yet understand functional foods. Consumers will have to be led gently into such products and the medical benefits not stressed. The relative naiveté of the consumer over the links between diet and health also is a major impediment to product development of functional foods. It may require a large producer, perhaps an alliance between a food and a drug company, to lead the way, explain what they do for you and how much you need. Great care is going to be necessary in involving consumers in such developments.

It is difficult to assign to any particular marketplace those markets that have been more successful than others in the production and consumer acceptance of technologically driven new products. The US companies have a greater degree of sophistication and success in the area of microwave foods whereas the Japanese market is much further advanced in relation to the development of functional foods. The New Zealand market is technologically more advanced in the area of dairy food ingredients whereas the European market is advanced in terms of consumer dairy products.

Clearly for companies to be successful and operate internationally the inputting of technological value into products is required. Companies will need to maintain a close watch worldwide, not only on consumer trends but on technological trends, and invest in research and development (Petroni, 1991). When new technologies are introduced into the market, it is not surprising that most people say they don't want them (as with food irradiation), as it is only the innovators who initially welcome it. Identifying the appropriate sectors (innovators, early adopters) is the key factor in the successful introduction of new products, based on technical innovation (Wright, 1990).

8.4
Consumer trends and product development

There is almost universal agreement that a successful new product comes out of the combination of a consumer need and a technical capability. Assessing the consumer need is of critical concern and an industry has grown devoted to tracking economic, social, political, demographic and other types of trends. Developments in the food sector highlight that the total food concept is no longer simply a series of ingredients but has benefits that go far beyond satisfying hunger. It can say both positive and negative things about consumers' lifestyles and attitudes, hence understanding these are critical to the development of successful new products. It is necessary to get out and immerse yourself in the market. Additionally, market research and market data are essential but are to be used astutely and in a combined manner.

Two consumer trends overall are affecting product development. The first is that old patterns of consumer demographics and psychographics are increasingly meaningless as people piece together component lifestyles. Increasingly psychographic profiles of segments are being produced which are rendering demographic classifications obsolete. The second is that few market successes in the past ten years have come from normal trend analysis. The marketers who have spotted real needs underneath the trends, and who have exploited what they saw in a truly entrepreneurial manner, are the ones who have won.

Critical food trends that are impacting new product development can be listed as shown in Table 8.3.

These trends are universal trends. How they are expressed in specific products, the use of the product, and level of understanding or degree of importance varies in different cultures. Some trends cut across all segments, such as the healthy trend, but others require targeting at a specific consumer segment.

It is essential to spot new trends early, verify that there is a trend, then develop possible scenarios with implications for the company's individual products or businesses. Clark (1991) contends that if we seriously wish to find market opportunities for new products then we need two things. These are, first, an insightful analysis and interpretation of the data (of which there is no shortage) and, second, a willingness to break new ground, a determination to be different, and a commitment to a winner. Care must be taken in ensuring that the trend is real. The emerging trend toward 'healthier' eating presented an ideal opportunity for the development of new fish products regarded as a healthy food in the United King-

Table 8.3 Critical food trends impacting on new product development

The ageing of the population and changing social structure with people wishing to express their desires and aspiration in food

The need for time, hence the need for convenience in foods

The need for adventure, hence the rise in ethnic foods

The concern for health and well-being reflected in the dramatic rise in healthy and 'lite' food products

The feeling for the environment and nature, leading to the rise in environmentally friendly and natural products

dom, but by only looking at the trends and failing to understand the consumer few foresaw the limited appeal of fish to the British public (Gofton and Marshall, 1992).

Relevant market research is a critical component in the development of new consumer products. Numerous papers list methods of how to undertake consumer research (e.g. Cooper, 1993; Davis, 1993). In the search for obtaining new ideas you can rarely expect to get them directly from consumers, just by asking them. Consumers think in terms of very minor improvements, a kind of enhancing of the product that they are used to using. Von Hipple (1986) notes that consumers' own real world experience inhibits their insights into new product needs and potential solutions. Consumers immersed in their own world are therefore unlikely to generate new product concepts that conflict with their now familiar products. A new product operation based on such a concept would produce largely minor improvements of little economic benefit. A research process should rather concentrate on searching for problems and solutions that consumers have not yet perceived, but would regard as important, once recognised.

Consumer problems are fundamentally different from wants and needs. Consumer problems deal with actual experiences of dissatisfaction producing more valid and meaningful data. Understanding from consumers about the dissatisfaction, drawbacks and annoyances associated with buying and using items in a specific product category will produce more meaningful ideas for product developments. Consumer focus groups, advisory panels and in-depth interviews are consumer-based qualitative techniques traditionally used for new product development, along with quantitative techniques such as multidimensional scaling or conjunct analysis. These quantitative and qualitative techniques are designed to provide consumer input into the new product development process. The techniques differ in terms of the type of consumer input desired, yet the literature has largely avoided the issue of what information is most useful from consumers for product development success (Fornell and Menko, 1981).

It must be remembered that market research only provides information about the consumers' present world, hence only some of the material on which the search for new ideas can be based. A high failure rate among new consumer food products can be traced partially to poor market research and poor marketing (Paschker, 1976; Smithburg, 1985). The serious drawback of many research methodologies is that they do not incorporate consumer problems. Problems are different from preferences and from perceived product attributes. Understanding consumer problems in combination with understanding their changing lifestyles will lead to better predictions of consumer behaviour, the development of better consumer products and marketing programmes. If market researchers change the perspective and the focus of the analysis, shifting from current needs, which generally are met well by existing products, to future needs, they would gain a competitive advantage in the satisfaction of emerging needs rather than a better satisfaction of consolidated needs. The basic problem is to recognise the so-called market weak signals indicating a current change so that the concept of the new product to be developed could be based on these signals. Bolongaro (1994) addressed this problem in the development of a long shelf-life food product using a delphi technique illustrating the steps needed to utilise such a research approach.

Lauglaug (1993) discusses the use of consumers to collaborate on developing products using a technical-market research (TMR) approach where TMR is a scientific approach to the collection, analysis and translation of consumers' needs, wants and expectations, focusing on shaping the consumers' thinking by creating the experience. It is a process that makes producers think as consumers and permits consumers to lead the product innovation process in a way that can significantly improve consumer satisfaction and market success. The TMR process explores concepts and products in 'antenna shops', pioneered by the Japanese. These represent a market place laboratory where new product concepts can be featured and explored with consumers. They represent a new approach to new product development in involving consumers in the innovation process. Antenna shops can also serve as a customer, market or competitor product intelligence collection technique (Lauglaug, 1993). The Japanese food industry has used this technique to a limited extent in the USA and Europe and extensively in Japan. The concepts of iced coffee and honey-lemon beverages were successfully introduced to the US market by the Japanese using this route (Mitarai, 1991).

New product development operations, well grounded in understanding the consumer, are in the best position to generate new ideas. Davis (1993) provides a good list of questions on consumer attributes that can be answered through market research (Table 8.4).

Table 8.4 Market research questions on consumer attributes

What is happening in the market?

What are the trends?

Who are the competitors?

How do consumers talk about the product in the market?

How do consumers use the products?

Are the needs being met by the current products?

How do consumers know that a product is working?

Do consumers use the product for acceptance or self-esteem?

Source: Davis (1993)

Successful new products must come out of the combination of a consumer need and a technical capability for a product developer. However, a careful balance is needed. O'Brien (1989) characterises the food industry as one being driven by markets rather than technology that produces products targeted at specific (and often volatile) markets. Since food products are low unit cost, high volume items, the speed with which they are developed and marketed is of considerable importance. This view, however, ignores that, for most products, good technology inputs remain invisible to the consumer and it is these technology changes that contribute incremental value additions to the food industry. Genetic selection to improve flavour and appearance value of produce is a good example.

The conflict with acceptance of products produced using emerging technologies is that they will often result in products largely new to the consumer and the producer must evaluate whether to launch the new product to gain a competitive advantage or undertake significant consumer tests before the launch of the product. Often the technology being implemented has societal implications, such as irradiated foods or bioengineered foods, and the resistance to their use needs to be overcome. Albrecht (1982) compared a list of emerging technologies with products introduced in the prior twelve years and concluded that most new products did not require a high-powered technology in their development, even classics such as Pringle's potato chips, but could be made with technologies readily available at the time of development. Is this true today? He cautions that new product marketing people, as well as technical people, can be easily seduced by emerging technologies. The prospect of talking to consumers about a revolutionary new technology being used to make an incredible new product is very attractive, and everyone is excited except the consumer.

An example of this is Quorn, the first big branded alternative to meat, now showing increasing consumer acceptance in the UK after initial consumer resistance to this (in consumers' eyes) radical new product. An application has been made to the FDA in the USA for approval for consumption as a food product (Milmo, 1994). This new-to-the-world product is a mycoprotein, low in fat, high in protein and fibre, containing no cholesterol and therefore having obvious consumer benefits. Originally designed for countries in the third world for protein deficiencies that never materialised, the product has had problems with no real features of its own, relying on soaking up flavours around it. The fact that Quorn is an engineered food worries consumers consciously trying to eat more healthily. With little mainstream consumer input into its development the product has suffered a faddish vegetarian image (Murphy, 1993). This product is now being totally relaunched, as a mainstream meat substitute, to expand the consumer base. The involvement and education of consumers in a major marketing effort will be critical to acceptance of Quorn as a mainstream product. Consumer involvement at an early stage of the development of this product and an assessment of consumer trends should have recognised the potential for this product.

Foods such as microwaveable foods, surimi based crabsticks, and 'lite' foods were all new to the market, but met consumer resistance in the early stages of their introduction. As the technology has improved so has product quality and consumers are recognising the benefits of these products which are now well established in the market. Such histories could be repeated with Quorn. Other products, such as low alcohol beer, still meet resistance in many markets owing to the poor image of the product.

Critical to the acceptance of new products, especially those based on new technologies, is their diffusion in the marketplace. This diffusion is communicated through certain channels over time, among members of a particular class (Frambach, 1993). Knowing the psychographic features of 'consumer innovators' who initiate markets by adopting novel items and then communicate them to the more conservative mass of consumers is particularly important during the product development process, since it allows the product features (and associated marketing material) to be geared to this primary market segment (Foxall, 1984). Considerable work has been undertaken on profiling innovators and adaptors of new food products (Foxall, 1993). Adaptors not innovators, albeit highly involved with the product field, are the highest purchasers of food innovations. Adaptors have a high level of commitment to a coherent behaviour pattern, such as healthy eating, and hence search within the appropriate product field for as many compatible items as are available. This is the consumer segment that must be targeted to obtain product acceptance.

Consumers have never before been exposed to the variety of foods now available, originating globally, and have never been so well educated regarding the quality of food. A company cannot place a food product in the market that it deems suitable and expect through persuasion to get consumers to buy it; the food product must have a demonstrable relative advantage, be compatible with existing social behaviour, not be overly complex, and be freely available on a trial basis. Company efforts should primarily focus on the relative advantage and observability of the product improvement to attract consumers to purchase. Trial availability has been shown to be useful in reducing the risks of converting non-users and encouraging existing users to use more, although brand loyalty still has a very considerable influence (Mitchell and Boustani, 1993). Clearly it is essential to understand the consumer and involve them in the development process at an early stage to identify the early adopters of a new product, especially one based on a technological development, and target them to accelerate the diffusion of the product in the marketplace.

Kirk *et al.* (1987) sum up the need for balancing technology development and consumer acceptance, for example in relation to nutritional products, in the statement: 'food manufacturers and marketers are not in a position to force feed the public, regardless of how good or nutritionally worthy they may perceive the product to be. While the food industry can contribute to moving things in a desirable direction, it does not create consumer needs and wants. On the contrary, if it is fortunate and skilful, it may identify and satisfy changes in the people's tastes. Timing is critical in the developing and marketing of a product.' Food manufacturers must deliver products that meet consumers' perceptions at a time when they are ready for them.

As consumers continue to look for food products that represent good value for money, have convenience and good nutritional value, it is food manufacturers who adroitly use technologies to create products that meet these criteria who will be the long-term winners.

8.5
Factors influencing product development

The food industry has major factors impacting product development that many other industry sectors do not have to contend with. In particular, in the USA increasing consumer concerns for safety and potential product liability claims from dissatisfied consumers have led to companies avoiding any risky, breakthrough concepts and remaining with tried and true line extensions. Inhibiting factors that have been brought about by governments and consumers are:

1   Government regulations

Governments have had major impacts in the areas of food safety, labelling and in nutritional requirements. Sweden, followed by the USA and the UK, have taken active steps to change people's diet. Public health policy, driven in many cases by consumer protection issues, is having a major impact on product development as various governments have produced nutrition guidelines for the populace, as well as enacting labelling legalisation, especially in the area of nutrient content claims.

Traditionally health claims have been banned in most industrialised countries, but public policy is changing. There is awakening a worldwide awareness by consumers of the prevention of chronic diseases. Legislation is being enacted to encourage consumers to make healthy food choices, impacting on product development. These developments are requiring producers to adopt a more holistic view of consumption rather than focusing on individual products and brands and to have objectives such as 'health of the nation' (as in the Heinz approach) for new product development (Dickie, 1993).

The commonality of country approaches is that the food component must be natural, not a supplement; this is a distinction that is becoming increasingly difficult to make. The usage and substantiation of health claims have been the subject of great debate and calls for increased government regulation to protect the product developer. Many companies will hold back on products carrying health claims while consumer interest in such products is high, forcing product developers into launching risky items from a labelling standpoint. Health responsible food development and marketing will be critical (Booth, 1989).

2 Packaging and environmental legislation

Key environmental issues impacting on the food industry are recycling, overpackaging and solid waste management. Biodegradability was ridden to a short death and recycling is not consumer friendly or is being forced on food producers or consumers such as in Germany and parts of Japan. With other countries considering even stricter measures, the trend is forcing considerable change in the approach to product development (Sloan, 1993). Although environmental concern is at an all time high and the majority of consumers are genuine in doing something about it, several counterbalancing trends will prohibit it offering a significant competitive edge in the next few years (Sloan, 1993). Despite 40% of consumers indicating a willingness to pay a little more for a product with an environmental edge, the reality is that the environment will not be competitive with other product attributes unless the environmentally friendly product is equally as convenient, is similar to the original product in every way, and offers a cost advantage. Developers need to justify every decision on positive grounds: safety, efficacy and quality.

3 Social/ethical issues

Consumers are influencing product development through consumer associations in Europe and the USA. The impact of messages such as 'additive free', 'sugar free', 'environmentally friendly' all have positive connotations and have led to a range of product opportunities. The critical element is for companies to anticipate the social impact of current and future issues and to bring their technological resources to bear to develop solutions to the problems such products and technologies will engender. Consumers are a product of their own environment and react to issues presented to them. For new food products of the more radical variety to succeed, such as functional foods, industry-based sources of information will need to be more credible.

The green consumers' concerns about the environment and their response in demanding environmentally friendly products generated a sense of consumer power that has now extended into social and ethical concerns (Adams, 1993). Associated with this is a rise in demand for organic products, albeit a limited market (Beharrell and Crockett, 1992).

Best (1993) proposes a fifth P of marketing, namely 'perspective', to master the intricate nesting of politics and science that confront food product developers today. Programmes are needed to position radical developments with societal implications beyond the immediate impact of the marketplace.

The implications of different governments' legislation in areas such as labelling, additives, nutritional requirements and packaging laws coupled with different socially acceptable norms pose immense complications to the product developer when developing a global product.

8.6
Emerging opportunities in product development

There are a number of interesting developments in the world that product developers must be aware of. Some are already a reality but offer considerable potential for the future:

1   Children's foods

Children-directed foods are a major growth category with meals, snacks and yoghurts already in the market specifically for children. Most of the products produced for children are designed with the idea of having them prepare them. Care, however, must be taken with such products. Children are valued and are more important to us today. Family sizes are falling, and the nurturing and education of children take longer, so they are dependent on adults for longer periods of time. Social values have shifted. Even though women spend more time in employment than ten years ago they actually spend more time looking after their children now. This has led, for example, to increased concern by consumers to ensure the nutritional value and correct labelling of children's foods.

2   Ethnic food product development

The ethnication of the diet has occurred in many countries as a result of travel and consumers wanting a sense of adventure. The primary ethnic foods are Chinese, Italian and Mexican but other ethnic foods are growing rapidly. Asia is likely to provide a considerable variety of these ethnic foods. Food products based on different ethnic cuisines are rapidly becoming global products and offer many opportunities for product development.

Crippen and Oates (1993) describe three types of ethnic segmentation trends – authentic ethnic, pseudo-ethnic and nouvelle ethnic. Certain consumers will demand authentic foods, but they will tend to be in the minority. Eating an ethnic food in a different market may not taste the way the authentic article is prepared but it gives the illusion or taste. Pseudo-ethnic foods find wide interpretation internationally. Hence manufacturers modifying foods for different markets must understand that modifications may be necessary to gain acceptance and that authenticity in many market segments is not essential. The nouvelle ethnic interpretation of many regional foods has been occurring internationally. These give the illusions of being ethnic but with a producer's own unique interpretation.

Understanding country and cultural norms is critical in developing ethnic food products or adapting products for particular markets. For example, an understanding of the dietary laws governing Moslems is critical for producers to provide products for that culture (Twaigery and Spillman, 1989).

3   Diet and 'lite' products

A glance at new product introductions tracked by *New Product News* in the USA clearly shows the dramatic climb in products bearing healthy or 'lite' claims. In 1988, 475 reduced/low-calorie/lite products hit the supermarket shelves; by 1991 the product category had increased by 155% to 1214. New areas of no cholesterol, very low fat, fat substitutes and similar 'healthy' products will ensure that this product category will continue to grow allowing consumers to indulge in food products without the feelings of guilt – extending markedly the range of food choice.

Improvements in taste, texture and mouthfeel, coupled with the development of new technologies such as supercritical extraction, will be the next big thrusts as manufacturers develop new ingredients to improve the attributes of these products. Simultaneously, many consumers are wanting fewer or no additives. These all combine together making it difficult to formulate a lasting healthy food product, equivalent to the original product.

4   Athletic or fortified foods

This is a rapidly developing food category, primarily today in the beverage area but poised to expand. In Germany the brand Isostar owns some 80% of the market. Gatorade is a major seller in the US market and Lucozade sports brand possesses some 75% of the UK market.

The original quest to provide a universal panacea for all athletes' needs has given way to a more rational approach. A runner's requirements under hot and humid conditions are now differentiated from someone who participates in a leisurely game of tennis. The physiology of the human body under conditions of stress or activity is now better understood. Informative labelling should guide consumers to the best choice for their needs.

A greater understanding of athletes' physiology will assist in developing new products in this category, and these will gradually become available to the average consumer. Such products are already available in a wide range of forms in Japan (Anon, 1990) but appear to be little understood by western consumers.

5 Mood and psychographic foods

Popcorn (1991) in her book sums up a number of interesting trends related to this area: 'Ergonomics hasn't hit the food arena yet – but it will.' Customised food will be next. It will be recognised that a man's nutritional needs are different from a woman's. We will be serving meals for mood – that reduce stress, enhance energy and induce sleep. A certain bread to quiet us and a special beef to stimulate us. Foods by age, by stage-of-life teen menus and menopause meals. Traditional medicinals to aid breathing or to boost us when we're nursing or have PMS. In the future we will see the extending of the area of foodaceuticals to the concept of psychoactive 'mood foods' (Smith, 1993) with processors seeing the allure and regulatory challenge of aphrodisiacs (O'Donnell, 1992).

With the ageing population there is the potential to consider the marketing of mood foods. Aromas have been shown to affect human brain patterns. Lemon and camomile for example deliver a relaxing effect. Others such as basil and peppermint stimulate. Folklore endows certain foods with psychological or mood-altering effects. As an ageing population accepts that diet affects physical and mental health, product developers will go the next step and continue to challenge what is allowable.

8.7
The consumer and product development of the future

As we enter the future, increasing global competition, rapidly advancing technology and changing consumers are all impacting on the undertaking of product development. It is those product developers that anticipate and respond promptly to new emerging consumer demands who will be the classic innovators developing products that meet consumers' expectations; it is those who risk investing, even though based on changes that may not as yet be clear. Trends in consumer spending on food and their increasing demands are already shaping a new feature. Three factors will impact future product development: the consumer, global competition and the organisation of the product development effort.

Future consumers will be driven by changing trends and a greater understanding of what they want. They will have a stronger voice in registering their preferences, and the ways their needs are met, to the food industry. Tomorrow's challenge is to create new value added foods that combine the attributes of taste, value, naturalness, nutrition and sensation with the benefit of convenience. The fact that the decision process is in the minds of individual consumers is something that product developers and marketers have to come to terms with.

The critical path to creativity requires skill in understanding the consumer, who should be the driving force behind the development efforts. The creative process has three elements – the consumer, technical research and marketing – each of which is dependent on the other. To understand consumers' needs, the issues that have to be addressed are the reason for product purchase, consumer perception of the product, what consumers want and know that they want in a product, their expectations of the product and what customers need or do not recognise that they need because of their limited technical knowledge of the product. This will require producers to focus to a greater extent on the role of the product in the consumer's choice of food, taking a holistic view of consumption rather than focusing on the individual product or brand and where that new product fits in the consumer's overall pattern of eating.

Consumers themselves may not always be clear and consistent as to what they want and will need help in accepting new products or to grasp the economic implications of their demands. The distinction between needs, wants and problems of consumers must be clear. The potential for misinterpretation by consumers will be considerable. Industry can play a role in influencing consumer choice, but a balance must be kept between modifying consumer behaviour and providing informational material along with a balance of products that informed consumers can select from.

Consumer eating habits can be viewed as a matter of culture, the products of social standards and relationships. The reflection is likely to be country-specific of changing consumer preferences and social trends, with wide variations in consumers and their needs in various regions of the world. The opportunities identified in consumer markets cannot today be exploited by mass marketing methods using the standard research methods to guide them: where a food fits in a consumer's overall consumption pattern will need to be determined. The development and marketing of new product variants is critical today for marketing to segmented consumer markets. State of the art market research is inadequate for this purpose, partly because of the impact of food attributes; e.g. health being quite idiosyncratic, depending for example on age, sex, work and leisure lifestyles and details of eating habits (Booth, 1989). The cost-effective measuring of consumer decision processes in an actual market situation such that the results can be aggregated to the operational level for the marketing and technical functions is the research problem to be addressed (Adams, 1993). This is the challenge to market researchers for the marketplace of the future.

Those product developers who bring consumers into the new product development process at an early stage, providing them with greater opportunity to input new ideas and be involved in the innovation process, are likely to succeed in the competition to win with new products. Today many companies get most of their information from direct customers (retail chains, agents and industrial customers) and only rarely, and somewhat at random, directly from consumers.

From a global perspective the food industry is facing an exciting but confusing future with increasing internationalisation of producers and consumers, compelling companies to look closely at their global strategies, products and future consumer markets. Companies that identify which products will succeed cross-culturally across borders and which need modifying to local market tastes are those that will succeed in new global markets. This gives huge advantages to companies such as Nestlé, BSN and Unilever which operate across several continents. Their long-standing presence in many global markets gives an unparalleled depth and breadth of development and market experience. Asian giants such as Ajinomoto of Japan, Taiwan's President Foods and Chaeron Pokphand of Thailand will be powerful companies in the future. It is the smaller and medium-sized companies who will carefully have to decide which markets (usually niche markets) to enter and their entry strategies, if they are to succeed in global markets (Buisson and Garrett, 1994).

With the ever-continuing increase worldwide in own-label brands and the emergence of pan-continental retailing groups, retailers will have an increasing role to play in the development of new products and must be prepared to invest in research. The food categories that have seen the most new product launches in recent years are those where the own-label share is low (Anon, 1993a) and the most line extensions and me-too products where the share is high. Own-label products continue to assume increasing importance in retailers' food business, for example about one-third in total in the UK and one-quarter in Germany. By turning own labels into brands in their own right, retailers are thinking like developers and marketers as well as purveyors. Further, the information that they obtain on consumer buying habits provides them with a competitive edge over manufacturers in deciding which new products to launch and then testing them on consumers. Retailers are thus taking an increasingly important role in product development and must accept a greater responsibility in developing new products if consumer expectations of improved products are to be met. Retailers also exercise considerable influence over producers through stocking policies, hence the need for retailers and producers to work more closely together if the consumer is to get products that offer a true advantage and extend food choice.

Increasing diversity of consumers requires producers to have a need for an accurate global view of the food market as a whole. Globally, though, cannot any more mean undifferentiated: it is easy to exaggerate the power of market analysis and demographic data that are standard in consumer research, especially in an increasingly fragmented, albeit international, market like food (Cannon, 1985). Knowledge is needed of the culture and the country or markets intended to do business in. This requires a thorough understanding of the finer points of the food chain, from producer to consumer. Individual market evaluations required are a major challenge. This is not to say that smaller producers will not succeed in niche markets, but usually such successes have been where the market has come to the producer.

Product modification versus product adaptation has emerged as a key product development issue. Mass production should allow for attractive pricing, with advocates suggesting that consumers will modify their wants and purchase for the standardised product because of such pricing (Levitt, 1983). Few food products have succeeded without modification. Food preferences are very culture bound while markets consist of many local and regional markets moulded through cultural, societal, religious and ethnic elements. This is particularly true of the Asian and European markets when dealing with an emotive product such as food. To attain the consumer acceptance desired in these markets necessitates product modification or new product development. Product modification is usually the method selected to speed products into new international markets. When modifying existing or developing new products a thorough understanding of your market is essential. Even with the increasing acceptance of ethnic foods internationally, creativity is required in modifying country-specific products, with their national connotations, to international markets. Appropriate geographic entry point and strategy selection for an international market are critical, with detailed consumer analysis and product testing (usually in market) being needed for a successful consumer food product.

The combination of changing consumers, changing marketplaces and the need to have a consumer-orientated product development process will require producers to evaluate how to organise the product development process. New product development bears an integral relationship to a company's strategy, helping it to define the company's choices in order to have a competitive advantage in the marketplace and to open up new markets. Critically, it is essential to look at product development from a global perspective, evaluating the new and vital markets that food consumers worldwide offer, and food products available globally, to extend food choice in existing markets. In some companies, product development is strategic and well organised but too often the picture in smaller food companies is of incremental short-term product development. Informal planning and no explicit product development strategy exemplify this, contrary to recommended practice. For companies to be successful in product development this must change.

Strategic product development requires both long and short range planning in the new product development cycle. To develop an international market, short- and long-term product and product development strategies should be implemented, instead of the tactic of product extension and adaptation that results in products of little benefit to consumers which is so prevalent in today's markets. Those companies that bring consumers into the development process and analyse how they bring new products to market, and who understand success and failure factors and how they achieve them, will optimistically improve their successes.

The use of cross-functional teams is on the increase to speed products to market. Marketing and Research and Development need to work together, and with consumers, seeking concepts internationally to define products with a high chance of success. Along with consumers, Research and Development must evaluate relationships between a product's physical and visual attributes, while Marketing simultaneously evaluates and selects the key target markets to bring the three creative elements of product development together. New technologies will allow producers to develop new products that extend consumer food choice. These will only succeed if the product offers demonstrable benefits to the consumer and is not so radical as to create consumer distrust of the product. Astute collection and usage of relevant market research data are required. This may mean organisational changes in many companies, with greater formalisation of the new product development process and development of new relationships with consumers.

Global product development will produce many challenges for companies, the strategy for this being dependent on the market targeted. Crippen and Oates (1992) warn that new product development must be considered in the light of cultural differences and ensuing food preferences. Diverse cultures do not share the same approach to product development. In the west much attention is often given to market testing, and getting to market with strong advertising support takes a long time, often with questionable consumer acceptance. In Japan many products are put into the marketplace to let consumers determine what products they like. Location of product development 'in market' may be necessary in certain parts of the world, but undoubtedly product testing will need to be, if consumer input is to be an integral part of the development and acceptance process.

The characterisation of new products in the food industry as largely improvements and revisions to products, which are perceived by consumers as offering little additional advantage with only a small proportion focused on new products, must change. If consumer choice is to be expanded, the current focus of product development needs to take a broader perspective, considering a holistic view of consumption and taking into account the expectations of benefits to the consumer in the new food product. Such an approach will require carefully designed market research and critically involve consumers in the whole development process from idea generation to commercialisation if expanding consumer choice and meeting consumer expectations are to be realised. Global products have the potential to significantly expand food choice if a proactive approach is taken to cross-cultural product development, interacting elements of different cultures' food products, and where necessary adapting the product to local tastes.

The design and implementation of a new product development process, that involves the developer (manufacturer and retailer) and consumers in the development of the product and its subsequent evaluation, should be considered a top priority for a company whose objectives are to develop products that expand consumer choice in world food markets, to be at the forefront of developing new products that benefit the consumer, and to be a successful global food company.

References

Adams, M. (1993) *Seeing Differently: Improving the Ability of Organisations to Anticipate and Respond to Constantly Changing Needs of Customers and Markets*, Report No. 93-103 Market Research Society, Cambridge, Massachusetts.

Adams, R. (1993) Green consumerism and the food industry: Further developments. *British Food Journal*, **95** (4), 9–11.

Albrecht, J. J. (1982) Technology's role in product development. *Food Technology*, **36** (9), 73–6.

Anon (1990) Doctored to taste. *Marketing*, April 15, 22–4.

Anon (1991) New products, new realities. *Prepared Foods New Product Annual*, 14–18.

Anon (1992a) Europe's 'Pharm Foods' opportunities. *Prepared Foods*, March, 75.

Anon (1992b) Ultrahigh pressure processing. *Prepared Foods*, August, 125.

Anon (1993a) Top 20 new product companies. *Prepared Foods*, Nov, 24–26.

Anon (1993b) Survey – the food industry. *The Economist*, Dec, 4, SS1–18.

Antony, M. T. and McKay, J. (1992) From experience: Balancing the product development process: Achieving product and cycle time excellence in high-technology industries. *Journal of Product Innovation Management*, **9**, 140–47.

Barkema, A., Drabenstott, M. and Welsh, K. (1991) The quiet revolution in the US food market. *Economic Review*, May/June, 25–40.

Beharrell, B. and Crockett, A. (1992) New age food! New age consumers! with or without the technology fix please. *British Food Journal*, **94** (7), 5–13.

Best, D. (1993) Food technology faces the future. *Prepared Foods*, May, 32–42.

Bingham, F. G. and Quigley, C. (1990) A team approach to new product development. *Journal of Marketing Management*, **6** (1), 47–58.

Bolongaro, G. (1994) Delphi technique can work for new product development. *Marketing News*, **28** (1), 11.

Booth, D. A. (1989) Health responsible food marketing. *British Food Journal*, **91** (6), 7–14.

Booze, Allen and Hamilton (1982) *New Product Management for the 1980s*, Booze, Allen and Hamilton Inc., New York.

Buisson, D. H. and Garrett, T. C. (1992) The Asia-Pacific food market opportunities and challenges. *The Food Technologist*, **22** (4), 10–15.

Buisson, D. H. and Garrett, T. C. (1994) The impact of the single European market on export and distribution of New Zealand's food products. *The Food Technologist*, **24** (3), 30–37.

Cannon, T. (1985) Marketing problems of the food chain: An overview. *Food Marketing*, **1** (1), 3–18.

Clark, A. M. (1991) 'Trends' that will impact new products. *Journal of Consumer Marketing*, **8** (1), 35–40.

Cooper, R. G. (1984) New product strategies. What distinguishes top performers. *Journal of Product Innovation Management*, **2**, 151–164.

Cooper, R. G. (1993) *Winning at New Products, Accelerating the Process from Idea to Launch*, 2nd edition, Addison-Wesley, Reading, Massachusetts.

Crippen, K. and Oates, C. (1992) Trends in changing far east food preferences and their implications to new product development. *Frontiers and Opportunities in Food and Agro-technology*, Times Conferences, Singapore, in press.

Crippen, K. and Oates, C. (1993) The changing Asia-Pacific food market, opportunities and challenges. In T. S. Sim and K. P. Yeow (eds), *Proc. Symposium of Food Quality and Safety From Manufacturers to Consumers*, Singapore Institute of Food Science and Technology, Singapore. pp. 244–54.

Czinkota, M. and Kotabe, M. (1990) Product development the Japanese way. *Journal of Business Strategy*, Nov/Dec, 31–6.

Davis, R. E. (1993) From experience: The role of market research in the development of new consumer products. *Journal of Product Innovation Management*, **10** (4), 309–17.

Dickie, N. (1993) Improving public health by means of good food: The Heinz approach. *British Food Journal*, **95** (5), 33–5.

Dornblaser, L. and Friedman, M. (1993) New product numbers down: Slow down or postponement. *Prepared Foods*, Feb, 55–6.

Duniaf, G. E. and Khoo, C. S. (1986) Developing low and reduced sodium products: An industrial perspective. *Food Technology*, **40** (12), 365–9.

Fornell, C. and Menko, R. D. (1981) Problem analysis: A consumer based methodology for the discovery of new product ideas. *European Journal of Marketing*, **5** (5), 61–72.

Foxall, G. R. (1984) *Corporate Innovation: Marketing and Strategy*, Croom Helm, London.

Foxall, G. R. (1993) The influence of cognitive style on consumers' variety seeking among food innovations. *British Food Journal*, **95** (9), 32–6.

Frambach, R. T. (1993) An integrated model of organisational adoption and diffusion of innovations. *European Journal of Marketing*, **27** (5), 22–41.

Friedman, M. (1990) Twenty-five years and 98,900 new products later. *Prepared Foods New Product Annual*, 24–5.

Gerrity, R. M. (1991) Globalisation – its impact on world trade: American agriculture and you. *Vital Speeches*, **58** (2), 52–8.

Gofton, L. R. and Marshall, D. W. (1992) Deconstructing sensory preferences; social factors influencing the demand for dark fish. In J. R. Burt, R. Hardy and K. J. Whittle (eds), *Pelagic Fish: The Resource and its Exploitation*, Fishing News Books, London.

Gorman, B. (1991) New products, new realities. *Prepared Foods New Product Annual*, 14–18.

Gruenwald, G. (1988) *New Product Development, What Really Works*, NTC Business Books, Lincolnwood, Illinois, Chapter 3, The hierarchy of new products. pp. 24–45.

Harmsen, H. (1994) Tendencies in product development in Danish food companies. *MAPP Working Paper No 17*, The Aarhus School of Business, Denmark.

Hughes, D. (1994) Forces driving partnerships and alliances in the European food industry. In D. Hughes (ed.), *Breaking with Tradition: Building Partnerships and Alliances in the European Food Industry*, Wye College Press, Wye.

Hunt, J. R. (1994) Nutritional products for specific health benefits food, pharmaceutical, or something in between. *Journal of the American Dietetic Association*, **94** (2), 151–3.

Kirk, J. R., Khoo, C. and Duniaf, G. E. (1987) Industry's role in translating dietary recommendations to food selection. *American Journal of Clinical Nutrition*, **45** (5), 1407–14.

Larson, M. (1991) Taste and value drive microwave foods. *Packaging*, Feb, 33–6.

Lauglaug, A. S. (1993) Technical – market research – Get customers to collaborate in developing new products. *Long Range Planning*, **26** (2), 78–82.

Lehr, H. (1991) New products from Europe and the world. *Food Engineering International*, April, 43–7.

Levitt, T. (1983) The globalisation of markets. *Harvard Business Review*, **61** (3), 92–102.

McPhee, M. (1993) Healthy, wealthy and wiser. *Prepared Foods*, Feb, 13–16.

Meenderink, K. (1992) When time is of the essence. *Marketing*, October 29, 31.

Meyer, R. S. (1984) Eleven stages of successful new product development. *Food Technology*, **38** (7), 71–8.

Milmo, S. (1994) Zeneca betting on its bioproducts. *Chemical Marketing Reporter*, **245** (10), 9.

Mitarai, S. (1991) The Japanese beverage market. *Prepared Foods New Product Annual 1991*, 30–32.

Mitchell, V.-W. and Boustani, P. (1993) Market development using new products and new customers: A role for perceived risk. *European Journal of Marketing*, **27** (2), 17–32.

Murphy, C. (1993) Doubts mushroom over Quorn future. *Marketing Week*, **16** (32), 22–3.

O'Brien, D. R. (1989) Managing for optimisation of technical creativity. *Food Technology*, **43** (2), 82–7.

O'Donnell, C. D. (1992) Aphrodisiacs, *Prepared Foods*, March, 69–70.

O'Reilly, A. J. F. (1991) Leading a global strategic change. *The Journal of Business Strategy*, **12** (4), 10–13.

Paschker, M. (1976) How to guarantee new product failure. *Sales and Marketing Management*, July 12, 40–42.

Petroni, G. (1991) New directions for food research. *Long Range Planning*, **24** (1), 40–51.

Popcorn, F. (1991) *The Popcorn Report: Revolutionary Trend Predictions for Marketing in the 1990s*, Doubleday, USA.

Selwyn, M. (1991) The new food chain. *Asia Business*, Dec 17, 26–34.

Shakla, T. P. (1992) Nutraceutical foods. *Cereals Food World*, **37**, 665–666.

Sloan, A. E. (1993) Consumers, the environment, and the food industry. *Food Technology*, **46** (8), 72–75, 91.

Smith, A. P. (1993) Meals, mood, and mental performance. *British Food Journal*, **95** (9), 16–18.

Smithburg, W. D. (1985) The development and marketing of new consumer products: some successes and failures. *Journal of Consumer Marketing*, **2** (3), 55–58.

Spaulding, M. (1992) Meeting consumer demands. *Packaging*, 37 (6), 31.

Swientek, B. (1993) Multiple forces shape a single EC market. *Prepared Foods*, July, 61–62.

Takeuchi, H. and Nonaka, I. (1986) The new product development game. *Harvard Business Review*, **64** (1), 137–148.

Twaigery, S. and Spillman, D. (1989) An introduction to Moslem dietary laws. *Food Technology*, **43** (2), 88–90.

VanderMenwe and L'Huillier M. A. (1989) Euroconsumers in 1992. *Business Horizons*, **32** (1), 34–40.

von Hipple, E. (1986) Lead users: A source of novel concepts. *Management Science*, **32**, 791–805.

Weller, E. (1992) Forget what the consumer is saying. *Progressive Grocer*, 128.

Wheelwright, S. C. and Clark, K. B. (1992) *Revolutionising Product Development: Quantum Leaps in Speed, Efficiency and Quality*, The Free Press, New York.

Wolf, C. (1994) A taste of tomorrow's foods. *Futurist*, **28** (3), 16–20.

Wright, G. (1990) Understanding the UK food consumer. *Journal of Marketing Management*, **6** (2), 77–86.

Cooking

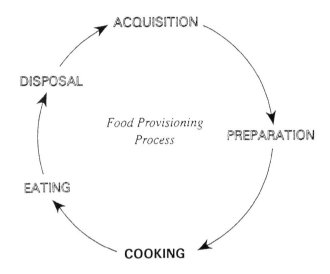

ACQUISITION

DISPOSAL

PREPARATION

*Food Provisioning
Process*

EATING

COOKING

Raw, cooked and proper meals at home 9

Anne Murcott

It is Sunday morning. Quite soon, the family will sit down to a meal of roast beef, potatoes, cabbage, beans and peas. In the kitchen, preparations have reached the final stages:

> This is the best part of the dinner, the gravy ... I'm a wonderful gravy maker though I say it myself. Fat, from the meat, as you can see. All the juices are there as well which will add to the flavour of the gravy. I also like to add a beef cube as well for added flavour. Flour in now. Coming to a nice consistency now. Mix the flour well in. I know it looks a bit of a ghastly sight at the moment, but I can assure you it'll be 'alright on the night'. Pour the potato water in. Add my beef cube, nice and crumbled in. There. Have a little mix again – mix it all together. Then add some of my gravy browning. Not too much. Leave that boil up now – nice and gradual. Wipe the gravy browning bottle down not for dabs to be all over the place. ('Beyond The Gravy', BBC Radio 4, September 25, 1987)

To anyone familiar with 20th century British eating, such a Sunday is immediately recognisable. Certainly some might ponder on the nutritional advisability of a meal that could mean consuming more saturated fats than dietary guidelines of the 1990s recommend. Others again may well observe that the Sunday roast and all its accompaniments is jockeying for position in British eating patterns with manufactured convenience foodstuffs, everyday dishes unimagined by our grandparents, family Sunday lunches at the pub, or eating on the wing – snacking and grazing – fitted round weekend outings, leisure, and shift-work. Sunday roast might just be due to become a matter for reminiscence, if not nostalgia. Certainly, menus and meal patterns are on the move in much of the Western world. Cooking the foods we eat, however, is not.

Cooking is one of the distinguishing features of the human species. While somehow or other all living organisms need to 'refuel', human beings do so in a fashion that marks them off from even their nearest evolutionary neighbours. As Boswell recorded in *The Journal of a Tour to the Hebrides with Samuel Johnson, LL.D.*:

> ...I had found out a perfect definition of human nature, as distinguished from the animal ... My definition of *Man* is, 'a Cooking Animal'. The beasts have memory, judgement, and all the faculties and passions of our mind, in a certain degree; but no beast is a cook. (Quoted in Kuper, 1977: p. v)

Moreover, cooking appears to be all-pervasive across the species. It is probably fruitless to ask how and why human beings came to adopt the habit of cooking their food. We are unlikely ever to be able to know for sure. But we can ask why and how, once cooking had been invented, the practice has persisted, and, moreover, why so many different versions of the practice have continued to flourish.

Modern laboratory sciences provide one set of answers. McGee provides a non-technically expressed summary, derived from physics and chemistry, of four basic reasons for cooking meat:

> to make it safe to eat, easier to chew and to digest (denatured proteins are more vulnerable to our enzymes), and to make it more flavorful. (McGee, 1984: p. 105)

He reports much the same of the effect of cooking on plant foods: it generally improves safety and edibility; it means harmful compounds are either leached from the tissues or inactivated. As for the taste, flavour is intensified since high temperatures make aromatic molecules more volatile and thus more readily detected (McGee, 1984: pp. 174–5).

Other laboratory sciences – the biological in general, psychophysiological in particular – detail the workings of human sensory mechanisms. In an equally non-technical statement, Rogers and Mela (1992) introduce the three sensory systems and consider both innate and learned likes and dislikes in relation to biological mechanisms controlling intake. While their discussion does not directly address reasons for cooking, it picks up McGee's point about flavour, providing a basis for considering a further set of answers as to possible preferences for cooked rather than raw foods. People like the taste.

The answers such sciences provide do not, however, exhaust the matter. They do not explain, for instance, why, even on a hot summer's day, women will still spend a couple of hours in the kitchen cooking a full Sunday roast rather than making something more quickly prepared: 'I don't know ... it's not right, not having a Sunday dinner on a Sunday' ('Beyond The Gravy', BBC Radio 4, September 25, 1987). Nor do they explain why in her *Every-day Cookery* book (as distinct from her better known *Household Management*) Isabella Beeton's midweek menus for late Victorian 'plain family dinners' included dishes involving a good number of different cooking procedures:

> Tuesday. – Cod fish and oyster sauce. – Curried beef. – Rabbit pie, sprouts, potatoes. – Bread and butter pudding.
> Wednesday. – Macaroni soup. – Roast fowls, ham, bread sauce, potatoes. – Jam tart.
> Thursday. – Rissoles made from cold chicken. – Leg of mutton, potatoes, mashed turnips. Plain plum pudding. (Beeton, c. 1895, p. 241)

And they do not tell us why the Yoruba have not only devised different ways of cooking yams, the staple food of that part of Africa (making them into bread, deep-fried fritters or steaming them), but will ridicule someone eating '(yam) loaf alone or with any food other than stew' (Bascom, 1977; p. 84).

To find answers to the full range of questions as to why human beings cook their food, we also have to turn to the modern social sciences, especially to social anthropology and sociology. This chapter, then, is devoted to an introduction to the contribution the combination of the latter two disciplines have made to understanding how and why people cook (together with occasional, albeit strictly amateur, excursions into social history). In the process, the conceptual apparatus of social anthropology and sociology – signalled by terms such as culture, social organisation, division of labour – will be shown both to apply to, and can be illustrated via, cooking.

One key point of departure for the discussion that follows needs to be stated right away. What human beings think, do and say goes well beyond biological requirements for survival – in eating as in anything else. Profound intraspecies variation in the systems of thought, the patterns of the deeds and words, requires analytic attention independent even of human psychology, to investigate those systems and patterns as phenomena in their own right. Though social anthropologists and sociologists quarrel about the meaning of the term culture, something like it remains indispensable to their investigations. For, not only do different human groups end up nourishing themselves by very variant routes, at the same time they perennially incline to invest quite complex meanings in what and how they eat, meanings that tell them – and us – something about the nature of the social group in question.

This chapter divides into three sections. The first establishes the ordinariness of cooking by pointing to its extensiveness – part and parcel of what sets the nature of human life apart from that of other forms of existence. The second picks out the cultural significance of cooking as part of a broader question of the definition of food itself – a 'food has no meaning if a man does not give it a meaning by willingly eating it' (Tremolières, 1970). The third considers the task of cooking, its social organisation and the import of technological developments in the domestic kitchen.

9.2
On the ubiquity of cooking – hearths, recipes and cookbooks

This section dips into the evidence about the place of cooking in human affairs. Here the point is not simply to illustrate that cooking is commonplace, but to indicate that it is evident in much of what human beings *in any case* do and say, make and invent. Technological innovations encompass the kitchen range, pots and pans – that much is obvious. Beyond that we find that one or other aspect of cooking is incorporated, rather as a matter of course, in long-term historical developments – from the creation of specialist terminologies to reading and writing. It is also engrained in the stream of routine, day-to-day activities in which members of one or other social group are engaged. It is integral to human society and its significance is to be analysed as such.

For the moment, cooking can be defined in terms of the application of heat. McGee reminds us of the physics of the process: 'the transfer of energy from a heat source to food' (McGee, 1984: p. 610), and that the various cooking techniques draw on three different principles of heat transfer: conduction, convection and radiation.

Anthropologists remind us of the social and cultural aspects of the process enshrined in the vocabularies of cooking that have been created to distinguish and define different methods human beings have invented for applying heat to food. Farb and Armelagos (1980, p. 107) report that the (American) English language, for instance, has at least thirty-five words for ways of cooking. Some terms (e.g. coddle), they point out, are rarely used, others are compounds (e.g. pan-fry, oven-bake) and some are used for particular dishes (e.g. to plank meat, or to shirr eggs – i.e. break into a greased and crumbed dish and bake (Whitfield, c. 1956: p. 126)).

Other bases for distinctions English speakers make include the amount of time needed for a cooking process (stewing is boiling at lowish heat for a long time, par-boiling much shorter on higher heat, and blanching even faster); the amount of cooking medium used (deep- as distinct from shallow- fried); the placing of food in relation to the heat source (spit-roasting compared with broiling – grilling on the European side of the Atlantic – where the food is placed below the heat source and grilling – as in 'barbecuing' – where it is placed above). By contrast, other languages signal different social categorisations of the procedure. Farb and Armelagos (1980) refer, for example, to an Ethiopian language which has words distinguishing the boiling of solids and liquids. And they identify the French language's distinction between *rôtir*, to roast, and *griller*, to grill, as one based on the shape of the food being cooked. As these authors point out, what is telling about cooking vocabularies is not so much that some languages make distinctions that are absent in others, but rather that one or other language entails reference to certain possibilities which are then arranged 'conceptually in ways that are culturally indicative' (Farb and Armelagos, 1980: p. 108). In other words, the names given to a cooking technique enshrine methods of applying heat to food culturally typical of one human group as distinct from another, or which within a group are culturally conventional for a certain occasion or for certain types of person.

If vocabularies of cooking can be treated as 'vehicles' for a social group's cultural expression of different techniques for the application of heat to food – a people's means of encapsulating a shorthand reference for memorising and passing them on to future generations – then the group's distinctive cooking practices are reflected in the utensils they create and the technologies they invent for making use of the fuel available to them. Once again, the point is that the invention of some sort of cooking pot and one or other type of hearth, oven and bakehouse registers the human ubiquity of cooking and, simultaneously, the cross-cultural diversity within the species at different periods of history. To take only a handful of haphazardly selected examples, we have, for instance, Rosemary Firth's (1977) account of the domestic interior of the bamboo and palm thatch house which she and her husband, Raymond, had built for them when they settled in a small village on the east coast of Kelantan in Malaya in 1939 to conduct a social anthropological study of the local fishing economy:

> It had a typical village-style kitchen at one corner of the house, with a sanded and raised platform for a wood fire, over which cooking-pots of iron rested on metal grids – while the smoke found its way out of a hole at the side of the roof. (Firth, 1977: p. 183)

She makes no mention of an oven. Similarly, amongst the LoDagaa of northern Ghana, Goody notes that almost all the food women cook for routine, domestic eating is boiled, for there are no ovens. A hearth is made of three stones on which pots are balanced, a separate hearth per pot (Goody, 1982: p. 70).

Likewise, according to Anderson (1988), Chinese cooking in private houses is primarily boiling and steaming, with stir-frying a minor adjunct, with the big stove, both literally and metaphorically, representing the traditional Chinese home. Apparently unchanged since 200 BC, the stove is 'an impressive creation', designed such that 'even a tiny amount of very poor fuel will suffice to cook a lot of food' (Anderson, 1988: p. 182). His description of the Chinese *batterie de cuisine* provides the basis for an intriguing comparison with modern western kitchen implements such as those detailed in Susan Campbell's (1980) practical manual aimed at the self-consciously interested sector of the British mass market.

By contrast, baking and roasting are far more prominent in other cuisines. The Romans for instance, imported their own bread ovens into Britain. Usually, Wilson reports, these were designed 'with a domed roof of rubble and tiles, and a flue in front. Wood or charcoal was burnt inside for a time and then raked out so that bread and cakes could be baked in the heated chamber' (Wilson, 1976: p. 210). It needs to be remembered, however, that even by the Mediaeval period in England, a house with a baking oven was not the norm; villages might have had communal ovens, or a specialist baker (see also Mennell, 1985). As for roasting, she documents the development of the spit in England from a simple Mediaeval item rotated by a lowly household menial to the complex mechanical devices of the Tudor period – now operated by a dog, or, in another design, by gravity (Wilson, 1976). In parallel she records the gradual four hundred year 'move', that began in the 12th century, of the fireplace from the centre of a room or hall to its placement against a side wall.

As a footnote to this comment on diversity of means of applying heat to food, the human capacity for inventiveness is illustrated by the British Army's improvisation of stoves using petrol-soaked sand in the North

African desert, and of tank crews tying a can of water to the hot exhaust into which tea could be thrown as soon as a lull in the action allowed (Ellis, 1990: p. 287). Parallel resourcefulness is illustrated in anecdotes of neat (American) tricks of setting wrapped and sealed food on the engine to cook during a long car journey, or in the dishwasher – an intriguing instance where clearing up after one meal overlaps with preparations for the subsequent ones. Both are reminiscent of the ingenuity in placing meringues to dry out in the airing cupboard. Put another way, we are reminded that heat sources serve more than one human purpose. Sometimes, as in the examples just cited, people make a virtue out of heat as a by-product. At others, equipment is deliberately designed to be multipurpose. An Aga complete with 'constant' hotplates and ovens may serve not only to heat the room but also fuel the central heating and hot water supply of the whole house – never mind providing a plate-warming compartment which, in farm kitchens, can double as emergency incubator for orphan new-born lambs.

Vocabularies of cooking on the one hand, and technological innovations on the other, come together in the production of cookery books. The idea of a recipe – the mainstay of cookbooks – had, however, to be invented first. Recipes in Europe are a great deal older than the use of the word in English to refer to cooking practices suggests. Defined as a 'statement of the ingredients and procedure necessary for the making or compounding of some preparation, especially of a dish in cookery', the *Oxford English Dictionary* gives the earliest culinary usage as 1743, adopting a term already in use to describe the preparation of medicines. Statements of quite this kind are not, however, found amongst all cultural groups – Lévi-Strauss, for instance, does not consider that the 'coarse cooking techniques' of the South American Indian tribes amongst whom he worked could be called recipes (Kuper, 1977).

Inventing recipes, or indeed the very idea of a recipe, is entwined with, and expressed via, another human invention, that of literacy. If the creation of vocabularies of cooking, or indeed anything else, allows some sort of shorthand communication, fixing it as a written text makes it available for discussion, comment and dissemination in a way that oral communication cannot. Some of the earliest known written records contain references if not to cooking certainly to foods and their agricultural production (for China of the 5th century BC see Anderson, 1988). A work, translated (Flower and Rosenbaum, 1958) under the title *The Roman Cookery Book* is attributed to Apicius,[1] a gourmand who lived in

[1]In his notes to his translation of Juvenal's *The Sixteen Satires* Peter Green comments that Apicius was 'renowned both for his gourmandise and his extravagance. It is said that after he had spent the rough equivalent of £1 000 000 he balanced his books, found he had no more than 100 000 gold pieces left, and poisoned himself, on the grounds that no gourmet could be expected to live on such a pittance. The cookery-book associated with his name is a late compilation' (Green, 1967, p. 111).

the reigns of Augustus and Tiberius during the 1st century AD, and a few European manuscripts survive from the late Middle Ages. Cookery books feature among the very earliest of printed books – the first known was published at Nuremberg in 1485 (Mennell, 1985).

For the social historian, cookery books are a valuable source of information about cooking. But just like any other document, their interpretation needs caution. It cannot be assumed that they are a straightforward representation of what actually took place in kitchens in the past, any more than they are in kitchens of the much more recent present. So, for instance, when Crawford reported that among the results of a British Food Enquiry carried out in the late 1930s, 'Mrs Beeton's Book' was far and away the most popular, he was careful to note that it was likely 'she receives more lip service than regular consultation' (Crawford, 1938: p. 113). These caveats notwithstanding, the point remains that the invention of printing amplifies the effects of literacy, significantly enlarging the potential readership.

9.3 Cooking, proper meals and the definition of food

Having pointed not just to the ubiquity of cooking in human society, but also the way in which it is integral to it, we now turn to take a closer look at its cultural significance. To do so, we need to attend to two things. First, we must revisit the definition of cooking. One general meaning deals solely with the application of heat, by whatever method. Another would be more inclusive, encompassing the opposite, the lowering of temperature, and extending to the creation of dishes by mixing different combinations of foodstuffs together, neither heating nor cooling them. Following on from this, we next need to backtrack somewhat to think about food more broadly. For one way of considering the cultural implications of cooking is to view it in the same light as contrasting eating amongst human beings with the refuelling required for the survival of any organism. Accordingly, we must address the very definition of food, culturally speaking.

Though human beings are omnivorous, it is well known that they do not eat everything available that is both nutritious and safe. The British will not eat earthworms, horses or Brunel the pet guinea pig. But Kazakhs, Belgians and the French will eat horsemeat and the Yoruba do eat guinea pigs. As for earthworms, a zoology student made the headlines (*Oxford City Courier*, June 5, 1986) with his fund-raising stunt of downing 300 of them, duly boiled, precisely because so doing is culturally deemed to break the rules for acceptable eating – evoking the human physiological response of revulsion and directing the human psychological capacity for disgust accordingly. What is to count as food, then, is culturally not biologically defined. Before human beings eat something, they have to have called it food. In other words, they invest the item with a meaning, thereby classifying it.

The cultural classification of things into food and non-food takes various forms. It may be a relatively simple matter of sorting items more or less as they stand into one category or the other. But it can also be arrived at by attention to some process of transforming a particular item. Transformations may be natural, ones that happen whether or not human beings are around. Fruits become definable as food at a certain point in what we name the ripening process; asparagus and fiddle head ferns are effectively only food at early stages of the shoots' growth. Or transformations may be artificial, deliberately effected by human intervention: converting green tomatoes – unripe, non-food – into chutney. Once modes of transformation enter the picture, the simple two-fold classification into food and non-food elaborates, allowing subdivisions into different types. 'Natural' processes, such as putrefaction, may be harnessed (as in cheese making) along with fermentation (in, for instance, making vinegars that are then used in pickling), drying, salting, freezing and smoking to alter the state of an item, thereby defining it as a different type of food. And then, of course, there is cooking. In these terms, it represents a further mode of transformation converting raw items, still called ingredients, into a dish, at last becoming food to be put into the mouth.

By now it is probably only too obvious that cooking as a mode of transformation can serve as a basis for transmuting the meaning of something from non-food to food. From cattle grazing in a field, via carcass as the product of the abattoir, pieces duly butchered and wrapped in plastic film on the chilled supermarket shelf, renamed yet again as a joint of beef in the kitchen, it is roasted in the oven to arrive on the platter ready to be carved and served as the meat for Sunday dinner – almost bringing us to the point at which this chapter began. Cooking has literally and metaphorically transformed the raw flesh, translated it culturally into food.

Sociologists and social anthropologists might be happy enough with the account presented so far. But they diverge in pursuing the analysis of cooking's cultural significance further. One line of enquiry is represented by the contribution of the French anthropologist, Claude Lévi-Strauss (e.g. 1966). Very crudely stated, his argument runs something like this.

Human beings present a paradox; they are simultaneously animal and not-animal. We have biological functions in parallel with other animals, but we also have intelligence, language and the capacity for memory, inventiveness and abstract thought that sets us wholly apart from other animals. In this, we are creatures of both 'nature' and 'culture'. As a corollary of the latter comes a neat twist. Human beings are well aware of the fact of their dual character. But with it, Lévi-Strauss argues, comes a subliminal tension which constantly requires resolution. We have perpetually to cope with the fact that we are simultaneously animal but not-animal.

In his view, the regular need to eat repeatedly faces us with that paradox, and cooking is one means via which we continually express our

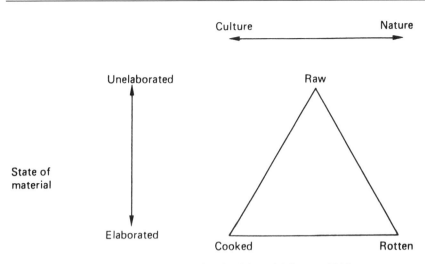

Figure 9.1   The culinary triangle. (After Lévi-Strauss, 1966)

attempts to deal with it. Though foodstuffs are of the natural world, reflecting the animal aspect of ourselves, by cooking we convert them to the cultural realm, reminding us of our human selves. The transformation, or elaboration, effected by heating raw items signals that we are 'cultured' in the way animals cannot be. Lévi-Strauss displays his argument via a scheme he calls the 'culinary triangle' (Figure 9.1).

Indeed in cooking, he says, we go further to represent additional distinctions, this time amongst various groupings of human beings and the activities in which they engage, such that different methods mark those distinctions. For instance, roasting, Lévi-Strauss claims, is a cooking technique associated with high prestige, masculinity and special occasions in both industrialising and industrialised societies. Boiling, on the other hand, is more prosaic, used for everyday dishes, for those of humbler birth, for women and children. This he shows in the developed form of his culinary triangle (Figure 9.2).

Taking a long and, it has to be admitted, a tortuous route, Lévi-Strauss ultimately insists that the way we transform our food, including the fact that we cook it as well as have different techniques for doing so, is ultimately to be analysed as a kind of language which can be shown, without our realising it, to communicate the very structure of human society, any and everywhere.

Fellow anthropologists have criticised Lévi-Strauss for methodological inadequacies, gaps in the argument, never mind conveniently selective attention to the evidence (for a review of some of the debates see Murcott, 1988). But it is not just for the sake of completeness that his work receives continued attention. Despite the criticisms that it deserves, it is

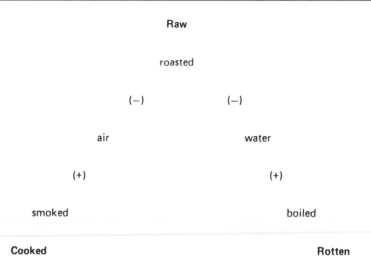

Figure 9.2   The culinary triangle developed. (After Levi-Strauss, 1966)

nonetheless hailed as seminal. For, along with the equally influential contribution of the eminent English social anthropologist Mary Douglas (e.g. Douglas, 1972, 1984, 1987), it demands we take seriously that cooking, as a typically human feature of eating, has an *analysable* set of cultural meanings – which brings us back to the Sunday roast.

Not only is the definition of food a matter of cultural classification rather than solely a biological necessity (with cooking a characteristic that can form part of that definition) so too is another human invention, the idea of a meal. People in all societies have clear notions of what is to count as a meal (cf. Mintz, 1992) and of the cooking techniques that are essential to the definition as one. In Britain, research during the 1980s reveals that the idea of a proper meal at home is (still) epitomised by the special Sunday version[2] (Charles and Kerr, 1988; Murcott, 1982, 1983a). Commonly – and tellingly – known as a 'cooked dinner' people can detail its character quite precisely, revealing clear 'rules' for its composition and preparation. It is not a whole menu, but one course presented on a single plate. It centrally consists of meat – flesh, not offal (e.g. liver), fresh, not preserved (e.g. sausages). Poultry is acceptable, fish is not. As essential to its definition are the accompanying vegetables, potatoes and at least one in addition that has to be green. It is not complete until gravy is added to the whole assemblage.

What is to make the dinner 'proper', however, lies not just in the essential elements of which it is composed. The rules go further, specifying the

---

[2]See Marshall, chapter 11, for extended discussion of eating occasions.

cooking techniques integral to its definition as this particular type of meal. The meat is to be roasted (baked) or grilled, the vegetables are to be boiled. The Sunday variant is special not only in the greater number of accompanying vegetables but also in that both meat and potatoes are roasted. The gravy is interesting. Its ingredients are an amalgam of the cooking mediums of the other items, fat from the roasting pan plus the meat juices, the water in which the potatoes (or green vegetables) were boiled. As such it reflects and recapitulates the techniques required for cooking the other items. To this extent it incorporates a combination of the rest of the dinner, over every item of which it is then to be poured.

The rules prescribing cooking techniques are thrown sharply into relief when seen alongside the preparation of other types of meal. The research reports comparison with alternative cooking styles firmly considered not 'right' for a proper meal. Making chips (French fries) rather than roast or boiled potatoes will not really do – this is 'cheating'. Making a stew or casserole of meat and vegetables all in one pot in the oven does not count as a cooked dinner, nor does quickly frying the meat in a pan on the hob. A real cooked dinner is a meal (plateful) that requires more than one cooking technique, separate preparation of the various elements, all needing regular attention over a long enough cooking time.

Analysis of the cultural significance of the meal and its associated prescribed cooking techniques has to go further to see in what way the whole affair is, as it were, seamlessly built into ordinary family life at home. In the ideal, the social convention is that women cook for men and children – a convention reported in society after society. More generally, women are charged with responsibility for the health and welfare of their families, including the smooth running of the household. As the geographical as well as cultural location of attending to the basic day-to-day necessities of sleeping, washing, dressing, eating, etc., the Western household represents some kind of anchor in people's lives and their social identities. Home is a place to return to, a place for refreshing oneself. The British cooked dinner is recognised as the meal proper to the home; on weekdays it is to be ready for the return home after work or school; on Sundays the superior variant marks that particular day of the week as different and special. And in a proper home, proper meals are properly cooked by the woman – a point to which the next section will return.

The example of the 'cooked dinner' illustrates one way in which the cultural significance of cooking can be analysed. It is not, nor is it intended as, a comprehensive account of the significance of every aspect of all day-to-day cooking at home. Additional research must examine further sets of meanings human groups in Britain, as well as elsewhere, invest in cooking and meal preparation, including the meanings, if any, attributed to people's own sense of changes in eating patterns. And care is needed to collect evidence based on direct observation of what people routinely actually do as well as that based on what they routinely say about their activities.[3] The main point remains: to cook food at all, to employ one cooking method or another has implications well beyond using heat to make foodstuffs tastier and safe to eat.

9.4
The social organisation of cooking: divisions of labour, tools, techniques and technology

The kind of sauce Britons call gravy can obviously be made in many different ways – with 'potato-water' as in the quotation with which this chapter opened, bean- or cabbage-water, stock or wine; with or without flour; with or without a beef cube, 'granules' or an 'instant' mix of flavouring, colouring and cornflour combined. Learned at home as a child, following instructions in a cookery book, incorporating commercially available additions or substitutes, they represent different orthodoxies and conventional wisdoms of different historical periods.

To be sure, the trend away from laborious and lengthy home preparation of dishes from basic ingredients to the use of manufactured 'pre-prepared' items is well known, widespread and extensively documented (e.g. in Mexico City, Vargas and Casillas, 1992; the Netherlands, van Otterloo, 1990; the USA, Levenstein, 1988). Viewing the social significance of this trend in cooking runs from commentators' warnings against the loss of even the simplest culinary skills and against the demise of good taste (cf. Driver, 1983) to whole social movements (e.g. those committed to whole foods or adopting 'exotic' cuisines) based on a revolt against what is perceived as an undue artificiality and superficiality of modernity (cf. Atkinson, 1983; van Otterloo, 1990). In addition there are thoroughgoing sociological analyses such as Mennell's elegant sociological thesis (1985; see also Goody's (1982) different, but equally compelling, social anthropological account) proposing that it is to grand-scale cultural shifts we must look, shifts which entail reducing differentiation between the culinary tastes associated with socio-economic variation but increasing greater width of choice for the whole populace.

Here, we shall be content with swiftly reviewing the socio-cultural significance of cooking with preliminary reflections on technological developments that penetrate the domestic kitchen in more or less the same way they have penetrated most spheres of life. To do so, we must, yet again, revisit the definition of cooking. For we need to take into account not just one other method of applying either heat or cold to foods, but the social organisation of its accomplishment, introducing questions such as who does what, how and for whom. Presenting this extended style of defining cooking, Schwartz Cowan (1989) points out the task of transforming raw ingredients into dishes entails tasks which are linked to others which they do not at all resemble. The procurement of foodstuffs,

---

[3]For an important cautionary tale with potentially serious practical policy implications, see Laderman's (1983) work revealing the intellectual inadequacy of analysis that neglects actual practice and is over-reliant on data about ideas about food.

access to storage, the acquisition of fuel, utensils and equipment, never mind their maintenance and cleaning, are self-evidently necessary to, but by no means the same as, actually attending to converting the ingredients into a meal. Implicit in what Schwartz Cowan calls this 'work process' is a sort of algebraic equation of time, pairs of hands, tools, energy source and equipment – an equation to be kept in mind as we proceed.

Among the many features marking professional from domestic cooking is a persistent gender differentiation. As already noted, in most societies women are very closely associated with ordinary domestic cookery. The development of more technically elaborate cooking methods and the emergence of socially prestigious cuisines carry with them an imperative to become distanced from the everyday cooking of the majority of the population. By and large, Mennell (1985) argues, this accounts for 'male chefs' but 'women cooks' across the western world. Cooking professionally as a chef or caterer crystallises as an occupation; cooking domestically, every day as part of the ordinary round of life, remains as a more diffuse cultural convention, part of a woman's social role. Modern attempts are made now and then to teach home cooking to schoolboys to prepare them to share the task with their future wives. In parallel, forecasts are made of increasing marital egalitarianism in the kitchen – though it has to be said that such forecasts run rather against the evidence. Cooking at home – in its widest sense – nonetheless persistently continues to be conventionally associated with women not men. Indeed, an analysis of nuances in the everyday usage of the verb 'to cook' suggests just such associations (Murcott, 1983b).

Explanations – not to mention histories of these innovations, the primary sources on which they depend and the manner in which they are marketed (cf. Murcott, 1983c) – for trends in the technological penetration of the kitchen typically turn on this 'gendering' of domestic cooking (cf. Mennell et al., 1992). The themes of efficiency and economy, time- and labour-saving, run through any number of accounts recording and analysing the implications for domestic culinary practices of historical transformations of women's social position – from mistress of a grand household, employment as cook, kitchen maid or factory worker of earlier eras to late 20th century full-time housewife and mother or 'busy working wife'. At the same time, there are counter trends, albeit on a smaller scale and socially far more selective. The second half of this century, at least, has seen periodic resurgences of domestic based interest in complex, novel as well as venerable, cooking skills, technologies and recipes. Books on the exceptionally labour-intensive nouvelle cuisine of the 1970s, on woks or tempura, along with pasta-making machines and sorbetières are to be found on the kitchen shelves of middle class Australian, American and British homes.

The former trend may need to be seen as tied firmly to the culturally prescribed refuelling necessary for generally getting by day-to-day. Correspondingly, the latter, countervailing, trend may need to be analysed not just as a selective self-conscious expression of good taste but more as the province of leisure for those who can afford to buy others' labour to deal with the mundane versions of kitchen work and housekeeping and thus release the time to engage in the recreation – and re-creation – at home of dishes needing potentially more laborious culinary skills. To move beyond such speculation, which patently does little more than restate the question, requires a good deal more work. A form of analysis is required that adequately comes to grips with a realm in which industrialised and craft modes of production coexist.

9.5
Conclusion

In a paper published as recently as 1994, Warde and Hetherington observe that 'the sociology of food preparation remains relatively unexplored, yet it deals with a range of socially symbolic and very time-consuming activities. It would amply repay further research' (Warde and Hetherington, 1994). As was implied at the end of the previous section of this chapter, conclusions that make confident statements about the cultural import and social organisation of domestic cooking are premature. There are far too few data for one thing. We shall have to move beyond stray, if intriguing, anecdotes and speculations of the kind presented here first. Further, some systematic analytic framework for thinking is needed, though we shall equally need to guard against any framework's becoming a mental straitjacket. At the least we might design a more coherent outline that brings together elements too commonly considered separately – the beginnings of which are sketched here by way of conclusion.

Since it is probably safe enough to assume that the ubiquity of cooking (remembering the broad definitions indicated above) will persist, we may as well use the journalists' questions 'where, when, what, who and how' as a starting point. For instance, we could start with 'where'. If we assume that the home will remain central to daily living arrangements, we can ask whether it will also continue to be as central a site of food preparation and eating. Provisioning, cooking and eating can, in principle, be detached from other home-based activities. Let us recall, for instance, that elements of hotel- and hospital-stay represent surrogates for the home. So, the unremarkable US convention for hotels to let rooms on the assumption that residents will find their own breakfast would raise a good many eyebrows in Britain. And in a great many countries of Africa and Asia it is automatically assumed that relatives, rather than the institution, will provide, cook and present meals to hospital patients.

Coming at the point from another direction, we might just see a revival of campaigns for kitchenless houses that were to be coupled with organising domestic work on a neighbourhood basis of the kind envisioned by 19th century utopian commentators, feminists among them (Hayden, 1978). Were such a move to take hold, pursued to its logical conclusion it would eventually result in the total disappearance of eating at home. Kitchenless houses are probably highly unlikely, but examining the significance for eating at home of trends towards eating 'out' has received scant social scientific attention.[4] We have little idea of whether eating in cafés, pubs and restaurants supplants or complements styles of domestic-based eating.

Posing questions about eating out not only addresses 'where' but also raises another two, 'when' and 'who'. Is eating out something of a special occasion or is it becoming part of an ordinary pattern that in the process relieves the member of the household usually charged with cooking from the work of doing so that day? And if it is to be the latter, will it be men, women or children who are relieved of the burden?

Staying with home-based eating leads us to consider 'how'. Just because food is consumed in the house need not mean it was cooked there. But a study of the social organisation of modes of home eating and their relation to 'niche' supplies of complete meals across the range to include take-aways, meals-on-wheels for the elderly (introduced in Britain during the 1939–45 war) and outside catering for parties, does not readily spring to mind. Even if provisioning and cooking is done at home, we are still faced with the lack of social scientific material and have to rely on limited access[5] to market research data. The glimpses we have of such findings are tantalising. For example, a forty-two per cent increase in the household ownership of microwave ovens is reported between 1985 and 1992 – households which almost all already had freezing as well as refrigerating equipment in one form or another. But social scientists know little of how these white goods are being used. Do they simply relieve the labour of, or even promote, patterns of home cooking with each meal prepared from basic ingredients, or is the freezer used as a bulk storage cabinet of meals purchased ready prepared that are simply defrosted/ cooked in the microwave oven? As provocative are findings carried in the July issue of the magazine *Good Housekeeping*, from a Gallup survey marking the seventieth anniversary in 1994 of the Good Housekeeping Institute. Two out of three people are reported to use convenience foods at least once a week as compared with 15% of adults who bake bread; 21% of adults judged pressure cookers the most useful kitchen gadget while freezers were described as 'your favourite non-essential domestic appliance'. But findings such as these cannot substitute for social scientific investigations of the significance of modern domestic equipment for food

---

[4]An important start is being made by Alan Warde whose study *Eating Out and Eating In* is one of the projects of the 1992–98 Economic & Social Research Council (UK) Research Programme *The Nation's Diet: the Social Science of Food Choice*, of which the present author is Director.
[5]Even though a proportion of such materials are publicly available, their commercial price is prohibitive to individual scholars and academic institutions alike.

preservation and preparation for the activities to which Warde and Hetherington (1994) refer.

Starting from the questions journalists pose is just a simple beginning, and a shorthand at that. The interconnections between the questions and their associated answers need to be part of the analytic effort. And the whole is ever to be integrated with related social scientific work on household structures, domestic divisions of labour, formal and informal economies, gender, age-grading and kinship and more besides. Only then will it be possible to bring the discussion introduced in this chapter up-to-date to demonstrate, for example, how microwaving is accommodated in the repertoire of cooking techniques and to consider its cultural impact on styles of eating and the social organisation of food production at home. Furthering the understanding of the cultural and social significance of food will, then, need to proceed in this direction, amongst others, in future.

Dedication   'Beyond The Gravy' was created and produced by Richard Thomas (1950–1994) to whose memory this chapter is dedicated. It was his flair, acumen and intellect that converted my 1982 paper, published in an academic journal, into a radio programme which successfully caught the imagination and interest of the listening public, yet simultaneously remained faithful to the analysis on which it was based.

Acknowledgements   I am grateful to Geraint Talfan Davies, Controller, BBC Wales, for his interest; to Iris Cobbe, Sound Librarian, BBC Wales, and Phil Strong for their helpful comments and advice; to Ben McManus, for talking about wars and the loan of his book; and to Taylor Nelson plc for sight of their report *Trends in Cooking Habits, 1985–93*.

References   Anderson, E. N. (1988) *The Food of China*, Yale University Press, New Haven and London.
Atkinson, P. (1983) Eating virtue. In A. Murcott (ed.), *The Sociology of Food and Eating*, Gower, Aldershot.
Bascom, W. (1977) Some Yoruba ways with yams. In J. Kuper (ed.), *The Anthropologists' Cookbook*, Routledge & Kegan Paul, London.
Beeton, I. (c. 1895) *Mrs Beeton's Every Day Cookery and Housekeeping Book*, new edition, Ward Lock, London.
Campbell, S. (1980) *The Cook's Companion*, Macmillan, London.
Charles, N. and Kerr, M. (1988) *Women, Food and Families*, Manchester University Press, Manchester.
Crawford, W. (1938) *The People's Food*, Heinemann, London.
Douglas, M. (1972) Deciphering a Meal. *Daedalus, Journal of the American Academy of Arts and Sciences*, **101**, 61–81.
Douglas, M. (ed.) (1984) *Food in the Social Order*, Russell Sage Foundation, New York.
Douglas, M. (ed.) (1987) *Constructive Drinking*, Cambridge University Press, Cambridge.
Driver, C. (1983) *The British at Table*, Chatto & Windus, London.

Ellis, J. (1990) *The Sharp End: the Fighting Man in World War II*, revised edition, Pimlico, London.

Farb, P. and Armelagos, G. (1980) *Consuming Passions: the Anthropology of Eating*, Houghton Mifflin, Boston.

Firth, R. (1977) Cooking in a Kelantan Fishing Village, Malaya. In J. Kuper (ed.), *The Anthropologists' Cookbook*, Routledge & Kegan Paul, London.

Flower, B. and Rosenbaum, E. (1958) *The Roman Cookery Book*, Harrap, London.

Goody, J. (1982) *Cooking, Cuisine and Class*, Cambridge University Press, Cambridge.

Green, P. (1967) *Introduction and Notes to Juvenal: The Sixteen Satires*, Penguin, Harmondsworth.

Hayden, D. (1978) Two Utopian Feminists and their Campaign for Kitchenless Houses. *Signs*, **4** (2), 274–290.

Kuper, Jessica (ed.) (1977) Introduction to J. Kuper (ed.), *The Anthropologists' Cookbook*, Routledge & Kegan Paul, London.

Laderman, C. (1983) *Wives and Midwives: Childbirth and Nutrition in Rural Malaysia*, University of California Press, Berkeley, CA.

Levenstein, H. (1988) *Revolution at the Table: The Transformation of the American Diet*, Oxford University Press, New York.

Lévi-Strauss, C. (1966) The Culinary Triangle. *New Society*, December, 937–940.

McGee, H. (1984) *On Food and Cooking: the Science and Lore of the Kitchen*, Allen & Unwin, London.

Mennell, S. (1985) *All Manners of Food*, Basil Blackwell, Oxford.

Mennell, S., Murcott, A. and van Otterloo, A. H. (1992) *The Sociology of Food: eating, diet and culture*, Sage, London.

Mintz, S. W. (1992) Die Zusammensetzung der Speise in frühen Agrargesellschaften. Versuch einer Konzeptualisierung. In M. Schaffner (ed.), *Brot, Brei und was dazugehört*, Chronos, Zurich.

Murcott, A. (1982) On the Social Significance of the 'Cooked Dinner' in South Wales. *Social Science Information*, **21**, 677–695.

Murcott, A. (1983a) 'It's a pleasure to cook for him...': Food Mealtimes and Gender in Some South Wales Households. In E. Gamarnikow *et al.* (eds), *The Public and the Private*, Heinemann, London.

Murcott, A. (1983b) Cooking and the Cooked in A. Murcott (ed.), *The Sociology of Food and Eating*, Gower, Aldershot.

Murcott, A. (1983c) Women's place: cookbooks' images of technique and technology in the British kitchen. *Women's Studies International Forum*, **6** (2), 33–39.

Murcott, A. (1988) Sociological and Social Anthropological Approaches to Food and Eating. *World Review of Nutrition and Dietetics*, **55**, 1–40.

Rogers, P. J. and Mela, D. J. (1992) Biology and the Senses. In National Consumer Council, *Your Food: Whose Choice*, HMSO, London.

Schwartz Cowan, R. (1989) *More Work for Mother*, new edition, Free Association Books, London.

Tremolières, J. (1970) A behavioural approach to organoleptic properties of food. *Proceedings of the Nutrition Society*, **29** (2).

van Otterloo, A. H. (1990) *Eten en Eetlust in Nederland*, Bert Bakker, Amsterdam.

Vargas, L. A. and Casillas, L. E. (1992) Diet and Foodways in Mexico City, *Ecology of Food and Nutrition*, **27** (3/4), 235–248.

Warde, A. and Hetherington, K. (1994) English households and routine food practices: a research note. *Sociological Review*, **42** (4), 758–778.

Whitfield, N. (c. 1956) *Kitchen Encyclopedia*, Spring Books, London.

Wilson, C. A. (1976) *Food and Drink in Britain*, Penguin, Harmondsworth.

Eating

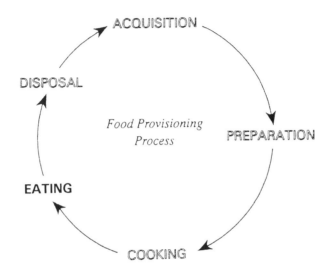

ACQUISITION

PREPARATION

COOKING

EATING

DISPOSAL

Food Provisioning Process

Are sensory properties relevant to consumer food choice? 10

Monique Raats, Béatrice Daillant-Spinnler, Rosires Deliza and Hal MacFie

10.1
Introduction

It is clear to many consumers that the taste of a food is a crucial parameter in determining its acceptability. Thus if questioned in the street or in a focus group, consumers will discuss at length how the flavour of a product has decreased, and how they will not buy cheaper foods because of poor sensory properties. In a study with husbands and wives, Schafer (1978) found that husbands rated taste followed by nutrition as being the most important determinant of food choice, whilst the wives rated nutrition followed by taste as being most important. McNutt *et al.* (1986) found taste after safety to be the most important factor. However, when one begins to examine indices of buying behaviour, it is equally clear to the researcher that taste is not the only crucial determinant, and in some cases is clearly well down the priority list.

At the moment of consumption the sensory experience is very direct and sensations from each of the senses reach the brain and are processed unconsciously as well as consciously. Factors at work include the direct sensory pleasure or disgust that the viewing, smelling, tasting, swallowing or hearing of the product imparts to its consumer. Sensory evaluation has been described as a scientific discipline used to evoke, measure, analyse and interpret reactions to those characteristics of foods and materials as they are perceived by the senses of sight, smell, taste, touch and hearing (Lawless, 1993). The measurement of these sensory properties provides a number of problems, including whether to measure the degree of sensory experience or simply the hedonic reaction to the food: these are discussed in the first section. Of interest in the current context is the effect of the consumer comparing his or her expectations of sensory properties with those actually delivered by the product. This may produce a synergistic improvement in enjoyment, or one of the two forces may dominate. This is a fairly recent area of study in food choice research and is reviewed in section 10.3.

Foods are used in many different contexts, and people's attitudes and beliefs regarding the foods will vary across these contexts. McKenzie (1976) argued that the consumption of food can have dramatically varied purposes for a single consumer. Efforts in the nutrition and food literature to relate attitudes to food consumption have been mainly ad hoc, not based on a theoretical model, and thus not predictive. These types of methods include a framework developed by Krondl and Lau (1978, 1982), the 'item by use' appropriateness method (Schutz, 1988) and the repertory grid method as described by Thomson and McEwan (1988).

At the moment of purchase the decision making process is purely cognitive and the sensory properties are those expected from previous memory, from direct claims on the package, inferred from the images and information, or indeed from handling the product itself. It is therefore appropriate to examine the role of sensory properties as just one element in a general attitudes and beliefs model. Krondl and Lau (1978, 1982) developed an approach in which the relative importance of various factors affecting food selection are studied: the factors included are taste, tolerance, prestige, price, convenience, health belief and familiarity. The method has been used to look at the relative importance of particular factors within groups of subjects (Lau et al., 1979, 1984; Reaburn et al., 1979) and differences between groups of subjects (George and Krondl, 1983; Hrboticky and Krondl, 1984). Many models describing the relationships between the many factors influencing food choice have been put forward (Booth and Shepherd, 1988; Kahn, 1981; Krondl and Lau, 1982; Pilgrim, 1957; Randall and Sanjur, 1981; Shepherd, 1985 (see Figure 10.1); Yudkin, 1956). Reviewing these models Shepherd (1989) concluded that, for the most part, these models are not quantitative and do not explain the relative importance of, and interaction between, the various factors. Shepherd and Sparks (1994) point out that, though these models are useful for indicating which variables need to be measured in the food choice area, they do not offer a framework within which to design studies or a basis upon which to build theories of human food choice. Methods in which the relationships between attitudes and beliefs are looked at in a more structured manner have been introduced from social psychology. These quantitative methods are examined in more detail in section 10.5 in the context of factors relevant to the consumption of milk.

Finally, the application of this research is to take this new understanding of the mechanisms that control food choice and acceptance and use it, for example, to convince consumers to eat more healthily if you are in health promotion, to lose weight if you are a dieting advisor, to buy your particular brand if you are a retailer or manufacturer for instance. The key is that you must communicate with consumers and in section 10.6 we look at some fairly recent results in this area.

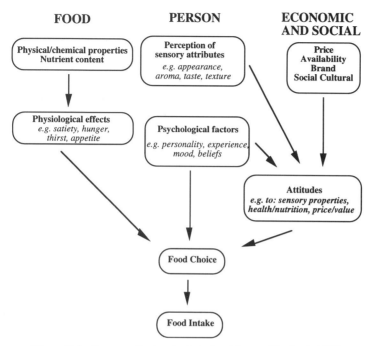

FOOD

PERSON

ECONOMIC AND SOCIAL

Figure 10.1   Factors affecting food choice. (*Source*: Shepherd, 1985)

Sensory preference is the hedonic dimension of acceptability (Pilgrim, 1957). It can be defined as the consumer's affective (i.e. related to pleasure or displeasure) response to a food product in a given context. As a consequence, the practical definition of sensory preference is the degree of sensory appreciation. Sensory preference can either be measured in a relative way (paired comparison, ranking) or in an absolute way, using an interval or ratio scale. The most commonly used is the nine-point hedonic scale proposed by Peryam and Pilgrim (1957).

Sensory preference according to ISO (1994) is an expression of the emotional state or reaction of an assessor which leads him/her to find one product better than another or several others. In general, hedonic measurement is both easy to carry out and not very time consuming, thus answers to specific questions can be obtained rapidly. However, the information obtained will always be the consumer's first impression; the level of liking determined by this type of measure could be different after an extended period of consumption. Indeed, sensory preference is an indicator of food acceptability which could or could not be a predictor of the consumer's behaviour. Many studies report the failure of hedonic measures to predict consumption. Cardello and Maller (1982) reported weak

**10.2
Predicting food acceptance from sensory preference?**

correlations between the preferred concentration of sugar or salt and the amount of sugar or salt added to a meal. Bellisle and colleagues (1987, 1988) observed discordance between hedonic responses and consumption behaviour. In both experiments, the preferred concentrations of sugar and salt were overestimated when determined by the nine-point hedonic scale. One explanation for these results could be that the amount of each sample eaten during the hedonic test was very small and did not correspond to real behaviour. Shepherd *et al.* (1991), using a relative-to-ideal scale, also report discordance between hedonic response and consumption measurement. On the contrary, Shepherd and Farleigh (1986) show that the distance from ideal is in accordance with the total intake of salt and the measurement of table salt use. According to these latter results, no obvious conclusion can be drawn about the predictive value of sensory preference measurements. Indeed, the lack of agreement between preference measurements and consumption measurements reported in some studies could be accounted for by the measure itself; or by biases in the experiments, i.e.:

- A protocol where consumers are asked to spit out the samples when they give their assessment is not natural (Pangborn and Giovanni, 1984) and can lead to preferences for sweeter samples than when subjects are allowed to swallow the samples (Mattes and Mela, 1986).
- A range with very high concentrations of sugar can lead the consumers to overestimate their preferred concentration (McBride, 1985).

Hedonic scales (like/dislike, agreeable/disagreeable) only take the consumer's affective response into account, that is to say the degree of pleasure given by the perception of the sensory properties. Part of this pleasure, however, results from the interaction between an affective and a cognitive dimension. For example, a consumer can associate the creaminess of an ice-cream with the fat contained in the product and if she/he is weight conscious, the pleasure given by the creaminess is reduced because of the threat of gaining weight. In the same way, one can give a low hedonic score to a very sweet product because one knows that this product contains lots of calories and this knowledge interacts with the pleasure as such. The question is, which scales take this interaction into account? The simple nine-point hedonic scale seems to take into account only the affective response of the consumer.

One alternative is the relative-to-ideal rating scale where the consumer is asked to give a preference for the level of a sensory characteristic, referring to his/her ideal. The scales are often line scales with end labels indicating 'not enough' or 'too much' of a particular attribute and a midpoint labelled 'ideal'. The relative-to-ideal scale has been used to determine the most preferred red colour of a strawberry jam (Griffiths *et al.*, 1984), preferred level of salt in a bread (Booth *et al.*, 1983) and soup

(Shepherd *et al.*, 1984). The population's most preferred level of sugar has been determined in lemon drink, chocolate, tomato soup (Conner *et al.*, 1986) and in plain yoghurt (Daillant and Issanchou, 1991). In general the relative-to-ideal scale works well when one can identify a particular sensory attribute, for example sweetness or creaminess, for the consumer to focus on. Daillant-Spinnler *et al.* (1994) have developed a more general index, termed the global relative-to-ideal rating, which is a folded scale which is labelled 'ideal for me' at one end and 'very far from my ideal' sausage/yoghurt/lemonade or whatever product is under test. These authors validate the scale against the ordinary relative-to-ideal rating and indicate that it does reveal a difference between two groups of consumers, high-fat and low-fat cream cheese eaters, when given information about whether the product was 'light' or 'normal'. In this case the performance of high-fat users is consistent with a previous study where no information was available, whereas the low-fat users modify their ideal point, presumably because some extra cognitive aspects, e.g. weight-consciousness, come into play.

10.3 Role of sensory expectations in food acceptance

The *Oxford English Dictionary* (1971) defines expectation as 'the action or state of waiting, or of waiting for (something); the action of mentally looking for something to take place; anticipation'. For instance, in marketing and consumer science language, Anderson and Hair (1972) described expectations as 'subjective notions of things to come' or 'type of hypothesis formulated by the consumer'. Olson and Dover (1979) considered consumer product expectations as pre-trial beliefs about the product. Thus, expectations appear very frequently in people's daily life, affecting their reactions and decisions, although sometimes unconsciously.

In the area of food consumption, expectation plays a very important role because it may contribute to improving or degrading the perception of a product, even before tasting it. Expectation is extremely related to consumer satisfaction or dissatisfaction, thus is often measured by the degree of disparity between expected and perceived product performance (Anderson, 1973). Two general types of expectation may be distinguished: (1) a sensory-based expectation, or (2) a hedonic-expectation. In the first case expectation affects the perception of sensory attributes and leads the consumer to believe that the product will possess certain sensory characteristics. The second expectation is related to like/dislike to a certain degree (Cardello, 1993).

Figure 10.2 shows the components of the process which starts at time *t*, when a particular product is chosen. This choice is based on several criteria involving expectation, beliefs, attitudes, and intentions. At *a posteriori* time *t* + 1, the product is used by the consumer and the product evaluation is made. As evaluation implies comparison of actual performance with some standard, the consumer may confirm or disconfirm his/

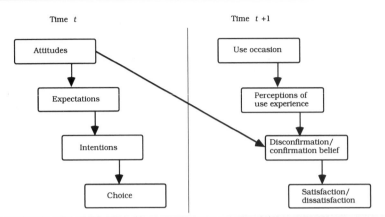

Figure 10.2   Conceptual model of disconfirmation of expectations process. (*Source*: modified from Cadotte *et al.*, 1987, and Oliver, 1980)

her expectation. In fact, a wide variety of issues and phenomena involve human expectations that occasionally may be disconfirmed. Confirmation occurs when performance matches the standard, leading to a neutral feeling. Performance better than the standard results in a positive disconfirmation, leading to satisfaction. Performance worse than the standard creates negative disconfirmation and then dissatisfaction (Cadotte *et al.*, 1987).

Four psychological theories may be considered to describe how the disconfirmation created by the expectations may influence product quality perception, namely: (1) assimilation; (2) contrast; (3) generalized negativity; and (4) assimilation–contrast (Anderson, 1973).

Assimilation theory (or cognitive dissonance) states that any discrepancy between expectations and product performance will be minimized or assimilated by the consumer, by changing his/her perception of the product to bring it more into line with his/her expectation. According to Festinger's (1957) theory of cognitive dissonance, when consumers receive two ideas which are psychologically dissonant, they tend to minimize the 'mental discomfort' created by an unconfirmed expectancy, by distorting their perception to make them more consonant.

Contrast theory assumes that the consumer will magnify the disparity between the product received and the product expected. When expectations are not matched by actual product performance, the consumer will evaluate the product less favourably than if she/he had no prior expectations about it. Contrast is thus the reverse of assimilation.

Generalized negativity supposes that any discrepancy between expectation and product performance will result in a generalized hedonic state, causing the product to have a less favourable rating than if performance had coincided with expectations. Thus, if a consumer expects a particular performance from a product but a different performance occurs, she/he will judge the product to be less pleasant than if she/he had no previous expectation. The assimilation–contrast approach assumes that there are limits of acceptance or rejection in consumer perception. If the disparity between expectation and product performance is sufficiently small to be in the consumer limit of acceptance, she/he will rate the sample based on assimilation theory, by putting the product more in line with expectations. However, if the discrepancy between expectation and actual product performance is so large that it falls into the zone of rejection, then the contrast effect comes into play and the consumer exaggerates the perceived disparity between the product and his/her expectation for it. Whether assimilation or contrast effect develops is a function of the degree of disparity between expectation and actual product performance.

Although expectation has not received very much attention in the sensory literature, it has been the focus of much greater attention in the fields of marketing and social psychology (Cardello, 1993). Carlsmith and Aronson (1963) used sweet and bitter solutions to show that when a person expects a particular event and a different event occurs, she/he will experience dissonance. In their study, when the consumer expected either a sweet solution and a bitter one came or when she/he expected a bitter solution and a sweet came, the solutions were rated more unpleasant. Thus, a bitter solution was rated more bitter and a sweet solution was rated less sweet. This study has demonstrated that under certain conditions the disconfirmation of a strong expectancy leads to negative effect.

Anderson (1973) found that there is a point beyond which consumers will not accept increasing disparity between product claims and actual performance, at least for a simple or easily understood product. More complex products may yield different results because consumers may tend to be more dependent on the information provided. This result has a very important implication in a marketing context because it showed that unrealistic consumer expectation generated by excessive promotional exaggeration can result in consumer dissatisfaction. He has emphasized that the results may vary depending on the product and recommended more research with a variety of products and services.

Cardozo (1965) investigated the factors that affected consumer satisfaction and indicated that satisfaction with the product under investigation (ballpoint pens) was influenced by the effort (which includes the physical, mental, and financial resources) expended to obtain the pen, and the expectations concerning to the product. His results suggested that satisfaction with a product may be higher when consumers expend considerable effort to obtain it than when they use only modest effort. This finding goes against the usual notions of marketing efficiency and customer convenience, although it is in line with cognitive dissonance theory (Festinger, 1957). This research also suggested that consumer satisfaction was lower when the product did not come up to expectations (product quality was lower than expected) than when the product met expectations. There was evidence of the contrast effect in these results; however, the author did not investigate whether or not a positive disconfirmation (product better than expected) also would produce disappointment, leading in this case to the assimilation model. The results of Cardozo have been contested by Anderson (1973) who pointed out methodological errors in his study.

Olshavsky and Miller (1972) studied the effects of expectancy for a relevant product to college students – a tape recorder – in both negative (product worse than expected) and positive (product better than expected) disconfirmations, under controlled laboratory conditions. Expectation was manipulated by changing the description of the product given to the subjects. The authors concluded that overstating the quality of a complex, multidimensional product apparently contributes to a more favourable evaluation and understatement to a less favourable evaluation.

In another study investigating the applicability of the disconfirmation model of consumer satisfaction/dissatisfaction with a restaurant, Swan and Trawick (1981) concluded that satisfaction increases as positive disconfirmation increases (as performance exceeds expectations), suggesting that the model may be useful in understanding customer satisfaction in retail service settings. In summary, they infer that satisfaction depends on how expectations are confirmed or disconfirmed, supporting the basic disconfirmation model of satisfaction.

Few studies concerning expectation and sensory evaluation are presented in the food science literature. However, the importance of the topic has been pointed out several times (Cardello and Sawyer, 1992; Fennema, 1985; Gacula *et al.*, 1986). Very often the sensory scientist deliberately eliminates these critical factors from consideration, because they are believed to be simply sources of bias in the data or are too difficult to measure and/or control (Cardello, 1993).

Olson and Dover (1979) presented one of the first studies on expectations and food. By giving information about the bitterness of ground coffee (the message emphasized that the coffee had no bitterness) before consumers tried the beverage, they evaluated how the expectations created by the information affected the assessment of the bitterness on a new brand of coffee.

This experiment had been planned to create an expectancy-disconfirmation (product more bitter than expected), therefore the coffee had been prepared using 50% more ground coffee than specified by the package directions. The authors criticized previous research on expectations for its methodological problems and for its failure to clearly conceptualize the expectation construct. In order to avoid such failures the authors used a control group (no information about the coffee bitterness) and they also measured the strength of consumers' belief expectancies about the attributes evaluated, demonstrating the creation of specific product expecta-

Table 10.1 Mean (with standard deviation in brackets) belief strength (scores can vary from 0 to 10) for the four attribute levels of the bitterness belief vector (adapted from Olson and Dover, 1979)

		Bitterness belief levels			
Group	Time of rating	Not bitter $(+2.18)$[a]	Slightly bitter $(+.10)$[a]	Fairly bitter (-1.39)[a]	Very bitter (-2.84)[a]
Experimental ($n = 20$)					
	Pre-trial (after ads)	6.70 (4.45)	2.70 (3.66)	0.55 (1.39)	0.05 (0.22)
	Post-trial	4.35 (4.65)	3.45 (4.01)	1.60 (2.96)	0.60 (2.26)
Control ($n = 18$)					
	Post-trial	2.06 (3.76)	3.72 (4.15)	3.39 (4.19)	0.83 (2.43)

[a]Mean evaluation of this belief level averaged over all experimental measurements (–3 to +3 scale, bad–good).

tions. Consumers who received the information rated the coffee as being less bitter (not bitter = 4.35) than did the control group (not bitter = 2.06). Thus, the relatively positive pre-trial expectations apparently limited the negativity of the post-disconfirmation product ratings. The Olson and Dover (1979) results are presented in Table 10.1.

This result supported a cognitive dissonance (assimilation) theory. Although the product by itself was very familiar to the subjects, the product brand was totally unfamiliar. According to the authors, rather different results may be obtained for disconfirmations of product expectations that have become established over a long period of actual use experience.

Using the fat-free and the regular versions of familiar products such as cake, crackers and cheese, Tuorila *et al.* (1994) investigated the effect of prior experience, dietary practices, label information and expectations on the sensory and hedonic ratings for the products. The authors varied the information about the amount of fat in the products by presenting three different conditions to the consumers: products which were unlabelled, products which were correctly labelled and incorrectly labelled products. They concluded that the label conditions had their desired effect in creating different hedonic and sensory expectations for the test products, supporting an assimilation model of the effect of disconfirmed expectations on food acceptance: ratings of liking change in the direction of expected liking.

Expectations have been indicated as a determining factor in the acceptance of novel foods. According to Cardello *et al.* (1985), many factors which may affect consumer expectations, such as product name, packaging, nutritional information, cost information and product presentation, had a significant effect on the consumer acceptance and stated likely purchase and use of novel foods. However, several of these variables do not have commensurate effects on familiar foods. These results followed the assimilation model, since the greater the degree a consumer's experience with the product matched his/her expectation, the greater was his/her liking of the product.

The temperature of the foods and beverages has been considered an important factor in determining liking, with the preference probably reflecting people's expectations about how much they will like the taste of a food at different temperatures. According to Zellner *et al.* (1988) these expectations could be based on direct experience with the taste, socially transmitted information about the taste or more general ideas of appropriateness.

A study has been carried out to investigate the effect on liking ratings of altering expectations concerning the appropriate temperature of unfamiliar beverages (goya brand guanabana and tamarind juices) (Zellner *et al.*, 1988). By splitting the subjects into two groups (group 1 = no expectations; group 2 = providing information about how the juices are normally consumed in the tropics), the results indicated that the liking ratings of the beverages were altered by changing subjects' expectations. The authors suggested that although sensory properties of foods are important in liking, human food preferences are often governed by factors outside the sensory aspect, as in this case, by expectations generated by cultural learning.

Cardello and Sawyer (1992) conducted three studies to investigate the effects of disconfirmed consumer expectations on food acceptability. In the first and second experiments, the authors used an unfamiliar product (experiment 1: water-soluble, edible film used to coat a sugar-free, salt water taffy carrier; experiment 2: pomegranate juice). The results for both studies followed an assimilation model; however, in the second one, when the expected bitterness was high, a contrast effect occurred.

The third experiment examined the effect of expectations on liking for a very familiar product with a high degree of brand loyalty: six brands of cola beverage. By manipulating the expectations of both directions (positive: consumers expected a cola they liked; and negative: consumers expected a cola they disliked), six different conditions of confirmation/disconfirmation were created. Table 10.2 shows that consumers who expected a worse product rated the product lower (e.g. group presented with large-positive disconfirmation: 6.4 post-test versus 7.8 pre-test) than they had in the pre-test (blind condition); whereas subjects who expected a better product rated it higher than in the pre-test (e.g. group presented with large-negative disconfirmation: 6.0 post-test versus 2.4 pre-test). An assimilation effect has been observed in this experiment.

It must be clear that at the point of purchase it is the expectations of the sensory properties that contribute to the choice decision. In the mainstream sensory literature there has been relatively little research into the

Table 10.2 Effect of expectation manipulations on mean acceptance ratings of cola beverages (adapted from Cardello and Sawyer, 1992)

Level of disconfirmation presented to group	Pre-test acceptability	Expected acceptability	Post-test acceptability	t-values post-test versus pre-test
Large-positive	7.8	3.8	6.4	-4.33^{**}
Large-negative	2.4	6.1	6.0	9.12^{**}
Intermediate-positive	7.8	5.3	6.1	-6.24^{**}
Intermediate-negative	2.2	5.3	5.5	9.83^{**}
Low/no disconfirmation	7.8	6.5	7.4	-2.46^{*}
Intermediate-negative[a]	2.2	5.2	5.6	9.20^{**}

[a]This experimental condition was intended to produce low/no disconfirmation, using a beverage with low pre-test acceptability.
$^{*}p < 0.05$; $^{**}p < 0.01$

mechanisms by which these sensory expectations are generated and further research is required.

The significance of the research into the way in which expectations enhance or degrade the perception and acceptance of a food or beverage is two-fold. First, it will direct the marketing sector in how to influence expectations. Second, it will guide the product developer and processor in setting specifications for the food product. The assimilation–contrast model is of particular interest in this context because if one can determine population tolerances within which deviations from expectations are not detected, then assimilation will occur. However, if products fall outside these limits the contrast effect will operate and the products will be perceived to have less of an expected sensory attribute than if no promise had been made.

10.4 Measuring the effect of context on sensory acceptance

A method to investigate consumer perceptions of the appropriateness of foods for particular uses has been developed by Schutz (1988) and co-workers. Using this method, food items are selected to represent a class of foods from a combination of literature research and by interviews with consumers concerning what they identify as items in that class. The uses, situations and attributes to be evaluated are obtained by eliciting information on when it would be appropriate to use such items. A representative sample of the desired population is then asked to rate the appropriateness of each food item for each of the various uses, attributes and situations in a matrix format on a self-administered questionnaire. The uses, situations and attributes are phrased in a use intention mode, i.e. they are typically stated as in 'when I want something I really like' or 'in a salad' or 'for children'. They can include time of day, occasion, where served, physiological states, how used, psychological characteristics, person served, physical characteristics and sensory characteristics. For analysis the mean appropriateness rating over the group or subgroup of subjects is determined for all 'item by use' combinations, and principal components analysis is conducted among all pairs of foods and for all pairs of uses, in order to determine the cognitive structure for the foods and for the uses.

A drawback of the previously described method is that the factors being studied are those generated by the researchers and are, in many cases, not specific to the food or population of interest. The repertory grid method, as described by Thomson and McEwan (1988), has been put forward as another method for gaining information on consumers' perceptions of foods. It allows individual consumers to determine their own, highly individual, range of contexts. The basis of the method is the elicitation of so-called 'constructs', ways in which two things are seen as alike and in the same way different from a third (Kelly, 1955). The repertory grid method has frequently been used in food-related research (e.g. Scriven *et al.*, 1989; Worsley, 1980).

Both the repertory grid method and the 'item by use' appropriateness method have been applied in a study on seven types of milk (Raats and Shepherd, 1991/92). The repertory grid method involved three phases: the elicitation of uses of milk, the elicitation of constructs (reasons or attributes which are used to distinguish the different types of milk) within the context of each use, and the ratings of the milks for appropriateness for use and for each construct. The 'item by use' appropriateness study included many of the use-attributes used by Bruhn and Schutz (1986), modified for the UK, with additional items added from other research and information on UK milk usage (Shepherd, 1988).

In the 'item by use' appropriateness study, higher-fat milk users did not use a particular use-attribute to distinguish between the milks; however, low-fat milk users used 'as a part of breakfast', 'when on a diet', 'when I want something that contains little fat', 'when I want something low in calories' to separate the milks. In the repertory grid study, the low-fat milk users mainly separated the milks on 'fat content', 'creaminess', 'richness', 'wateriness' and 'healthiness', whereas higher-fat milk users mainly separated the milks on 'taste/flavour', 'richness', 'wateriness', 'acquainted with', 'fresh/natural' and 'creaminess'. More variation in the data was accounted for using the repertory grid method than in the 'item by use' appropriateness method, indicating that the constructs used in the former approach were more relevant to the subjects in distinguishing between the milks studied. The repertory grid method was thus more suitable for drawing out information pertinent to understanding which sensory attributes are of importance to consumers.

The repertory grid method lends itself well to cross-country and -language comparison. Raats and Shepherd (1993) therefore used it to study the place milk occupies within the total diet, its perceived appropriateness and consumer beliefs concerning different types of milk (long-life versus fresh and different levels of fat content) in four European countries: Finland, France, the Netherlands and the UK. Twenty subjects (10 male, 10 female, 18–74 years) were interviewed in each country. Even though this small sample cannot be considered a random cross-section of the population of each given country, nor are they necessarily representative of the attitudes of their country as a whole, it would seem unlikely that the responses of the subjects fail to reflect the general attitudes of the people within these cultures.

The results indicate that consumers in each of the countries studied found the 'aroma, flavour or taste' and 'fat content' of the milks or products made with a milk to be the important distinguishing characteristics. On the whole consumers rate the milk they use most as highest in aroma, flavour or taste. This is possibly a form of hedonic assessment and is in agreement with previous work carried out by Tuorila (1987). British and Finnish consumers also used 'available, convenient or handy' and 'acquainted/familiar with or used to' to describe their most used milks.

Aside from the construct 'fat content', all groups of consumers except the Finnish used 'creamy' to denote the milks of a higher fat content. Other terms describing the milks of higher fat content were 'rich' (used by French and British consumers) and 'full' (used by French and Dutch consumers). A number of constructs were unique to French consumers; these could be summarised as 'body, filling, firm or heavy' and 'foamy, frothy or smooth', describing milks or products made with milk of a higher fat content. Finnish, Dutch and British consumers described the lower fat milks and products as 'healthy'. Dutch and British consumers also described these products as 'liquid, thin or watery'. French consumers used the construct 'digestible' to describe milks or products made with a milk of lower fat content. This work indicates that the sensory qualities associated with the sensations of 'fat content' are described with a range of terms that varies with country or language.

The repertory grid method is thus a powerful method for eliciting the role of sensory properties in a range of food products across many languages and ethnic backgrounds because it enables the subject to use a vocabulary that is directly relevant to them. This tool should be in the armoury of any researcher interested in investigating reasons for food selection.

10.5
Structured attitude and belief models

The theory of reasoned action (see Figure 10.3) as proposed by Ajzen and Fishbein (1980) is a development of their earlier expectancy-value model of attitudes (Fishbein and Ajzen, 1975). The theory of reasoned action is concerned with rational volitional behaviour, that is behaviours over which an individual has control. The theory suggests that rather than assessing attitudes towards a person or object (or in the case of food,

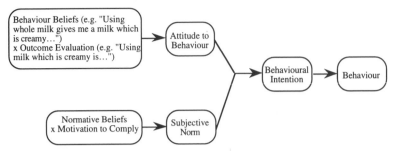

Figure 10.3 A schematic representation of Ajzen and Fishbein's (1980) theory of reasoned action.

choice of a food), attitudes should be assessed in relation to action, to the performance of particular behaviours, e.g. eating, drinking, or using a food or a drink. Behaviour is predicted by intention to perform the behaviour, which is in turn predicted by a person's own attitude towards the behaviour and the social pressure perceived by the individual to perform the behaviour ('subjective norm'). Attitudes towards a behaviour are predicted by a person's belief about the outcome of a behaviour, modified by the value the person attaches to those outcomes, e.g. beliefs about the degree of presence of a particular sensory property are modified by the subject's view of the desirability of that property in the final product. The resulting score can be added in to the final attitude index alongside all the other functional, ethical and social beliefs. We shall illustrate the use of the model for the case of milk consumption. All the components of the model can be assessed by suitable questionnaire items.

A number of studies have used this approach, and consistent relationships between attitudes and food choice have been identified (e.g. see Shepherd, 1989; Shepherd and Sparks, 1994; Stafleu *et al.*, 1991/92). In general, attitude and subjective norm are found to be good predictors of individual intention to choose a certain food item, although the attitude component has usually been found to have greater predictive power. Where belief items have been assessed these have shown good prediction of the attitude component.

The consumption of milk of different fat contents was studied by Tuorila (1987). Eleven belief items were studied, including sensory (e.g. tastes greasy, leaves greasy aftertaste), nutritional/health, suitability and price items; all but price were found to be highly correlated with the attitude score. Subjects were found to have more positive beliefs about the type of milk that they normally consumed, signifying the importance of beliefs in predicting behaviour. A modified version of the questionnaire used by Tuorila (1987) was used by Shepherd (1988) in a study of low-fat milk consumption. Shepherd (1988) concentrated on the belief-evaluations and the attitude component of the theory of reasoned action. Thirteen beliefs were assessed, including sensory, nutritional, functional and price items. Principal components analysis was carried out on the belief-evaluation products. The sensory and functional items were related and separated from the nutritional beliefs; the price items did not relate clearly to any other items. Tuorila (1987) found sensory belief-evaluation products to outweigh nutritional concerns in the selection of milks of different fat levels, whilst Shepherd (1988) found nutritional belief-evaluation products to outweigh sensory and functional concerns in the selection of low-fat milks.

Table 10.3 Mean scores (possible range: -16 to $+16$) for the individual belief-evaluation products ($b \cdot c$) and correlations (r) between individual belief-outcome evaluation products and attitude, with β-coefficients from regression analysis for whole, semi-skimmed, and skimmed milk. n varied from 252 to 256

$b \cdot e$	Mean score	Correlation with attitude (r)	Regression of attitude (β)
Whole milk (79% of the variance explained)			
Can use as a drink by itself	6.37	0.40***	0.17***
Can use for cooking	5.21	0.28***	0.02
Can use in tea/coffee	5.55	0.28***	−0.05
Can use with cereal	6.66	0.42***	0.11*
Has a high-fat content	−6.38	0.52***	0.17***
Helps control weight	−5.36	0.33***	0.03
Is creamy	−4.83	0.52***	0.11**
Is good for the heart	−7.18	0.51***	0.10*
Is healthy	−1.86	0.72***	0.43***
Tastes good	5.37	0.57***	0.27***
Semi-skimmed milk (76% of the variance explained)			
Can use as a drink by itself	3.10	0.55***	0.09
Can use for cooking	5.29	0.44***	−0.13**
Can use in tea/coffee	6.23	0.49***	0.06
Can use with cereal	5.60	0.50***	−0.01
Has a high-fat content	3.98	0.35***	0.02
Helps control weight	4.12	0.54***	0.05
Is creamy	1.45	−0.01	−0.01
Is good for the heart	6.09	0.65***	0.20***
Is healthy	6.49	0.71***	0.34***
Tastes good	2.64	0.72***	0.47***
Skimmed milk (78% of the variance explained)			
Can use as a drink by itself	−1.23	0.48***	0.01
Can use for cooking	3.70	0.52***	0.03
Can use in tea/coffee	3.66	0.65***	0.13**
Can use with cereal	1.59	0.67***	0.10*
Has a high-fat content	6.75	0.39***	0.01
Helps control weight	5.38	0.46***	0.08*
Is creamy	4.91	0.30***	0.02
Is good for the heart	8.28	0.53***	0.00
Is healthy	8.05	0.65***	0.37***
Tastes good	−1.22	0.74***	0.45***

*$p < 0.05$; **$p < 0.01$; ***$p < 0.001$

In a recent study in which 257 British consumers completed a postal questionnaire about the consumption of three milk types (whole, semi-skimmed and skimmed), designed according to the theory of reasoned action (Raats, 1992), the relative importance of the individual belief-evaluation products in predicting attitudes was investigated in three ways. Table 10.3 shows mean scores for individual belief-outcome evaluation product and attitudes scores; and the correlations between the individual belief-evaluation products and the attitudes scores, along with the multiple regressions of attitudes on the individual belief-evaluation products for whole, semi-skimmed and skimmed milk. Subjects had the most positive attitude towards semi-skimmed milk. 'Is good for the heart' was the most negative belief-evaluation product for whole milk, and 'can use with cereal' the most positive. For semi-skimmed milk, 'is healthy' had the highest belief-evaluation product and 'is creamy' had the lowest. The most negative belief-evaluation product for skimmed milk was 'use as a drink by itself' and the most positive was 'is good for the heart'.

For each of the milks, the two belief-outcome evaluation products contributing most highly to the multiple regression related to 'is healthy' and 'tastes good'. However, whereas 'healthy' had the highest β-coefficient in the equation for whole milk, 'tastes good' had the highest β-coefficient for semi-skimmed and skimmed milks. Other variables were less important in predicting attitude but several were nonetheless significant, e.g. 'can use as a drink by itself' and 'has a high-fat content' for whole milk, and 'is good for the heart' for semi-skimmed milk (see Table 10.3).

The structure of the behavioural beliefs was investigated by examining how the individual belief-evaluation products separated into groups, using a principal components analysis with varimax rotation. The results of these analyses are shown in Table 10.4. The results are similar for the three types of milks. Taste- and use-related items related to component 1 for each of the milks but, whereas for whole milk the items on fat and healthiness all loaded on component 2, for semi-skimmed and skimmed milks the healthiness items loaded on component 2 and the fat-related items on component 3 (see Table 10.4).

For each of the milks, the beliefs about 'is healthy' and 'tastes good' were most associated with attitude but the pattern varied between the different milks, with 'is healthy' being more important for whole milk and 'taste' being more important for semi-skimmed and skimmed milks. Part of the reason for this difference may lie in the different structure of beliefs concerning the types of milks. For whole milk, principal components analysis of the belief-evaluation products revealed two components, the first related to taste and use and the second to health and fat. For semi-skimmed and skimmed milk the solution was similar except that the health-related items separated from the fat-related items therefore giving a three-component solution. It would appear that there are similarities in the way in which people think about the two types of low-fat milks which differ

Table 10.4 Principal component (PC) loadings for the behaviour belief-outcome evaluation products (b·e) from the principal components analysis of whole (n = 250), semi-skimmed (n = 253), and skimmed milk (n = 246), with the percentage variance accounted for

b·e	Whole milk		Semi-skimmed milk			Skimmed milk		
	PC1 32%	PC2 26%	PC1 44%	PC2 14%	PC3 11%	PC1 42%	PC2 18%	PC3 10%
Is creamy	0.16	0.72	−0.12	−0.03	0.88	0.08	0.12	0.89
Has a high-fat content	−0.10	0.79	0.24	0.28	0.69	0.13	0.35	0.75
Is healthy	0.25	0.71	0.21	0.87	0.15	0.18	0.85	0.27
Is good for the heart	−0.02	0.82	0.26	0.83	0.18	0.09	0.88	0.14
Helps control weight	−0.08	0.66	0.25	0.78	−0.04	0.17	0.71	0.18
Tastes good	0.69	0.24	0.79	0.20	0.02	0.83	0.09	0.12
Use as a drink by itself	0.77	0.00	0.82	0.08	0.04	0.80	−0.05	0.09
Can use for tea/coffee	0.78	−0.01	0.76	0.28	0.03	0.72	0.32	0.10
Can use in cooking	0.73	−0.12	0.65	0.37	0.22	0.61	0.45	−0.10
Can use with cereal	0.82	0.08	0.74	0.25	−0.05	0.86	0.19	0.10

from the way in which they think about whole milk. Different types of concern may come to the fore in making decisions to consume (or not to consume) any particular type of milk.

In studies of foods other than milk, attitudes has been found to be better predicted by consideration of the belief-evaluation products concerning the taste of the food rather than belief-evaluation products concerning health consequences (Shepherd, 1987; Shepherd and Farleigh, 1986; Towler and Shepherd, 1992). The results for milk are somewhat more equivocal. However, in these two cases both health and sensory concerns were significantly related to attitude. The results from the present study show a similar effect of both health and taste concerns being the predominant influences on milk selection, although whether health or taste predominates varies between the different milk types.

The Fishbein and Ajzen model is based on the assumption of 'reasoned' action and is thus particularly appropriate for patterns of food consumption that have been consciously adopted such as choosing to eat a 'healthy diet', as in the case of choice of particular milks (Raats, 1992; Shepherd, 1988; Tuorila, 1987), eating meat (Richardson et al., 1993), and consuming organic foods (Sparks and Shepherd, 1992). The use of multiple regressions of attitudes on individual belief-evaluation products enables the role of sensory versus other factors to be prioritized. The technique is really a population model based on the principle of weighted utility, so subjects are assumed to weigh up all factors and sum them to give an overall score. This may be the case for people's decision making when sometimes selecting a particular food from a supermarket shelf or a menu. However, a weighted utility model may be inappropriate in some contexts. At the moment of consumption or purchase, the sensory experience or expectations may be intense and factors such as the unconscious or impulsive nature of some decisions come into play. The role of a 'reasoned' action model in determining sensory enjoyment, or food acceptability overall, can be questioned. Criticisms of the theory of reasoned action include that it is limited to volitional behaviour, it deals with very specific behaviours and it is too individualistic (see Kashima and Gallois, 1993, for a discussion of some of these issues). There are a number of modifications which may increase the theory's statistical predictions of behaviour, including perceived control (therefore forming the theory of planned behaviour (e.g. Ajzen, 1991)), self-identity, perceived moral obligation, variety seeking, and attitude variability (see Shepherd and Sparks (1994), and Sparks (1994), for a discussion of some of these topics).

From a food marketing point of view, a disadvantage of the Fishbein and Ajzen model is that it does not provide targetable segments of the population and prioritize the key factors controlling their food selection and enjoyment patterns. However, the model was never intended to be used as a marketing tool, but as an indicator of the key parameters that populations are using to make conscious decisions about certain forms of behaviour. It is therefore very useful to government departments, or large companies, interested in changing behaviour. If intention to change is the antecedent of behaviour, then the model will indicate which parameters to target in a healthy eating campaign. Some issues regarding communicating with consumers are discussed in the next section.

10.6
Communicating about sensory properties to consumers

Many individual subjects' characteristics, such as whether or not subjects are habitual users of the product in question, may have a significant impact on sensory assessment. Differences in attitudes towards products owing to knowledge of product identity (e.g. via packaging) can effect sensory ratings more than 'objective', physical differences between products (Aaron et al., 1994; Shepherd et al., 1991/92). Shepherd et al. (1991/92) found that when consumers were informed about product characteristics such as low-fat content, they thought products were less creamy.

Flavoured milk samples varying in sweetener (aspartame versus sugar) and fat content (low- versus full-fat milk) were used in a study in which one group of subjects was given no information about the samples, while others initially rated the samples with no information given and then again with the provision of information on the fat and sugar content of the samples (Shepherd et al., 1991/92). Analysis of the responses of all subjects together revealed that ratings of sweetness, body, healthiness and calorie content were influenced by the information whilst liking and likelihood of buying were not. However, differential effects were found when the subjects were divided on the basis of attitudes towards low- and full-fat flavoured milks. When looking at the type of sample towards which individuals had more positive attitudes, information increased both liking and likelihood of buying.

Information on use, attitudes and beliefs with regard to full- and reduced-fat spreads was elicited from a group of 101 consumers (Aaron *et al.*, 1994). Subjects were also asked to rate overall liking of a reduced-fat spread sample labelled as 'reduced-fat spread 40% fat' or 'full-fat margarine 80% fat' and liking of selected attributes. Analyses of hedonic and sensory ratings revealed no effect of label alone for the overall group. However, consistent interactive effects of label information with consumer attitudes, beliefs, and (to a lesser degree) with the type of spread normally used, were found. Label information had the overall effect of shifting sensory judgements in a direction more consistent with an individual's beliefs; i.e. hedonic responses from subjects with more positive attitudes towards reduced- or full-fat spreads were significantly greater in the respective label condition. Seemingly small effects of label/information alone may disguise sizeable, but contrasting, effects on selected population subgroups.

The food industry incorporates sensory evaluation in the quality assurance and quality control procedures including: incoming inspection, in-process control, final product inspection and product surveillance (Reece, 1979). Sidel (1988) suggested that industry go yet a step further and use sensory analysis to establish product sensory specifications. Sensory evaluation can also play a role in the marketing of food products through the development of a product vocabulary, the development and maintenance of sensory data bases, relating laboratory data to consumer data and decreasing the cost and time of bringing a product to the market (Pearce, 1980). Fishken (1990) described sensory quality as '...that complex set of sensory characteristics, including appearance, aroma, taste and texture, that is maximally acceptable to a specific audience of consumers, those who are regular users of the product category, or those who, by some clear definition, comprise the target market.'

Some producers tell the consumers by way of advertisements that they are paying attention to the aroma, flavour and texture of their products. Advertisements carrying photographs of sensory assessors and product profiles have been sighted over the past few years. Products carrying sensory descriptions on the packaging are also being marketed, the descriptions also being incorporated in advertising campaigns. When purchasing foods such as fresh produce, meat, bread and cheese in shops or almost anything at open-air markets, consumers receive little or no product information. Many product descriptions which include the usage of hedonic terms like 'good flavour', 'attractive' and 'pleasant' are very general and one wonders what technique was used to create these descriptions. The legal contesting of sensory claims is a newly emerging phenomenon and in future it will be necessary to ensure that proper sensory techniques are used to ensure that the information being given to the consumer is correct and meaningful.

Sensory profiling of products used by consumers in their own homes is another possible means of collecting a vocabulary of consumer relevant descriptors which can be used for subsequent advertising claims. This type of data-collection has been reported for Scotch whiskies (Guy *et al.*, 1989) and canned lagers (Gains and Thomson, 1990).

The idea of improving consumer perception of fruit and vegetable quality by improving the quality of information has been mooted in a number of studies. A study carried out by the CEMAGREF in France concluded that consumers were interested in getting more detailed information about the taste quality of fruit, but did not pursue the idea any further (Alavoine *et al.*, 1990). In a Dutch study (Cramwinckel *et al.*, 1989) on the consumer perception of the quality of potatoes, functional and sensory quality profiles were developed and tested, the hypothesis being that a choice made by the consumer with the aid of relevant product information contained in a quality profile will lead to a higher quality of perception than a choice based on very limited information. The study consisted of two parts, the first being the development of 'quality profiles'. A consumer panel was used to find out which product characteristics/attributes consumers were interested in when purchasing and/or consuming potatoes. It was concluded that consumers wanted information about ease of peeling and preparation of potatoes and also information about sensory or eating quality characteristics. 'Quality profiles' for eight potato varieties were then developed and a pamphlet produced with the eight 'quality profiles' in it and information on how to use them. No hedonic information was included in the profiles – it was left up to the consumer to decide what they personally considered a positive product characteristic.

The second part of the study consisted of the testing of the 'quality profiles'. Half the households in a consumer panel chose two potato varieties with the aid of quality profiles. The other half of the households in a consumer panel chose two potato varieties from a list of variety names. In the Netherlands, potatoes are sold under variety names. All respondents assessed the potatoes they received for preparation and/or eating quality.

The average score given by the group of people who received information in the form of 'quality profiles' was significantly higher than the average score given by the group that did not receive information in the form of 'quality profiles'. This difference was even more marked in the group of people who actually prepared the potatoes. Respondents were also allowed to make comments about the potatoes they received; the group who used the 'quality profiles' also made significantly more positive comments and less negative comments about the potatoes. Most of the comments made were about peeling and preparation. It was concluded that choosing potatoes with the aid of information about product characteristics/attributes can improve the quality as is perceived by the consumer. By clarifying the differences in product characteristics/attributes the perceived quality increases. Sensory profiling was found to be a useful tool for finding detectable differences between products and which words best described these differences; thus a vocabulary in which to describe differences could be developed. Product information must then be presented to consumers in a suitable form and through appropriate channels. Armed with information about product properties that are difficult to assess at the point of purchase, the eating quality, a consumer can then make an informed choice, therefore finding the product that best fits his or her needs.

10.7 Concluding remarks

It has not been possible in this short chapter to address all the mechanisms by which sensory preferences will influence food choice. For example, the pioneering work of Steiner (1977) has indicated the existence of innate sensory preferences for sweetness and innate dislike of bitterness and sourness. Saltiness, on the other hand, is a preference that is apparently learned soon after birth. The anthropological origins of such preferences are clear, but the fact that sensory preferences may be driven or modified following the repeated consumption of a food that is valuable to the body in some way, e.g. contains a pharmacologically active compound, has been shown in adults (see Forbes and Rogers, 1994). In this model sensory preference is modified through choice and then influences subsequent choices of foods containing similar sensory properties, but not the active compounds.

Our aim has been to examine the methodological problems facing the more applied researcher trying to prioritize and model the influences governing the selection and subsequent enjoyment of a particular food. We make a distinction between foods that are selected after detailed cognitive processing, which will involve the use of structured attitude and belief models, and those that are not. For the latter products it is more likely that sensory properties, correctly measured, may well form good predictors, although the mediating effect of context of use will still be important. For all products the pattern of disconfirmation or confirmation of the expected sensory properties of a product are likely to prove a critical factor in the decision to repurchase, and this aspect needs more research. Finally, the quality and nature of the information that is passed to consumers about sensory properties is an urgent research topic because of the increasing trend towards sensory claims.

References

Aaron, J. I., Mela, D. J. and Evans, R. E. (1994) The influences of attitudes, beliefs and label information on perceptions of reduced fat spreads. *Appetite*, **22**, 25–37.

Ajzen, I. (1991) The theory of planned behavior. *Organizational Behavior and Human Decision Processes*, **50**, 179–211.

Ajzen, I. and Fishbein, M. (1980) *Understanding Attitudes and Predicting Social Behavior*, Prentice-Hall, Englewood Cliffs NJ.

Alavoine, F., Crochon, M. and Bouillon, C. (1990) Practical methods to estimate taste quality of fruit and how to sell it to the consumer. *Acta Horticultura*, **259**, 61–5.

Anderson, R. E. (1973) Consumer dissatisfaction: the effect of disconfirmed expectancy on perceived product performance. *Journal of Marketing Research*, **10** (2), 38–44.

Anderson, R. E. and Hair Jr., J. F. (1972) Consumerism, consumer expectations, and perceived product performance. In M. Venkatesan (ed.), *Proceedings of the Third Annual Conference of the Association for Consumer Research*, Association for Consumer Research, Iowa City, Iowa.

Bellisle, F., Giachetti, I. and Tournier, A. (1988) Comment déterminer la salinité préférée d'un aliment: Comparaison de l'évaluation sensorielle et des tests de consommation. *Sciences des Aliments*, **8**, 557–64.

Bellisle, F. and Lucas, F. (1987) La measure de préférences alimentaires: Les tests brefs d'evaluation sensorielle permettent-ils de prédire la consommation? *Sciences des Aliments*, 7 hrs série **8**, 33–7.

Booth, D. A. and Shepherd, R. (1988) Sensory influences on food acceptance – the neglected approach to nutrition promotion. *British Nutrition Foundation Bulletin*, **13**, 39–54.

Booth, D. A., Thompson, A. and Shahedian, B. (1983) A robust brief measure of an individual's most preferred level of salt in an ordinary foodstuff. *Appetite*, **4**, 301–12.

Bruhn, C. M. and Schutz, H. G. (1986) Consumer perceptions of dairy foods and related-use foods. *Food Technology*, **40**, 79–85.

Cadotte, E. R., Woodruff, R. B. and Jenkins, R. L. (1987) Expectations and norms in models of consumer satisfaction. *Journal of Marketing Research*, **2** (8), 305–14.

Cardello, A. V. (1993) What do consumers expect from Low-cal, Low-fat, lite foods? *Cereal Foods World*, **38** (2), 96–9.

Cardello, A. V. and Maller, O. (1982) Acceptability of water, selected beverages and foods as a function of serving temperature. *Journal of Food Science*, **47** (5), 535–8.

Cardello, A. V., Maller, O., Masor, H. B., Dubose, C. and Edelman, B. (1985) Role of consumer expectancies in the acceptance of novel foods. *Journal of Food Science*, **50**, 1707–18.

Cardello, A. V. and Sawyer, F. M. (1992) Effects of disconfirmed expectations on food acceptability. *Journal of Sensory Studies*, **7**, 253–77.

Cardozo, R. N. (1965) An experimental study of consumer effort, expectation, and satisfaction. *Journal of Marketing Research*, **2** (8), 244–9.

Carlsmith, J. M. and Aronson, E. (1963) Some hedonic consequences of the disconfirmation of expectancies. *Journal of Abnormal and Social Psychology*, **66** (2), 151–6.

Conner, M. T., Haddan, V. and Booth, D. A. (1986) Very rapid, precise assessment of effects of constituent variation in product acceptability: Consumer sweetness preferences in a lime drink. *Lebensmittel Wissenschaft und Technologie*, **19**, 586–90.

Cramwinckel, A. B., van Mazijk-Bokslag, D. M. and Raats, M. M. (1989) Analytische en emotionele kwaliteit: Een experiment over kwaliteitsbeleving. *Voedingsmiddelentechnologie*, **22** (20), 141–5.

Daillant, B. and Issanchou, S. (1991) Most preferred level of sugar: Rapid measure and consumption test. *Journal of Sensory Studies*, **6**, 131–44.

Daillant-Spinnler, B., Issanchou, S. and Schlich, P. (1994) Influence of fat content on acceptability of cream cheese: ideal relative response, consumption and attitudes. (Unpublished manuscript.)

Fennema, O. (1985) The placebo effect of foods. *Food Technology in Australia*, **37** (11), 516–526.

Festinger, L. (1957) *A Theory of Cognitive Dissonance*, Row-Peterson, Evanston IL.

Fishbein, M. and Ajzen, I. (1975) *Belief, Attitude, Intention and Behaviour, An Introduction to Theory and Research*, Addison-Wesley, Reading MA.

Fishken, D. (1990) Sensory quality and the consumer: viewpoints and directions. *Journal of Sensory Studies*, **5** (3), 203–209.

Forbes, J. M. and Rogers, P. J. (1994) Food selection. *Nutrition Abstracts and Reviews (Series A)*, **64** (12), 1065–78.

Gacula Jr., M. C., Rutenbeck, S. K., Campbell, J. F., Giovanni, M. E., Gardze, C. A. and Washamii, R. W. (1986) Some sources of bias in consumer testing. *Journal of Sensory Studies*, **1**, 175–82.

Gains, N. and Thomson, D. M. H. (1990) Sensory profiling of canned lager beers using consumers in their own homes. *Food Quality and Preference*, **2**, 39–47.

George, R. S. and Krondl, M. (1983) Perceptions and food use of adolescent boys and girls. *Nutrition and Behavior*, **1**, 115–125.

Griffiths, R. P., Clifton, V. J. and Booth, D. A. (1984) Measurement of an individual's optimally preferred level of a food flavour. In J. Adda (ed.), *Progress in flavour research, Proceedings of the 4th Weurman Flavour Research Symposium*, Dourdan, France, May 9–11, 1984, Elsevier Science Publishers B.V., Amsterdam. pp. 81–90.

Guy, C., Piggott, J. R. and Marie, S. (1989) Consumer profiling of Scotch whisky. *Food Quality and Preference*, **1**, 69–73.

Hrboticky, N. and Krondl, M. (1984) Acculturation to Canadian foods by Chinese immigrant boys: changes in the perceived flavor, health value and prestige of foods. *Appetite*, **5**, 117–126.

ISO (1994) *Sensory analysis – vocabulary. Revised Canadian proposal.*

Kahn, M. A. (1981) Evaluation of food selection patterns and preferences. *CRC Critical Reviews in Food Science and Nutrition*, **15**, 129–153.

Kashima, Y. and Gallois, C. (1993) The theory of reasoned action and problem-focused research. In D. J. Barker, C. Gallois and M. McCamish (eds), *The Theory of Reasoned Action: Its Application to AIDS – Preventive Behaviour*, Pergamon Press, Oxford. pp. 207–226.

Kelly, G. A. (1955) *The psychology of personal constructs: a theory of personality*, Norton, New York.

Krondl, M. M. and Lau, D. (1978) Food habit modification as a public health measure. *Canadian Journal of Public Health*, **69**, 39–48.

Krondl, M. and Lau, D. (1982) Social determinants in human food selection. In L. M. Barker (ed.), *The Psychobiology of Human Food Selection*, AVI, Westport. pp. 139–151.

Lau, D., Hanada, L., Kaminskyj, O. and Krondl, M. (1979) Predicting food use by measuring attitudes and preference. *Food Product Development*, **13**, 66–72.

Lau, D., Krondl, M. and Coleman, P. (1984) Psychological factors affecting food selection. In J. R. Galler (ed.), *Nutrition and Behavior*, Plenum Press, New York. pp. 397–415.

Lawless, H. (1993) The education and training of sensory scientists. *Food Quality and Preference*, **4**, 51–63.

Mattes, R. D. and Mela, D. (1986) Relationships between and among selected measures of sweet-taste preference and dietary intake. *Chemical Senses*, **11** (4), 523–539.

McBride, R. L. (1985) Stimulus range influences intensity and hedonic ratings of flavour. *Appetite*, **6**, 125–131.

McKenzie, J. C. (1976) Food is not just for eating. In D. Hollingsworth and E. Morse (eds), *People and Food Tomorrow*, Applied Science, London. pp. 138–143.

McNutt, K. W., Powers, M. E. and Sloan, A. E. (1986) Food colors, flavors, and safety: a consumer viewpoint. *Food Technology*, **40**, 72–78.

Oliver, R. L. (1980) A cognitive model of the antecedents and consequences of satisfaction decisions. *Journal of Marketing Research*, **17**, 460–469.

Olshavsky, R. W. and Miller, J. A. (1972) Consumer expectations, product performance, and perceived product quality. *Journal of Marketing Research*, **9** (2), 19–21.

Olson, J. C. and Dover, P. A. (1979) Cognitive effects of deceptive advertising. *Journal of Marketing Research*, **15** (2), 29–38.

Oxford English Dictionary (1971) The compact edition of the *Oxford English Dictionary*, Oxford University Press, Oxford. p. 928.

Pangborn, R. M. and Giovanni, M. E. (1984) Dietary intake of sweet foods and of dairy fats and resultant gustatory responses to sugar in lemonade and to fat in milk. *Appetite*, **5**, 317–327.

Pearce, J. (1980) Sensory evaluation in marketing. *Food Technology*, **34** (11), 60–62.

Peryam, D. R. and Pilgrim, F. J. (1957) Hedonic scale method of measuring food preferences. *Food Technology*, **11**, 9–14.

Pilgrim, F. J. (1957) The components of food acceptance and their measurement. *American Journal of Clinical Nutrition*, **5**, 171–175.

Raats, M. M. (1992) The role of beliefs and sensory responses to milk in determining the selection of milks of different fat content. Unpublished doctoral dissertation, University of Reading, England.

Raats, M. M. and Shepherd, R. (1991/92) An evaluation of the use and perceived appropriateness of milk using the repertory grid method and the 'item by use' appropriateness method. *Food Quality and Preference*, **3**, 89–100.

Raats, M. M. and Shepherd, R. (1993) The use and perceived appropriateness of milk in the diet: A cross-country evaluation. *Ecology of Food Nutrition*, **30**, 253–273.

Randall, E. and Sanjur, D. (1981) Food preferences – their conceptualization and relationship to consumption. *Ecology of Food and Nutrition*, **11**, 151–161.

Reaburn, J. A., Krondl, M. and Lau, D. (1979) Social determinants in food selection. *Journal of the American Dietetic Association*, **74**, 637–641.

Reece, R. N. (1979) A quality assurance perspective of sensory evaluation. *Food Technology*, **32** (9), 37.

Richardson, N. J., Shepherd, R. and Elliman, N. A. (1993) Current attitudes and future influences on meat consumption in the U.K. *Appetite*, **21**, 41–51.

Schafer, R. B. (1978) Factors affecting food behavior and the quality of husbands' and wives' diets. *Journal of the American Dietetic Association*, **72**, 138–143.

Schutz, H. G. (1988) Beyond preference: Appropriateness as a measure of contextual acceptance of food. In D. M. H. Thomson (ed.), *Food Acceptability*, Elsevier Applied Science, London. pp. 115–134.

Scriven, F. M., Gains, N., Green, S. R. and Thomson, D. M. H. (1989) A contextual evaluation of alcoholic beverages using the repertory grid method. *International Journal of Food Science and Technology*, **24**, 173–182.

Shepherd, R. (1985) Dietary salt intake. *Nutrition and Food Science*, **96**, 10–11.

Shepherd, R. (1987) The effects of nutritional beliefs and values on food acceptance. In J. Solms, D. A. Booth, R. M. Pangborn and O. Raunhardt (eds), *Food Acceptance and Nutrition*, Academic Press, London. pp. 387–402.

Shepherd, R. (1988) Belief structure in relation to low-fat milk consumption. *Journal of Human Nutrition and Dietetics*, **1**, 421–428.

Shepherd, R. (1989) Factors influencing food preferences and choice. In R. Shepherd (ed.), *Handbook of the Psychophysiology of Human Eating*, Wiley, Chichester. pp. 3–24.

Shepherd, R. and Farleigh, C. A. (1986) Preferences, attitudes and personality as determinants of salt intake. *Human Nutrition: Applied Nutrition*, **40A**, 195–208.

Shepherd, R., Farleigh, C. A. and Land, D. G. (1984) The relationship between salt intake and preferences for different salt levels in soup. *Appetite*, **5**, 281–290.

Shepherd, R., Farleigh, C. A. and Wharf, S. G. (1991) Effect of quantity consumed measures of liking for salt concentrations in soup. *Journal of Sensory Studies*, **6**, 227–238.

Shepherd, R. and Sparks, P. (1994) Modelling food choice. In H. J. H. MacFie and D. M. H. Thomson (eds), *Measurement of Food Preferences*, Chapman and Hall, London. pp. 202–226.

Shepherd, R., Sparks, P., Bellier, S. and Raats, M. (1991/92) The effects of information on sensory ratings and preferences: The importance of attitudes. *Food Quality and Preference*, **3**, 147–155.

Sidel, J. L. (1988) Establishing a sensory specification. In D. M. H. Thomson (ed.), *Food Acceptability*, Elsevier Applied Science, London. pp. 43–54.

Sparks, P. (1994) Food choice and health: Applying, assessing and extending the theory of planned behavior. In D. R. Rutter and L. Quine (eds), *Measurement of Food Preferences*, Avebury, Aldershot. pp. 25–45.

Sparks, P. and Shepherd, R. (1992) Self-identity and the theory of planned behavior: assessing the role of identification with green consumerism. *Social Psychology Quarterly*, **55**, 388–399.

Stafleu, A., De Graaf, C., van Staveren, W. A. and Schroots, J. J. F. (1991/92) A review of selected studies assessing social-psychological determinants of fat and cholesterol intake. *Food Quality and Preference*, **3**, 183–200.

Steiner, J. E. (1977) Facial expressions of the neonate infant indicating the hedonics of food-related chemical stimuli. In J. M. Weiffenbach (ed.), *Taste and Development: The Genesis of Sweet Preference*, DHEW, Publication No. NIH 77-1068, US Government Publishing Office, Washington DC. pp. 177–188.

Swan, J. E. and Trawick, I. F. (1981) Disconfirmation of expectation and satisfaction with a retail service. *Journal of Retailing*, **57** (3), 49–67.

Thomson, D. M. H. and McEwan, J. A. (1988) An application of the repertory grid method to investigate consumer perceptions of foods. *Appetite*, **10**, 181–193.

Towler, G. and Shepherd, R. (1992) Application of Fishbein and Ajzen's expectancy-value model to understanding fat intake. *Appetite*, **18**, 15–27.

Tuorila, H. (1987) Selection of milks with varying fat contents and related overall liking, attitudes, norms and intentions. *Appetite*, **8**, 1–14.

Tuorila, H., Cardello, A. V. and Lesher, L. (1994) Antecedents and consequences of expectations related to fat-free and regular-fat foods. *Appetite*, **23**, 247–263.

Worsley, A. (1980) Thought for food: investigations of cognitive aspects of food. *Ecology of Food and Nutrition*, **9**, 65–80.

Yudkin, J. (1956) Man's choice of food. *Lancet*, **i**, 645–649.

Zellner, D. A., Stewar, W. F., Rozin, P. and Brown, J. M. (1988) Effect of temperature and expectations on liking for beverages. *Physiology and Behaviour*, **44**, 61–68.

11 Eating at home: meals and food choice

David Marshall

'This is it. At last, after all that shopping, preparing and cooking, it is time to eat.'

Eating occasions are part of what Braudel (1973) once described as 'the everyday activities of ordinary existence'. In developed countries, such as the UK or USA, the motivation to eat, as part of that ordinary existence, is rarely depletion driven and consumers satisfy more than simple physiological hunger needs when they eat, as the previous chapters have shown. One area that has been largely neglected is the influence of the occasion on food choice. The timing and form of eating occasions arise from a complex interaction of physiological feedback and socio-cultural factors which includes rules about cuisine and situation, as well as beliefs about nutrition and health (Birch, 1993).

Each culture distinguishes between eating occasions according to when and how often the occasion takes place (routinisation), what food is served (content), who is involved (participation), how the food is consumed (method) and the significance of the event (function) (Whitehead, 1984). This chapter focuses on that penultimate stage in the food provisioning process and examines the extent to which eating occasions, and in particular meals, shape food choice. It focuses on eating at home, not eating out, and looks at why meals remain important (social) events.

This chapter examines the importance of meal patterns and the implicit rules about how meals are structured in Europe and North America. Following from this the chapter goes on to look at how the daily menu and notions about proper meals are driving choice and to consider what is happening to the family meal, in response to broader social and economic changes. The method of eating focuses on table manners, and the question of who partakes in the meal highlights inequalities in access to food by gender and class. Finally, the concluding section offers a speculative framework to examine the impact of meal occasions of food choice.

Food choice is not some random process but a patterned activity exhibiting 'restrictive patterns of sequence and combination' that help to define social categories as each family, and individual, works out a regular pattern of mealtimes, ordinary and celebratory food. Furthermore, 'sociality' is encoded in meal patterns and structures which reflect the grammar of eating in a sort of food syntax (Douglas, 1972, 1976). Through the process of acculturation and socialisation consumers learn which foods are appropriate for which occasions; for example, children by the age of ten can recognise which foods belong to which meals (Birch, 1993).

Meal patterns tell us something about the eaters but also about what is appropriate. They, meal patterns, relate to the organisation of food and drink and can include the time at which the meal was served, the frequency of the event, where it was eaten, who was present, the order of the dishes and the food served at the meal (Roos *et al.*, 1993). The meal itself can be differentiated by the number of courses, the dishes in each course, the food presentation, meal duration, use of accoutrements and so on (Goode *et al.*, 1984). Goode and her colleagues identified a clear pattern of meals among an Italian-American community which followed seasonal, life stages and weekly cycles. Not only was there a clear distinction between meals served throughout the day, but midweek and weekend revealed distinct differences in the meal patterning (Table 11.1). This patterning reflected differences in work and leisure schedules within the community. At the weekend, breakfasts were more elaborate and Sunday dinner was considered unique with more courses, more variety, and an extended attendance list (Goode *et al.*, 1984). While effects of seasonality on availability, and consequently food choice, are less apparent today,

**11.1
Meal patterns and structured choice (routine)**

Table 11.1 Weekly meal cycles in an Italian–American community

	Weekday	Friday	Weekend	
A.M.	Breakfast or partial breakfast	Breakfast or partial breakfast	Elaborate breakfast	
Lunch	Full lunch or abbreviated	Full lunch or abbreviated	Late and abbreviated lunch	
Dinner	Gravy or platter	Fish or meatless Gravy or platter	Saturday 'Non-cooking' meal in home or celebratory eating out	Sunday Gravy and/ or whole meal
Post-	Club: simple party or Dessert and coffee	Late night breakfast	Late night breakfast	Dessert and coffee

Source: Goode *et al.* (1984), p. 178.

eating is still shaped by notions of time, cyclicality and tradition (Gofton, 1986a,b; Goode *et al.*, 1984).

In examining meal patterns as a means of understanding social relations, Douglas (1972) offered an interesting insight into the patterning of meals and menus via a complex framework which categorised eating according to the daily menu, meal, courses, helpings and mouthfuls. The daily menu, for example, can be broken down into 'early', 'main', 'light' and 'snack' occasions with breakfast and dinner as primary exponents of the first two occasions whereas lunch, high tea and supper are secondary exponents of light meals, and afternoon tea and nightcap secondary exponents of snacks.

The definition of a meal is centred on the idea that it is 'a "structured event", a social occasion organised by rules prescribing time, place and sequence of actions ... (and) ... is strictly rule bound as to permitted combinations and sequences (quoted from Nicod, 1979: 56–7; see also Douglas and Nicod, 1974). Indeed, in certain cultures an eating occasion can only be described as a meal if it includes a staple – potatoes in the UK, pasta in Italy or rice in S. India (Farb and Armelagos, 1980; Katona-Apte, 1975). Snacks, however, are unstructured, with one or more self-contained items. Meals can also contain very little where simple elements carry the ceremony of banquet, as in the Japanese tea ceremony, the Eucharist or Cha No Yu (Visser, 1992). Both meals and snacks have their place in the domestic pattern but meals can be further classified according to the course structure, and cooking practices, which reveals the extent to which food combinations are restricted by the meal format (see Appendix 11.1, which provides a structural framework for main meals adapted from Douglas (1972)).

The implicit rules for constructing and scheduling meals provide continuity in the food system and at the same time accommodate flexibility in the choice of food. They do not specify the use of particular foods but how to combine foods in distinct dishes that are recognised and appropriate for the occasion (Goode *et al.*, 1984). Goode and her colleagues found that their American-Italian community had reduced some meals in scale, and added and deleted formats as well as introduced new foods into meals that were less content-specific. As a consequence, it was possible to permit some flexibility in choice and accommodate individual preferences while producing a recognisable, and acceptable, meal.

While there is a broad range of content negotiation within the meal the format remains fixed, and it appears the structure *per se* is actually more important in the choice process than the contents, as Douglas claims:

Now I know the formula. A proper meal is *A* (when *A* is the stressed main course) plus 2*B* (when *B* is an unstressed course). Both *A* and *B* contain each the same structure, in small, *a* + 2*b*, when *a* is the stressed item and *b* the unstressed item in a course. A weekday lunch is *A*; Sunday lunch is 2*A*; Christmas, Easter and birthdays are *A* + 2*B*. (Douglas, 1976: 259)

Research undertaken by Douglas and Nicod (1974) into English working class eating habits revealed this tri-partite meal structure centred on a staple (potato or bread) plus a centrepiece (meat, fish or eggs with one or more vegetables) covered in a dressing (gravy). The progression through the second and third courses brought with it greater discretion to omit elements, dominance of the visual over the sensory, progression from savoury to sweet, hot to cold, potato to cereal and smaller quantities. Minor, or secondary meals, reflected this structural arrangement substituting bread for potato as the staple component in the meal and treating trimmings in the first course as optional. For Douglas the importance of the structure lay in the idea that 'the smallest, meanest meal metonymically figures the structure of the grandest, and each unit of the grand meal figures in the whole meal – or the meanest meal ... A meal stays in the category of meal only in so far as it carries this structure and allows the part to recall the whole' (Douglas, 1976: 257–258). While this research has been criticised for its failure to test these rules, to account for historical influences and for the undue weight placed on the symbolic aspects of consumption at the expense of material concerns (see Mennell *et al.*, 1992), it remains an important reminder of the need to consider how the structure of eating occasion affects what consumers eat. Other researchers have used some of these ideas to investigate food choice among the elderly, meal patterning among anorexics and to examine the relationship between meal formats and food choice (Hodes, 1991; Marshall, 1993; Roos *et al.*, 1993).

Perhaps the most interesting variation on the meal structure is the hamburger which offers a centrepiece (beef), staple (bread), trimmings (vegetables) and dressing (ketchup) in a compressed symbolic representation of the minor meal (Visser, 1992). Even the aeroplane meal reflects the temporal and structural characteristics of the meal, as the sequence of meals helps travellers to adapt to the time changes. The 'mild' unfamiliarity of the dishes resonates with the 'foreign' experience but the meal has a reassuringly familiar structure.[1]

Many of these rules for structuring dishes are encoded in recipes that have been learnt at home, derived from cookery books, friends, neighbours, magazines, food packages, recipe leaflets, etc.; the list goes on. Recipes are passed on through generations and across households and while they may vary in detail the basic form of the dish remains intact (Attar, 1990; Goode *et al.*, 1984). Yet recipes are never socially neutral, making various assumptions about the nature of the occasion, the

---

[1] The aeroplane meal usually offers an hors d'oeuvre, a main course (with its $a + 2b$), and a dessert followed by coffee and the complimentary chocolate, which adds that something special. Most passengers seem to follow the normal sequence of courses when eating, FMW with its $A + 2B$, although some anarchy is permitted at high altitudes, with only strangers to answer to.

number of people present, levels of skills, equipment, ingredients and resources (McKie and Wood, 1992; see Goody, 1982). Other than cookery books, there is relatively little documented evidence on the repertoire of recipes used and how often particular dishes are served, alternated and changed, or indeed how eating out, travel and the media influence what is served at home.[2] These recipes, the basis of the dishes served in the course of the meal, contain the necessary rules to transform the food into something culturally acceptable as 'edible' (Rozin, 1982). But knowing the recipe is not enough, one has to know where it fits into the overall pattern.

Much of this evidence on meal structures is restricted to European and American culture but Mintz (1992) proposes a variation on this structural theme which he claims is portable across cultures and can accommodate a range of meals. Focusing on plants, not animals, he identifies three food 'categories': the 'core' or starchy centre (bread, rice, potatoes, tortillas, couscous, boiled and mashed yams, taro, manioc and so on), the 'fringe' or flavour-giving foods and the 'legume' or protein-giving foods. The core–fringe–legume is reflected in the meal pattern and while foods differ across cultures the pattern remains. Mintz highlights the differences between core and fringe and contrasts the lighter coloured, less flavoured, soft chewy texture of the core with the colourful, more flavoured, solid texture of the fringe. He claims that this simple 'core, fringe, legume pattern' (CFLP) pattern allows one to make comparisons between cuisines. It is based on the idea that the parallels in cuisine emerged after mastery over animal and plant husbandry. Yet, the CFLP pattern has been breaking down as a direct consequence of sugars and fats gradually replacing complex carbohydrates at the centre of the meal and literally eroding the structure of the meal itself (Mintz, 1982, 1992). For Mintz, 'the interests of the planters, the slave traders, and the merchants; the taxing and military power of the state; the medical and political/economic arguments that were made for sugar' (1992: 18) all played their part in its meteoric rise. At the same time eating becomes reduced to 'grazing' on bits of fried protein and soft drinks provided by the food industry as the structural 'paradigmatics' and 'syntagmatics' of the meal begin to break down (Mintz, 1985). Add to this the demise of traditional matrilocal culinary patterns, a product of broader social changes, and eating moves towards a state of 'gastro-anomie' (Fischler, 1979). For Symons it seems that 'the modern food marketer pursues the abolition of this "grammar" of meals, perhaps in the name of "freedom of individual choice" and requiring time

[2]Alan Warde is currently researching the impact of eating out on household food provisioning under the 'Eating Out Eating In' project funded by the 1992–1998 Economic and Social Research Council (UK) Research Programme 'The Nation's Diet: the Social Science of Food Choice'.

constraints on eating to be considered an obstruction' (Symons, 1991: p. 210–211).

The growth of snacking has been identified as a recent phenomenon which has been stimulated by social and economic changes within our 'time scarce' society and facilitated by a food industry ready to supply an abundance of more convenient ready-made portions and snacks to accommodate our modern lifestyles (see Gofton, chapter 7). Falk (1994) essentially claims that the meal ritual has shifted its focus from 'communion to communication' as the basis for sociality. He claims that the focus of the modern meal is on conversation, stimulated by the sensory experiences. The desire for individual expression renders more formal eating less appropriate and as a consequence eating has become less of a ritual. Moreover, the meal has become marginalised in favour of non-ritual 'oral side involvements', which include snacks (Falk, 1994). Further support for this view appears to lie in the growth of 'snack' products and the increased trend towards eating out. So where does this leave the meal?

The breakdown of the meal may not be as dramatic as it seems and recent evidence suggests that, in the UK at least, the growth in snacking has stabilised. Snacking accounts for around 19% of eating occasions in the home, which means that the majority of food is still consumed as part of a 'formal' meal occasion (Taylor Nelson, 1990, 1993a). Furthermore, because meals tend to contain more food than snacks, they account for a higher share of the food consumed.

The structural approach offers a different perspective on the thinking behind what shapes food choice by focusing attention on meal patterns and structures. While the trend towards informality is apparent, and convenience is appealing, the rules about what is appropriate remain an important consideration for the consumer.

11.2 What's on the daily menu (content)?

Given the continuing importance attached to the domestic meal, this section looks at market research data from the UK to examine the relationship between meal occasions and food choice at home. The daily meal pattern, which forms a 'skeletal structure' for the day's activities, is highly dependent on the employment status of women and, in households composed of nuclear families, views on sexual division of labour (Charles and Kerr, 1988; Goode et al., 1984; see section 11.4).

Data from the Taylor Nelson Family Food Panel shows that 80% of all the food consumed was still eaten as part of a 'formal meal'. The number of in-home eating occasions grew by around 5% between 1987 and 1993 and there was some evidence of meal polarisation towards the beginning and end of the (working) day. The decline in the number of lunch and tea-time meal occasions was matched by an increase in evening meals (Taylor Nelson, 1990, 1993a).

Breakfast was regarded as an important eating occasion and one of the most frequently occurring meals. A high proportion of people consumed breakfast every day and both children (under 11) and adult men (over 45) were key participants at this occasion (Taylor Nelson, 1990). Cooked breakfasts were more common at the weekend when the households were more likely to eat together (see also Goode *et al.*, 1984). Consumer expectations about what constituted a 'proper' breakfast were changing, with convenience and health driving food choice at this eating occasion. Most breakfasts involved little more than toasting or grilling. The recent launch of products such as 'Pop Tarts' and 'Breakfast Bars'[3] reflect the fact that breakfast, with limited time to eat and prepare food, is more characteristic of a snack than a meal occasion (King, 1985; Nicod, 1979). Breakfast was dominated by a small number of products, notably breads and rolls, breakfast cereals, yellow fats and marmalade (Taylor Nelson, 1990; see Appendix 11.2).

Despite a decline in the number of lunch occasions, a consequence of more women working full-time, this was the second most frequently eaten meal. Midweek lunches at home were characteristically light meals that involved little planning, preparation or cooking and much of the food was eaten cold. Lunch at home included more women who were not in full-time employment and young children (under 6). It was more common among older households (over 45). This age group are also more likely to regard this as their main meal of the day and follow a meal format that resembles dinner (Tilson *et al.*, 1993).

The choice of food for lunch was complicated by the differences between midweek and weekend lunches. Overall the traditional lunch (which bore more of a resemblance to dinner) was being replaced by light, cold and increasingly 'healthy' products. Midweek lunches were less formal and sandwiches, soup or 'complete meals' were popular. Yoghurts, canned soups, cold meats, prepared savoury dishes, bread and rolls and fresh fruit and pâté were supplanting hot carcass meats. More people were reported to be eating lunch outside the home as a consequence of changing work patterns (Taylor Nelson, 1993a).

Afternoon tea, which in Britain usually meant a cup of tea, sandwiches and cakes, was less common. This, however, is an interesting meal in the sense that it can take on the guise of a snack, a light meal or even a main meal if used for children's early evening meals. (In looking at meal structures, Douglas (1972) made a clear distinction between the afternoon tea which is a snack and the high tea which is classified as a light meal.)

---

[3]'Pop Tarts' are manufactured by Kelloggs and consist of a fruit-flavoured filling sandwiched between two layers of cereal based outer, which is designed to be toasted prior to eating. 'Breakfast Bars' are a muesli type bar produced by Uncle Toby's in Australia and described as 'a quick and convenient way to enjoy breakfast on the go, that combines great taste and nutrition in every bite'.

Dinner, or the evening meal, may not be the most frequently consumed meal, but it was the occasion at which most food was consumed (Taylor Nelson, 1990). The majority (90%) of evening meals were described as main meals although lighter meals were more common after 8.00 p.m. This trend towards eating the main meal in the evening is reflected throughout Europe and North America in response to changing working patterns. In one British survey 81% of the sample claimed to eat their main meal between 6.00 and 9.00 p.m. This was even more marked among the under-54 age group, in households with children, and in households with more than three people (Mintel, 1993).

In the UK, savoury foods dominated these evening meals and fresh potatoes, bread and rolls, vegetables (fresh, frozen and salad), yellow fats, sauces and dressings, gravy products, beef and poultry, plus cakes, tarts and pastries were important components. Pork and lamb was being substituted for poultry, fresh and prepared vegetables were increasing in popularity, and consumers were beginning to experiment with ethnic foods such as rice, pasta and other accompaniments (Taylor Nelson, 1990).

The main meal at the weekend is more likely to take place in the middle of the day. Sunday dinners have always had a special place in British culture, as occasions when the family are most likely to eat together, converse, and even argue. According to Warren (in the 1950s): 'Sunday dinner in the English sense is strictly a family matter and its roots lie very deep, sociologically and psychologically it is a very important symbol' (1958: 67). While its importance has diminished, it remains the main family meal of the week (Taylor Nelson, 1990). The distinction between weekend and weekday is marked then by a commitment, at least in principle if not always practice, to a more formal meal to mark the climax of the week's activities.[4]

Besides the daily patterning of meals there is further evidence of how food selection is related to the occasion and ideas about what is appropriate to eat[5] (Marshall, 1993; Marshall and Bell, 1995; see also Schutz, 1988, 1994). The diversity of (even domestic) eating occasions makes classifying meals difficult, but research in the USA has shown that American consumers associate hamburgers and hot sandwiches with 'regular meals', roasts and fresh cooked vegetables with 'special meals' (Figure 11.1).

---

[4]In Australia, it seems that this 'special' meal occasion is moving to the start of the week as a direct consequence of individuals playing more sport on the weekends. Monday night is the new high point of the week for many Australian families.

[5]See Belk (1975) for a discussion of the influence of situation of choice. He has argued for an objective definition of situation to include physical surroundings, social surroundings, temporal perspectives, task definitions and antecedent states, and in research into snack products and meat has shown how situational factors, taken along with product and person characteristics, helped to explain choice.

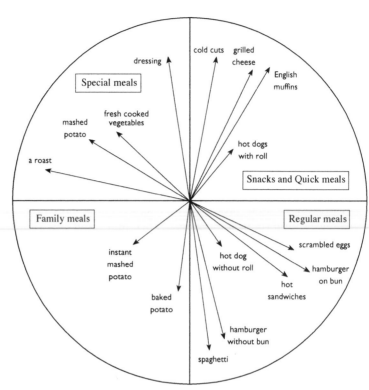

Figure 11.1 Positioning products by meal occasion. (*Source*: Adapted from Assael (1992), p. 545)

Market research conducted in the UK found that cereals, marmalade and dried fruit were almost exclusively breakfast foods, slimming products, soup and game products were mainly used at lunch, whereas rice, pasta and fresh continental fruit and vegetables were heavily reliant on evening meal occasions (Taylor Nelson, 1990; Appendix 11.3). But while certain foods are exclusively tied to certain occasions, many foods are used across a range of meals; this appears to be related to more than simply the type of food. Unfortunately, there is limited information on how the foods are combined to construct meals. The results from a Swedish study showed a general lack of agreement among a sample of school children (10–17 years) on what constituted a meal, resulting in a high percentage of unique meals among the sample. There was, however, a high degree of similarity between the grouping of items and meals on the school lunch menu (Ahlström *et al.*, 1990).

An analysis of meals recorded as part of a diary study in the North East of England in 1987 (unpublished) confirmed that dinner could include a wide variety of foods and follow several formats. In Table 11.2, households in the sample were clustered according to the types of food

Table 11.2 Clustering of 102 N. E. England households at dinner by proportions of foods consumed

	F^*	1	2	3	4
Households (n)		7	63	25	4
Items[**] (n)		105	85	96	49
Foods (%)					
coffee		6.0	3.5	2.5	3.3
tea	(b)	8.7	6.6	12.0	4.7
no alc[1]	(c)	9.6	0.9	1.2	1.5
alcohol[2]		0.7	0.3	0.3	1.3
milk	(a)	1.7	0.4	1.9	1.5
desserts	(b)	2.7	4.2	1.9	1.2
biscuits	(c)	5.5	6.3	12.9	2.4
bread	(b)	7.9	5.8	9.0	5.0
preserves	(c)	1.5	0.7	2.4	0.0
cereals	(c)	3.1	0.2	0.4	0.0
cheese	(b)	2.7	1.9	4.0	0.6
eggs		2.2	2.2	2.6	1.9
fats		1.4	0.5	1.6	1.2
margarine		0.1	0.0	0.2	0.0
fruit	(a)	2.0	4.2	5.6	0.3
rice	(c)	0.4	1.1	0.5	6.5
chicken	(c)	2.1	2.0	0.9	8.1
sheep	(c)	0.1	0.6	0.1	2.7
pig	(b)	1.6	2.7	1.4	3.6
beef	(c)	1.2	3.6	1.5	2.5
other meat		5.4	7.0	5.8	3.7
fish	(c)	2.6	2.8	1.4	6.6
sauces		1.1	1.2	0.7	0.3
misc[3]	(b)	3.4	4.2	6.9	3.0
vegetables	(b)	18.3	26.1	19.0	18.3
potatoes	(c)	8.0	10.9	3.1	18.0
pizza		0.2	0.2	0.3	0.0
Condition (%)	(c)				
frozen		5.5	13.0	5.4	17.2
mwave[4]		2.1	5.5	1.4	5.1
rprep[5]		29.2	16.7	20.6	20.5
hot		52.1	62.3	35.7	82.9
cold		39.7	28.3	50.8	12.8

Source: Gerhardy, Hutchins and Marshall (unpublished)

[a]$p < 0.05$; [b]$p < 0.01$; [c]$p < 0.001$.

[*]F = F-test.

[**]Items = number of foods during two-week period

[1]no alc   = no alcoholic drinks
[2]alcohol   = alcoholic drinks
[3]misc   = miscellaneous
[4]mwave   = microwaved
[5]rprep   = ready prepared

served at 'dinner' over a two-week period. The majority of households (62%) were contained within cluster 2. They followed the 'meat and two veg' format consuming higher proportions of beef, other meat, vegetables and potatoes, all served hot. This group also consumed proportionally more desserts. A smaller number of households (4%) in cluster 4 were more likely to serve fish, chicken, pork and rice as well as a lot of potatoes and proportionally more frozen food. In cluster 3 (25% of households) dinner meant tea, bread, jam and biscuits. In cluster 1, households consumed proportionally more tea, coffee and cereals. It is difficult to draw any firm conclusions from this analysis; suffice to say that there was a high degree of variation on the types of foods used across what the participants described as dinner, and no significant demographic differences were apparent across the household clusters. An analysis of breakfast and lunch meals also revealed a range of food being eaten. Despite the variation across the meals the 'meat and two veg' pattern appeared to dominate in this region of the UK. Further analysis of meals containing fish provided support for the tri-partite structure suggested by Douglas and Nicod (see 11.1) (Marshall, 1993).

As Mintz (1985) reminds us, foods are interchangeable across the basic structure of the meal and these changes often reflect changing social, economic and political circumstances (see Marshall, chapter 1). But while meal structures are relatively enduring, the nature and type of foods which are permitted at each point in the annual, weekly, or daily cycle, or even within the meal, are less fixed, as we continually redefine what is acceptable and appropriate. In addition, preparation and cooking can play an important role in determining how products are used, transforming a product and deeming it acceptable in another part of the meal structure. Part of that explanation lies in an understanding of what is proper and how that is changing.

11.3
Redefining 'proper' meals (more content)

In the UK, eating 'properly' means eating at least one 'proper meal' per day (an issue explored in chapter 9 by Anne Murcott). This usually means a 'cooked dinner' comprising meat, potatoes and vegetables which is eaten at the table,[6] not in front of the television, served on plates and with appropriate knives, forks, spoons, glasses and napkins provided for the dinner (Allen, 1973; Murcott, 1983a). Table manners are essential (see section 11.5) and there is usually a seating order for the family. Moreover, 'proper meals' are planned, involve 'forethought' and commitment to times, places and other people. They are for family, close friends or guests, although proper meals can be individual affairs. Responsibility for preparing the 'proper meal' usually falls on women in the household, a

[6]Over half of the respondents in a Mintel survey claimed to still eat at the table. This was more pronounced in the over-55 age group and families (Mintel, 1993).

feature not confined to western societies, and this reflects labour and power differentials within the family (Charles and Kerr, 1988; Cline, 1990; Murcott, 1982, 1983a; see also Murcott, 1986, for a discussion of the Tallensi meal). The ubiquitous 'tv dinner' introduced in the USA in 1953 offered an alternative to cooking for one, but this attempt to mimic the home-cooked 'proper meal', like many of the ready meals currently on sale, if anything, signified an implicit desire to eat 'properly'. It offered an alternative to eating out, or snacking, rather than an alternative to a 'proper meal' (Jerome, 1975).

The midweek evening dinner meal bears the hallmark of a 'proper' meal, but there appears to be some modification of the three-course structure and some divergence from the meat and two vegetable format. Only 3% of UK adults surveyed had a full three-course meal midweek, 49% claimed to have two courses, and the remainder one course (Taylor Nelson, 1990).

The growing interest in 'ethnic' food and the introduction of 'ethnic' dishes into the domestic meal repertoire seem to raise some questions about the ubiquity of 'meat and two veg' but it is worth further investigation. The recent success of rice[7] and pasta, making gains on the traditional potato in the United Kingdom, and the growing interest in cook-in-sauces might be taken as evidence that 'meat and two veg' is a thing of the past (Taylor Nelson, 1993a). Cook-in-sauces, for example, are almost exclusively for 'ethnic' dishes, bought primarily by DINKIES[8] and over 35s who would like to 'eat properly' but do not have the time (*Marketing*, 1994). The use of these sauces, however, advocates change in other parts of the meal structure and usually means substituting traditional staples such as potatoes for rice to retain 'authenticity'. The success of these products owes much to the commodification of 'ethnic' eating as the food industry continues to bring new flavours and foods to the supermarket shelves.

The 'internationalisation' of domestic cuisine may not be as rapid or extensive as it might appear from examining the supermarket shelves, reading newspaper and magazine food columns or watching the growing number of cookery programmes on television. Despite the success of ethnic cuisine in Australian restaurants and food malls, it has met with limited success in the home. One survey revealed the traditional 'British' meal, i.e. 'meat and two veg', was alive and well amidst a multicultural panoply of ethnic dishes (Ricketson, 1992; Shoebridge, 1992). Suburbia, it seems, was less quick to adopt the new ethnic dishes, preferring to stay with the tried, tested and accepted meals. Only 4% of Australians had

[7]In British cuisine rice was, until comparatively recently, something which one had for pudding. Its acceptance as a staple in the evening meal demands a different variety, cooked in a different way, and served with an entirely new combination of other foods.
[8]DINKIES is a marketing term used to describe couples with dual income and no kids.

cooked Mexican, Thai, Greek or Indian food at home, although Italian (24%) and Chinese meals (18%) were more popular.[9] Exactly where these meals were being incorporated into the domestic cooking routine was less clear-cut. If using pasta constitutes an Italian meal, are Australians really adopting the cuisine or simply adapting their own? In Australia, where many of the indigenous foods were rejected in favour of imports, there appears to be less of a problem with producing a foodstuff and declaring it Australian if it is produced in the country. Despite such claims, many 'ethnic' dishes remain clearly identified. As such the ever-successful pasta, sauce and pizza remain Italian, or in the case of the latter, increasingly American.

It will be interesting to see how the present generation of young people, currently exposed to a broad range of 'ethnic' cuisines, are changing their ideas about what constitutes a proper meal. Young Australians, for example, are reformulating their own ideas about 'proper meals' and 'dinner'[10] as an essential part of a wider process which involves off-loading their cultural baggage in the search for an Australian cuisine (Ripe, 1993). In Scotland young people, under 35, appear to be reformulating their ideas about what makes for a proper meal: 'spaghetti bolognaise' and 'chilli con carne' are considered proper, and 'meat and two veg' may be losing some ground to chicken stir fry.[11] In Britain there appears to be something of a trend towards savoury dishes, notably casseroles, pasta and 'ethnic dishes', that reflects a willingness to try new foods (Taylor Nelson, 1990):

> despite the much vaunted growth in informal eating and 'grazing', consumers' perceptions of meals have changed in line with shifts in the types of food consumed at each meal. Thus, while a pizza may have been regarded as being a light or snack food in the early 80's, as the product has become absorbed into the mainstream British diet so it is now consumed at main or substantial meals. (Taylor Nelson, 1990: 9)

While these meals often include a staple, meat and vegetables the 'flavour principle', different cooking styles and unusual ingredients identify them as non-indigenous.[12] Proper meals are then to some extent dynamic

---

[9]Interview with J. Peters Young and Rubicam, Melbourne, Australia, 17, December, 1993.

[10]This is based on numerous discussions and several focus groups conducted with a sample of Australian undergraduates at the University of Queensland, Brisbane, during sabbatical leave at the end of 1993.

[11]These preliminary findings arose from interviews with a number of young Scottish people recruited to take part in 'The Marriage Menu' study funded as part of the 1992–1998 Economic and Social Research Council's (UK) Programme 'The Nation's Diet'.

[12]Meat may be chopped into small pieces, removing the need to carve and serve the joint, itself an important ritual, while the vegetables are either absent from the occasion or incorporated into the meal as part of the sauce. For example, the addition of 'Chicken Tonight' creamy curry sauce leaves no doubt as to the main ingredient of the meal and when it is to be served. Combining vegetables and chicken in a sauce and incorporating 'foreign' ingredients represents a formidable transgression of the rules and renders the meal acceptable, but only in certain circumstances. Serving chicken curry with additional vegetables, or even a salad, is one solution to the structural problem but this is unlikely to elevate the dish to major meal status.

although the notion extends far beyond the food on the plate, and while new foods and flavours are being adopted, they must find their place within an existing pattern of meals where proper still means 'meat and two veg'. Pizza, spaghetti bolognaise, Chinese stir fry are more widely accepted as main meals but they are borrowed, the rules for combining foods are different, and they are to a large degree 'ready-made' which makes them less than proper. Proper meals not only say something about what food is appropriate – they also say something about the relationship between those eating together.

11.4
Eating at the family table (participation)

Despite the decline in the nuclear family, that is, mother, father and children living together (currently 25% of households in the UK, USA and Australia), most people spend some part of their life attached to a family in its various forms (CSO, 1995; Olsen *et al.*, 1983; Senauer *et al.*, 1991). The family has a pervasive effect on food choice particularly in the formative years and many ideas about 'proper eating' are passed on through this consumption unit. Furthermore, there is a suggestion that meals where the family eat together are in decline in both the UK and the USA (Beardsworth and Keil, 1990; Levenstein, 1988; Milburn, 1994) and changes in patterns of work and leisure make it increasingly difficult for the family, or household, to eat together on a regular basis.

Research evidence from the UK (section 11.2) and Australia (Craig and Truswell, 1988), however, suggests that a high proportion of households still attach considerable importance to the meal occasion. In a UK survey 57% of adults claimed to eat together every day or almost every day, with a further 16% eating together most days. This was more likely to be the case with two-person households than larger households. Eating together was reportedly higher among the elderly or retired, and lower socio-economic groups (Mintel, 1993).

Research by Charles and Kerr (1988) revealed the importance still attached to the 'proper meal' in nuclear households and the unequal distribution of food across family members. Notions about proper meals played a key role in food choice and there were variations on what was considered proper for men (high status meat, cake and alcohol), women (fruit, biscuits, sweets) and children (beefburgers, sausages and fish fingers). The best cuts of meat (and largest slice of cake) were reserved for men – a legacy from the days when women were dependent on the male breadwinner that reflected differential status and power within the family unit. Male preferences had a profound influence on what was served at mealtimes in these households (Charles and Kerr, 1988; see also Pill and Parry, 1989). Despite being largely responsible for food shopping and preparation, women's own preferences were often subordinate to those of their family (Charles and Kerr, 1988; CSO, 1990; Dare, 1988; Gillon *et al.*, 1993; see also Davis and Rigeux, 1974; Putnam and Davidson,

1987). One major difficulty was trying to satisfy all of the individual likes and dislikes and at the same time complying with cultural mores about what was appropriate and proper.

The changing nature of women's domestic role (Gershuny and Jones, 1987; McKenzie, 1986; Pahl, 1984) and supposed liberation from the kitchen (sink) is evident in more liberal attitudes among younger housewives towards using processed food (see Gofton, chapter 7, for a fuller discussion of the impact of these changes on food choice). As individual family members begin to take on more responsibility for their own food choice, the meal occasion is likely to see a decline in the idea of one communal dish for all. Different working patterns mean that it is more difficult for the family to eat together at the same time. What is not apparent is whether members who are not present at the family meal eat the same food as the rest of the family or consume different meals to suit their own tastes and time schedules. Higher socio-economic groups were more likely to only eat together at weekends among Backett's (1992) sample of middle-class Scots where the 'family meal' was largely affected by time schedules of family members, views on the appropriate socialisation of children, the eating preferences of individuals within the group and the importance which the family placed on this occasion as an opportunity for communication. Regardless of the frequency of eating together, family meals, when they do occur, offer a 'daily assurance of family identity and propriety' (Murcott, 1986).

Children are exposed at an early age to foods consumed by the rest of the family (Kyhlberg et al., 1986), and once in the high chair they are brought to the table to eat with other members of the household; although there is little research into the manner in which parental meal practices influence children's eating habits. Charles and Kerr (1988) stress the manner in which parents regard it as part of their responsibility to teach children how to eat 'properly' and this extends beyond the selection and recognition of food to table manners, etiquette and comportment (see Elias, 1978). The young and adolescent are renowned for their rebellion against established and regimented eating patterns. The dislike of meal-times as a restriction on freedom, and an intrusion into playing time, is an understandable reaction from children who see little point in learning table manners. At least the old adage that children should be 'seen and not heard' is giving way to one which encourages children to converse at the table, albeit under adult 'surveillance'. Any wonder that adults regard children's eating habits as rebellious and the antithesis of the structured approach adults adopt towards eating (James, 1981; see also James, 1990).[13]

---

[13]Research is currently under way as part of the 1992–1998 Economic and Social Research Council's (UK) Programme 'The Nation's Diet' which suggests that peer group influence is an important factor in shaping food choice among children.

New working patterns, new household structures with an increasing number of single-person households and cohabitation, limited time availability, leisure commitments, more eating out, health concerns, financial constraints, changing attitudes of the food preparer towards mealtimes and eating together, and tradition are all playing their part in the way we eat 'meals' as a 'family'. If anything there is a trend towards 'lighter' meals but 'grazing' and 'snacking' do not, as yet, appear to be supplanting domestic meals (Ekström, 1991; Taylor Nelson, 1993b).

11.5 Table manners (methods)

Burping, belching, wiping one's mouth on the table cloth and spitting are not generally regarded as acceptable forms of behaviour at the European table. Table manners, or the lack of them, are a marker of 'civilised conduct' (Elias, 1978) and act as a system of civilised taboos which condition our behaviour at the table. In the process they reduce tension and protect people from one another's knives and baring of teeth (Visser, 1992). Eating with knives (weapons) and forks, or chopsticks, is something many of us take for granted.[14] Eating with the hands is generally discouraged in the west, and in cultures where it is practised it becomes an art form in its own right signifying informality and social intimacy. Using the fingers to eat necessitates warm, as opposed to hot, food and where the fingers must be used, for example eating chicken legs, particular protocol applies, in this case eating with the thumb and forefinger. Society does make exceptions for certain members such as the young, the old and the infirm but most individuals put a safe distance between themselves and their food when eating at the table (Elias, 1978).

These unspoken rules about how food should be eaten are found across cultures and they can be a source of embarrassment for the uninitiated – eating is also about etiquette. In western cultures it is generally regarded as impolite to lick one's knife, or fingers, return chewed food to the plate and start eating before everyone else is served. In the USA it is acceptable to start eating once you have been served your food. Second helpings may be frowned upon at formal dinners where to ask for more implies that there was not enough in the first place, or it could be construed that one enjoyed the meal so much that it is a compliment to ask for more. In Denmark, for example, it is customary to ask for more, in Ireland second helpings are a must, in Bulgaria you are warned to take small portions and will be expected to eat several. In China the diner should not eat *cai* (meat, fish or vegetables) before being served rice: to do so is considered greedy. Furthermore, second helpings of *cai* can be taken only if accompanied by more rice (Visser, 1992).

---

[14]The first recorded use of the fork was by the wife of the Venetian Doge, Domenico Selvo in the eleventh century (Visser, 1992, on Elias).

11.6
Meals as markers
(function)

While manners tell us something about the eater, the meal can mark off cultural and social boundaries. Every culture has its own distinctive cuisine[15] characterised by a range of basic foods, frequent use of a set of recipes, flavourings, specific processing characteristics, as well as rules about how those foods are combined and what is appropriate for the occasion (Rozin, 1982; Rozin and Rozin, 1981; Sokolov, 1991; Symons, 1993; Visser, 1992). Indeed the English Sunday roast beef, the Australian barbecued steak, Chinese stir-fry, Malaysian beef satay, the American hamburger and chips have all become symbols of national cuisine. While ingredients are subject to change, as new food varieties are introduced, exotic ingredients are added and traditional practices wane, many of the rules still remain intact. As Sokolov (1991: p. 15) reminds us: 'old recipes never die, they add new dishes and ingredients to old recipes and slough off the losers, the evolutionary dead ends'. This ability to accommodate change can be seen, for example, in Europe's adoption of many New World foods such as maize, chillies and sweet potatoes in the 16th century (Sokolov, 1991).

Meals are unifying in the sense that they bring cultures and groups together both physically and symbolically. Through national dishes we identify cuisine, through meals we can relate to how one group of people organise their eating occasions spatially, temporally and structurally. At the same time meals highlight social and economic inequalities: at one extreme the soup-kitchen symbolises the loss of social dignity and dependence on charity; at another level the substitution of poorer quality foods, even particular brands of food, are seen to be lower status. In Britain, industrialisation, the growth in urban life and economic inequality contributed to new social distinctions which emerged amidst the myth of 'progress' in living standards for all (Burnett, 1979; Gofton, 1986a). Booth (1902) contrasted the diets of the rich and poor, highlighting the inequalities which existed at the turn of the century. While Rowentree's (1901) poor existed on a diet of bread, bacon, tea and sugar, Palmer's (1952) middle class feasted on an abundance of beef, fowl and fish to the extent of opulent waste (see Veblen, 1899). Despite the blurring of class boundaries in Britain there remain distinct differences between working- and middle-class food purchases (Tomlinson and Warde, 1993), but differences between meals remain undocumented.

Bourdieu's (1979) analysis highlights the fact that class differences exist as much in the manner in which food is eaten, and the formality surrounding the occasion, as in the types of food eaten. In French working-class households the freedom from a collective progression of the family

---

[15]'Cuisine' can be regarded as an arrangement on food-related customs, which are culturally elaborated and transmitted.

through the courses, or the use of the same plates across courses, or the shared spoon may be regarded as slovenliness and such short-cuts are only permissible when one is 'at home' with the family. In contrast, the bourgeoisie is required to eat with 'all due form' – the conviviality of the working-class meal is replaced by a strict sequence of dishes and an adherence to form and function in the home.

11.7 Pulling the menu together

Meal patterns and structures offer an interesting insight into the social relations between people who habitually eat together, which was the primary motivation behind the structural analysis of meals (Douglas and Nicod, 1974), but they also tell us something about food choice. Indeed, it has been argued elsewhere that food has simply been regarded as a medium through which to study other phenomena, and not really treated as a subject worthy of study in its own right (Beardsworth and Keil, 1990).

A closer examination of meals reveals the complex interaction between (1) the food, (2) the degree of processing and (3) the cooking method in shaping ideas about what is appropriate for different types of meal occasion. Figure 11.2 depicts the interaction between these factors and proposes a conceptual framework for understanding how consumers construct meals at home. As one moves out from the centre of the figure from 'special meals' towards 'snacks', eating occasions become more

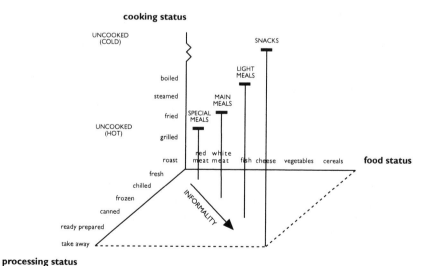

Figure 11.2 A framework for examining food choice across eating occasions in the United Kingdom and North America

informal, more individualistic and more unstructured, which is the direction some believe food consumption is moving in response to broader economic, social and political changes (Falk, 1994; see Gofton, chapter 7; Mintz, 1985). Less time is required for planning, shopping, food preparation and cooking and this is reflected in the type of products served and the duration of the meal. Consequently, the rules regarding content are less rigid and 'light meals' or 'snacks' are more likely to accommodate new foods (Douglas, 1976). These less formal eating occasions are temporally mobile and not tied to specific places, times of the day, week or year. Unlike 'special' meals, or to a lesser extent 'main' meals, they can be eaten in front of the tv, or on the move.

As one moves back into the centre of the figure, the converse of the above is true and meals become more formal, involve more people, and become more structured, along the lines suggested by Douglas and Nicod (1974), with specific times and places for eating. 'Special' and 'main' meals take on the appearance of 'proper meals'.

The 'food status' dimension reflects the social status which consumers attribute to food. Charles and Kerr (1988) found 'foods were ranked hierarchically in terms of their social status, and their distribution within the family reflected the relative power and status of family members' (1988: p. 36). They alluded to the existence of a food hierarchy in which 'red meat is the most highly valued food followed by poultry, fish, eggs, cheese, fruit leaf vegetables and cereals in that order' (Charles and Kerr, 1990: p. 37). This was reflected in what was chosen for celebratory and 'proper' meals and the 'food status' dimension reflects this hierarchical distribution. Moreover, it focuses on the centrepiece of the meal (the '*a*' in Douglas's '*a* + 2*b*' structure), or in the case of snacks the individual food item. The location of meat at the top of this hierarchy echoes the revered position of meat in developed economies (Fiddes, 1991; Twigg, 1983). Despite the growth in sales of white meat, mainly poultry, fuelled by falling prices, increased convenience and concerns about health, it carries less status than better cuts of red meat. The same is true of fish although exotic species, or whole fish, carry high status. Increased exposure to vegetarianism is changing consumer perceptions of vegetables but, in a society of meat eaters, vegetables are more likely to be seen as outside the mainstream and at the side of the plate, accompanying meat. They are not commonly perceived as centrepieces in their own right, except for particular groups like vegetarians (Marshall *et al.*, 1994). In the west no meal would be classified as such without the staple, be it potatoes, pasta, rice or cereal, but these foods on their own, without meat or fish, are regarded as lower status centrepieces.

This dimension is problematic because of the diversity of products within any one food category. The status of meat, for example, depends on the cut and degree of processing, with steak and chops revered as high status followed by mince, stewing meat and liver, while sausages and beefburgers remain at the bottom (Charles and Kerr, 1990).[16] This is not to suggest that the higher status foods are excluded from lower status occasions, for example a hot roast beef sandwich is a perfectly acceptable snack. The food status hierarchy, presented in Figure 11.2, reflects a general notion that those foods at the bottom are less likely to be accepted at the centre of more formal meals (nut roast for Christmas dinner). Ideally, hierarchies should be constructed for individual product groups.

The 'processing' dimension reflects the differential status attached to the product form, that is whether food is fresh, chilled, frozen, canned, etc. Fresh foods carry the highest status that reflect the time and skill required to prepare them as well as perceptions about superior taste. Freshness in foods is still highly valued and this is reflected in greater use of fresh foods in special meals. A higher proportion of processed products are more likely to be found in less formal meals such as 'light' meals and 'snacks'. In the right place they are perfectly acceptable, but a greater proportion of the food served in special meals will be fresh, whereas light meals and snacks are likely to contain more processed foods which require little or no further preparation. Some evidence is provided in the Family Food Panel which showed that a higher proportion of evening meals, the main meal of the day, contained fresh produce (Taylor Nelson, 1990). Furthermore, as Gofton shows in chapter 7 of this volume, the success of processed convenience products reflects broader social and economic changes but these products still have their place in the meal system. The acceptance, for example, of highly processed foods is less likely in special meal occasions where guests are present. It seems that a degree of 'culinary competence', and investment in time and effort, are essential for successful entertaining.

The cooking dimension, which draws upon Lévi-Strauss (1966),[17] reflects European and North American views whereby hot cooked foods have attained higher status than cold uncooked foods. Different cooking methods can, in turn, confer status on the food, and consequently the meal (see Murcott, chapter 9). This is reflected in Figure 11.2; however, unlike Lévi-Strauss it is not presented here as a universal structure for analysis, but more a reflection of practice (Lehrer, 1972). Whereas roasting confers high status, frying or boiling imparts lower status on the food. The centrepiece is more commonly roasted or grilled, while staple and trimmings are usually boiled, or occasionally fried.[18] While evidence

---

[16]In the case of fish, fresh shellfish, exotic species, cod, haddock and plaice are at the top followed by ling and coley, in the middle are frozen and processed products with highly processed fish fingers and fish burgers at the bottom (Gofton and Marshall, 1992).
[17]Cooking, according to Lévi-Strauss (1966), distinguishes man from nature, and in his treatise on the categories of raw and cooked he proposed that categories of cooking were appropriate symbols for social differentiation (see Leach, 1970).
[18]The Tuscan dinner follows the ascending order of boiled (lesso), fried (fritto), stewed (umido), and roast (arrosto) (Visser, 1992).

suggests that a high proportion of foods served at meals are uncooked (Taylor Nelson, 1990), this status of the meal is judged by the centrepiece and the cooking method applied. The highlight of the special meal is inevitably a hot '*pièce de résistance*'.

The framework illustrates the complexity of trying to incorporate each of the different elements when one approaches the problem of food choice from a 'meal perspective' as opposed to a 'product perspective'. The first decision for the consumer is not 'What food do I buy?' but 'What is the occasion?', 'Who will be present?', 'What type of meal is befitting?' and then 'What food do I serve on this occasion?' This, in turn, can drive decisions about whether fresh, frozen or canned products are acceptable and how the food is to be cooked. The status of the meal, or individual course, is reflected in the food, its form and how it is cooked. The appropriateness, for example, of fish in the meal reflects its limited place in the British diet, but different forms of the product are acceptable in different types of eating occasion. The whole fish is suitable for entertaining, but the humble processed boil in the bag version is more likely to be served for children's tea. While there are strong functional arguments centred on taste, flavours, appearance and the presence of bones, the nature of the occasion also appears to affect the decision on what to use (Marshall, 1993).

The domestic meal pattern may have its origins in the sequencing of the formal meal but everyday, or ordinary, is not elaborate, as Visser reminds us:

> Day to day eating is regular, much less copious than feasting, and done with a few people whom we know more intimately than anyone else, largely because we often meet and eat with them. Everyday meals are not intended to surprise or impress, or challenge us. The expected is what we look for; we achieve it through customary behaviour but low decorum, and through order … Daily meals in turn order our day, providing occasions for meeting friends and family and for resting from work. There are fewer rules of politeness than we find at formal feasts but rules exist and are observed. (Visser, 1992: p. 345)

11.8
Concluding remarks

Despite the increase in snacking, meals remain an important influence on food choice. They offer some degree of continuity (Roos *et al.*, 1993) in an ever-changing world of food. In accommodating that change, however, new food products have to find a place among what already exists and comply with the cultural folkways (see Buisson, chapter 8). Fischler (1988) has shown that desire for variety (neophilia) has to be balanced with the conservatism in choice (neophobia). This conflict he called the 'omnivores paradox' and nations of neophobics are uneasy with anything too unorthodox when it comes to food (Fallon and Rozin, 1983). Food, however, may be rejected because it is inappropriate as well as distasteful, dangerous or disgusting (Rozin, 1982). Inappropriateness is based on cultural convention and food mores which offer some guidance on accep-

EATING AT HOME   285

table food combinations which meet the needs of the participants and the occasion. These implicit rules about how meals should be constructed put food choice into context (Marshall, 1993).[19]

Unfortunately, much of the current debate about what people eat stops after the point of purchase and there is relatively little discussion about the nature of eating occasions and how consumers construct meals. Beyond classic recipe dishes, or manufactured versions of 'ethnic' dishes, we have little information on domestic meals, where brands fit into the meal pattern, or the extent to which individuals are taking on more responsibility for their own meals. The private nature of domestic eating presents enormous difficulties in conducting such research.[20]

Convenience and health (see Gofton, chapter 7, and Anderson *et al.*, chapter 5) are increasingly driving choice but the 'deconstruction' of the meal may not be as dramatic as once claimed (Taylor Nelson, 1993b). The notion of what constitutes a proper meal, the number of courses, its structure, and where and when it should be eaten still persist. Sometimes we cook ourselves 'proper' meals with three courses to remind ourselves that we are not alone. The patterning and structure of meals are culturally specific and for each society this arrangement of foods reasserts, reflects and continually reconstitutes our social identity. The appeal of sitting down to a recognisable combination of foods, with a familiar structure, is testament to the fact that we are concerned with more than feeding, and with meals, not simply food. Meals remind us of where we came from (the family), where we are (absent from the family), where we have been (travel experiences) and where we hope to be (back as part of a group). They offer us an identity, but they also require time, skills, access to cooking facilities and usually other people to share it with. The selection of food, like much consumption, is not simply the outcome of some deterministic process based on taste preferences and biological needs but it is socially constructed and learned (Bocock, 1993). This applies as much to food as any other form of consumption. Now, whose turn is it to do the washing up?

Acknowledgements

Thanks are due to Taylor Nelson AGB, in particular Giles Quick, for providing market research data from the Family Food Panel Special Reports (1990, 1993) and their permission to use this information in section 11.2. Thanks also to Richard Hutchins and Hubert Gerhardy for their assistance in producing Table 11.2.

---

[19]The idea of eating occasions driving choice may be more of a British than a European trait influenced, among other things, by access to food supplies and storage facilities (Renner, 1944).

[20]Despite the criticisms, Nicod's research was enterprising in that he actually lived with his sample for the period of fieldwork, rather than relying on second-hand accounts, and could observe what was actually happening in the home (Douglas and Nicod, 1974).

References   Ahlström, R., Baird, J. C. and Jonsson, I. (1990) School Children's Preferences for Food
Combinations. *Food Quality and Preference*, **2**, 155–165.

Allen, L. G. (1973) In M. Douglas (ed.), *Etiquette: Table in Rules and Meanings*, Penguin,
London. pp. 219–220.

Assael, H. (1992) *Consumer Behaviour*, 4th edn, PWS-Kent Publishing Company, Boston.

Attar, D. (1990) *Wasting Girls' Time: The History and Politics of Home Economics*, Virago,
London.

Backett, K. (1992) Do As I Say Not As I Do. Paper presented at *ESMS/BSA Med. Soc.
Joint Conference*, September 1992, Edinburgh.

Beardsworth, A. and Keil, T. (1990) Putting Menu on the Agenda. *Sociology*, **24** (1), 139–
151.

Belk, R. W. (1975) Situational Variables and Consumer Behaviour. *Journal of Consumer
Research*, **2** (December), 157–164.

Birch, L. L. (1993) Children, parents and Food. *British Food Journal*, **95** (9), 11–15.

Bocock, R. (1993) *Consumption*, Routledge, London.

Booth, C. (1902) *Life and Labour of the People in London*, First Series: *Poverty*, Vols. I and
II (reprint 1970). AMS Press, New York.

Bourdieu, P. (1979) *La Distinction*, Minuit, Paris. (English translation: (1986) *Distinction: A
Social Critique of the Judgement of Taste*, Routledge and Kegan Paul, London.)

Braudel, F. (1973) *Capitalism and Material Life, 1400–1800*, Weidenfeld and Nicolson,
London.

Burnett, J. (1979) *Plenty and Want: A Social History of Diet in England from 1815 to the
Present Day*, Scolar Press, London.

Charles, N. and Kerr, M. (1988) *Women, Food and Families*, Manchester University Press,
Manchester.

Charles, N. and Kerr, M. (1990) Gender and Age Difference in Family Food Consumption.
In J. Anderson and M. Ricci (eds), *Society and Social Science: A Reader*, The Open University, Milton Keynes.

Cline, S. (1990) *Just Desserts: Women and Food*, André Deutsch Limited, London.

Craig, P. L. and Truswell, A. S. (1988) Changes in Food Habits when People Get Married:
Analysis of Food Frequencies. In A. S. Truswell and M. L. Wahlqvist (eds), *Food Habits
in Australia, Proceedings of the First Deakin/Sydney Universities Symposium on Australian
Nutrition*, René Gordon, Victoria. pp. 94–111.

CSO (1990) *Social Trends*, HMSO, London.

CSO (1995) *Social Trends*, HMSO, London.

Dare, S. E. (1988) Too many Cooks? Food Acceptability and Women's Work in the Informal Economy. In D. M. Thomson (ed.), *Food Acceptability*, Elsevier, London. pp. 143–
156.

Davis, H. L. and Rigeux, B. P. (1974) Perception of Marital roles in Decision Processes.
Journal of Consumer Research, **1** (June), 5–14.

Douglas, M. (1972) Deciphering a Meal. *Daedalus*, **101** (1), 61–81. (Reprinted in *Implicit
Meanings: Essays in Anthropology*, Routledge and Kegan Paul, London. pp. 249–75.)

Douglas, M. (1976) Culture and Food. In *Culture; Essay on the Culture Programme, Russell
Sage Foundation Annual Report 1976–77*. pp. 51–58. (Reprinted in M. Freilich (ed.)
(1983), *The Pleasures of Anthropology*, Mentor Books, New York. pp. 74–101.)

Douglas, M. and Nicod, M. (1974) Taking the Biscuit: The Structure of British Meals. *New
Society*, **19** (30 December), 774.

Elias, N. (1978) *The Civilising Process* (2 vols). Vol. I: *The History of Manners*. Vol. II: *State
Formation and Civilisation*, Basil Blackwell, London.

Ekström, M. (1991) Class and Gender in the Kitchen. In E. Fürst *et al.* (eds), *Palatable
Worlds: Sociocultural Food Studies*, Solum Forlag, Oslo.

Falk, P. (1994) *The Consuming Body*, Routledge, London.

Fallon, A. E. and Rozin, P. (1983) The Psychological Bases of Food Rejections by Humans.
Ecology of Food and Nutrition, **13**, 15–26.

Farb, P. and Armelagos, G. (1980) *Consuming Passions: The Anthropology of Eating*,
Houghton Mifflin Company, Boston.

Fiddes, N. (1991) *Meat: A Natural Symbol*, Routledge, London.

Fischler, C. (1979) Gastro-nomic and Gastr-anomie: Sagesse du Corps et Crise Bioculturelle
de l'Alimentation Moderne. *Communications*, **31** (Automme), 189–210.

Fischler, C. (1988) Food Self and Identity. *Social Science Information*, **27** (2), 257–92.

Gershuny, J. and Jones, S. (1987) The Changing Work/Leisure Balance. In J. and T. Home (eds), *Sport, Leisure and Social Relations*, Routledge and Kegan Paul, London. pp. 9–50.

Gillon, E., McCorkindale, L. and McKie, L. (1993) Researching the Dietary Beliefs and Practises of Men. *British Food Journal*, **95** (6), 8–12.

Gofton, L. R. G. (1986a) Rules of the Table. In C. Ritson, L. Gofton and J. McKenzie (eds), *The Food Consumer*, John Wiley and Sons, London. pp. 127–154.

Gofton, L. R. G. (1986b) Market Change Social Change. *Food and Foodways*, **1**, 253–277.

Gofton, L. R. G. and Marshall, D. W. (1992) *Fish: Consumer Attitudes and Preferences – A Marketing Opportunity*, Horton Publishing, Bradford.

Goode, J. G., Curtis, K. and Theophano, J. (1984) In M. Douglas (ed.) *Food in the Social Order: Studies in Food and Festivities in Three American communities*, Russell Sage Foundation, New York. pp. 143–218.

Goody, J. (1982) *Cooking Cuisine and Class: A Study in Comparative Sociology*, Cambridge University Press, Cambridge.

Hodes, M. (1991) Food for Thought: Issues for the Social anthropology of eating Disorders. Unpublished ms. Cited in S. Mennell *et al.* (1992) *The Sociology of Food: Eating, Diet and Culture*, Sage, London.

James, A. (1981) Confections, Concoctions and Conceptions. In B. Waites (ed.), *Popular Culture: Past and present*, Redgrave Publishing Company, New York.

James, A. (1990) The Good, the Bad and the Delicious: The Role of Confectionery in British Society, *The Sociological Review*, **34** (4), 666–688.

Jerome, N. W. (1975) On Determining Food Patterns of Urban Dwellers in Contemporary United States Society. In M. Arnott (ed.), *Gastronomy: The Anthropology of Food and Food Habits*, Moulton Publishers, The Hague, Paris. pp. 91–112.

Katona-Apte, J. (1975) Dietary Aspects of Acculturation in South Asia. In M. Arnott (ed.), *Gastronomy: The Anthropology of Food and Food Habits*, Moulton Publishers, The Hague, Paris. pp. 315–326.

King, S. (1985) The British Meal. *Admap* (March), 160–167.

Kyhlberg, E., Hofvander, Y. and Sjolin, S. (1986) Diets of Healthy Swedish Children 4–24 Months Old. *Acta Paediatricia Scandinavica*, **75**, 937–946.

Leach, E. (1970) *Lévi-Strauss*, Fontana Modern Masters, London.

Lehrer, A. (1972) Cooking vocabularies and the Culinary Triangle of Lévi-Strauss. *Anthropological Linguistics*, 14.

Levenstein, H. (1988) *Revolution at the Table: The Transformation of The American Diet*, Oxford University Press, New York.

Lévi-Strauss, C. (1966) The Culinary Triangle, *New Society*, December, 937–940.

McKenzie, J. C. (1986) An Integrated Approach – With Special Reference to Changing Food Habits in the UK. In C. Ritson, L. Gofton and J. McKenzie (eds) *The Food Consumer*, John Wiley and Sons, London.

McKie, L. J. and Wood, R. C. (1992) People's Sources of Recipes: Some Implications for Understanding Food Related Behaviour. *British Food Journal*, **94**, 2, 12–17.

Marketing (1994) Top of the Shops. August 25, 14–17.

Marshall, D. (1993) Appropriate Meal Occasions: Understanding Conventions and Exploring Situational Influences on Food Choice. *The International Review of Retail, Distribution and Consumer Research*, **3** (3), 279–301.

Marshall, D. and Bell, R. (1995) The Relative Influence of Situation and Meal Occasion on Food Choice. In T. Worsley (ed), *Food Choice*, Adelaide: Food Choice Conference, Adelaide (forthcoming).

Marshall, D. W., Anderson, A., Forster, A. and Lean, E. J. (1994) Healthy Eating: Fruit and Vegetables in Scotland. *British Food Journal*, **96**, 7, 18–24.

Mennell, S., Murcott, A. and van Otterloo, A. H. (1992) *The Sociology of Food: Eating Diet and Culture*, Sage, London.

Milburn, K. (1994) Food Choice and the Domestic Situation: The use of Qualitative Methods. Paper presented at *Food Choice and Eating Habits: Research and Practice Issues Workshop, Victoria Infirmary, Glasgow* organised by Scottish Colloquium on Food and Feeding (SCOFF) and BSA Sociology of Food Group.

Mintel Leisure Intelligence (1993) Cooking and Eating Habits, **4**, 1–29.

Mintz, S. (1982) Choice and Occasion: Sweet Moments. In L. M. Barker (ed.), *The Psychobiology of Human Food Selection*, AVI Publishing, Westport.

Mintz, S. (1985) *Sweetness and Power: The place of Sugar in Modern History*, Penguin, New York.

Mintz, S. (1992) A Taste of History. *The Times Higher Educational Supplement*, May 8, 15–18.

Murcott, A. (1982) On the Social Significance of the 'Cooked Dinner' in South Wales. *Social Science Information*, **21** (4/5), 677–695.

Murcott, A. (1983a) It's a Pleasure to Cook for Him: Food Mealtimes and Gender in some South Wales Households. In E. Garmarnikow *et al.* (eds), *The Public and the Private*, Heinemann, London.

Murcott, A. (1983b) Cooking and the Cooked: A note on Domestic Preparation of Meals. In A. Murcott (ed.), *The Sociology of Food and Eating*, Gower, Aldershot. pp. 178–185.

Murcott, A. (1986) You are What You Eat – Anthropological factors Influencing Food Choice. In C. Ritson, L. Gofton and J. McKenzie (eds), *The Food Consumer*, John Wiley and Sons, London. pp. 107–127.

Nicod, M. (1979) Gastronomically Speaking. In *Nutrition and Lifestyles, Conference Proceedings of British Nutrition Foundation*, Applied Science Publishers, London. pp. 53–65.

Olsen, D. H., McCubbin, H. I. *et al.* (1983) *Families: What Makes Them Work?* Sage Publications, Beverley Hills.

Pahl, R. E. (1984) *Divisions of Labour*, Blackwell, Oxford.

Palmer, A. (1952) *Moveable Feasts*, Oxford University Press, Oxford. (Reprinted 1984.)

Pill, R. and Parry, O. (1989) Making Changes – Women, Food and Families. *Health Education Journal*, **44**, 3, 158.

Putnam, M. and Davidson, W. R. (1987) *Family Purchasing Behaviour: II Family Roles by Product Category*, Management Horizons Inc., Price Waterhouse, Columbus, Ohio.

Renner, H. D. (1944) *The Origins of Food Habits*, Faber and Faber, London.

Ricketson, M. (1992) Dishing Up the Past in a New Australia. *Time*, June 22, 52–59.

Ripe, C. (1993) *Good-bye Culinary Cringe*, Allen and Unwin, Sydney.

Roos, G. M., Quandt, S. A. and DeWalt, K. M. (1993) Meal Patterns of the Elderly in Rural Kentucky. *Appetite*, **21**, 295–298.

Rowentree, B. S. (1901) *Poverty: A Study of Town Life*, Macmillan, London.

Rozin, E. (1982) The Structure of Cuisine. In L. M. Barker (ed.), *The Psychobiology of Human Food Selection*, AVI, Westport, pp. 189–203.

Rozin, E. and Rozin, P. (1981) Culinary Themes and Variations. *Natural History*, **90** (2), 6–14.

Senauer, B., Asp, E. and Kinsey, J. (1991) *Food Trends and the Changing Consumer*, Eagan Press, Minnesota.

Schutz, H. (1988) Beyond Preference, Appropriateness as a Measure of Context Acceptance in Food. In D. M. Thomson (ed.), *Food Acceptability*, Elsevier, London.

Schutz, H. (1994) Appropriateness as a Measure of Cognitive-Contextual Aspects of Food Acceptance. In H. J. H. MacFie and D. M. H. Thomson (eds), *Measurement of Food Preferences*, Blackie Academic and Professional, Glasgow.

Shoebridge, N. (1992) Nouveau What? I'll Go Meat and Three Veg. *Sport in Business Review*, May 28, 72–75.

Sokolov, R. (1991) *Why We Eat What We Eat*, Summit Books, New York.

Symons, M. (1991) Eating into Thinking: Explorations in the Sociology of Cuisine. *Ph.D. thesis* (unpublished), Flinders University of South Australia.

Symons, M. (1993) *The Shared Table: Ideas for Australian Cuisine*, Australian Government Press Publication, Canberra.

Taylor Nelson (1993a) *Family Food Panel Management Summary*, Winter/Spring, August, Taylor Nelson House, 44–46 Upper High Street, Epsom, Surrey, KT17 4QS.

Taylor Nelson (1993b) Light Meals: The Growth of Informal Eating Occasions. *Family Food Panel Special Report*, November, Taylor Nelson House, 44–46 Upper High Street, Epsom, Surrey KT17 4QS.

Taylor Nelson (1990) What's for Breakfast, Lunch, Tea-time, Evening Meal, *Family Food Panel Special Report*, February, Taylor Nelson House, 44–46 Upper High Street, Epsom, Surrey KT17 4QS.

Tilston, C. H., Neale, R. J., Gregson, K. and Tyne, C. H. (1993) Food Consumption Patterns: An Elderly Population in Leicester Receiving Meals on Wheels. *British Food Journal*, **95**, 2, 15–20.

Tomlinson, M. and Warde, A. (1993) Social Class and Change in Eating Habits. *British Food Journal*, **95**, 1, 3–10.

Twigg, J. (1983) Vegetarianism and the Meanings of Meat. In A. Murcott (ed.), *The Sociology of Food and Eating*, Gower, Aldershot.

Veblen, T. (1975) *The Theory of the Leisure Class*, 3rd Edition, George Allen and Unwin, London. (First published in 1899, Macmillan, New York.)

Visser, M. (1992) *The Rituals of Dinner: The Origins, Evolution, Eccentricities, and Meaning of Table Manners*, Viking, London.

Warren, G. C. (1958) *The Foods we Eat: A Survey of Meals, Their Content and Chronology by Season, Day of the Week, Region, Class and Age, conducted in Great Britain by the Market Research Division of W. S. Crawford Ltd*, Cassell, London.

Whitehead, T. L. (1984) Sociocultural Dynamics and Food Habits in a Southern Community. In M. Douglas (ed.), *Food in the Social Order: Studies in Food and Festivities in Three American Communities*, Russell Sage Foundation, New York.

Appendix 11.1 A structural framework of categories to describe the main meal

	Primary structures	Primary elements	Primary classes	Subclasses
M	MW	FIRST	1. antipasta	
A	MWV	SECOND	2. fish	2.01 grilled fish
I	MVW	MAIN	3. entrée	2.01 fried fish
N	FMW	SWEET	4. dessert	2.03 poached fish
	FMWV	SAVOURY	5. cheese	
	FMVW			
	FSMW			
	FSMWV			
	F(S)MW(V)			

	Secondary structures	Secondary elements	Secondary classes	Subclasses
M	Various involving	Fa	1.1 soup	1.11 clear
E	secondary elements	Fb	1.2 hors d'oeuvres	1.12 thick
A	Fa...d, Ma.b., Wa.c.	Fc	1.3 fruit	
L		Fd	1.4 fruit juice	
		Ma	3.1 meat dish	
		Mb	3.2 poultry dish	
		Wa	4.1 fruit	
		Wb	4.2 pudding	4.11 steam
		Wc	4.3 ice cream	4.12 milk

Source: Adapted from Douglas (1972)

Douglas proposed this structural framework for main meals and illustrated it with reference to dinner which comprised 'First', 'Second', 'Main', 'sWeet' and 'saVory'* courses which mirror the order of service, at formal eating occasions. Primary 'dinner' structures included MW,

*In *Deciphering a Meal* the 'savory' element is denoted by a 'Z' and not 'V' as used here.

MWV, MVW, FMW, FMWV, FMVW, FSMW, FSMWV, FSMVW (conflated as (F(S)MW(V)). 'Antipasta', 'fish', 'entrée', 'dessert' and 'cheese' are the respective primary classes of these courses, and each one of them can be further subdivided in subcategories; antipasta can be subdivided into 'soup', 'hors d'oeuvre' and 'fruit'; 'entrée' into subcategories of 'meat' and 'poultry'; while 'dessert' reveals secondary elements consisting of 'fruit', 'pudding' or 'ice cream'. Further subclassification reflects cooking practices: for example, only specific types of cooked meat can be used as the meat and poultry entrée. Using this structural framework it is possible to explain the limited combination of courses in the meal, or foods within individual courses.

Appendix 11.2 Most important foods across meal occasions in UK households (% of occasions on which food was served)

Breakfast	Lunch	Tea-time	Evening meal
Bread and rolls 60%	Bread and rolls 55%	Bread and rolls 51%	Fresh potatoes 49%
Breakfast cereals 66%	Yellow fats 35%	Yellow fats 42%	Bread and rolls 32%
Yellow fats 51%	Fresh potatoes 29%	Cakes, tarts, pastries 34%	Prepared (frozen) vegetables 28%
Marmalade 32%	Salad vegetables 22%	Fresh potatoes 24%	Fresh root vegetables 27%
Eggs 13%	Fresh medium fruit 21%	Salad vegetables 23%	Salad vegetables 24%
Fresh medium fruit 10%	Cheese 20%	Prepared (frozen) vegetables 15%	Yellow fats 24%
Bacon 9%	Fresh root vegetables 16%	Cheese 14%	Table sauces and dressings 18%
Jam 8%	Cakes, tarts, pastries 14%	Fresh root vegetables 12%	Cakes, tarts and pastries 17%
Sweet/savoury spreads 5%	Prepared vegetables 14%	Table sauces and dressings 11%	Fresh medium fruit 16%
Salad vegetables 2%	Table sauces and dressings 13%	Fresh medium fruit 10%	Gravy products 16%
Sausages 2%	Gravy products 12%	Eggs 10%	Prepared savoury dishes 16%
Diet bread and crispbread 2%	Prepared savoury dishes 12%	Prepared savoury dishes 10%	Beef 14%
Dried fruit 2%	Fresh leaf vegetables 11%	Cold meats 9%	Fresh leaf vegetables 14%
Cheese 2%	Cold meats 10%	Jam 8%	Ice-cream and mousse 12%
	Eggs 9%	Baked beans 7%	Poultry 12%
	Beef 8%	Ice-cream and mousse 7%	Cheese 12%
	Poultry 7%	Beef 7%	Eggs 9%
	Yoghurts 6%	Poultry 6%	Fresh soft fruit 7%
	Fresh soft fruits 6%	Sweet biscuits 6%	Baked beans 7%
	Canned soups 6%	Fresh leaf vegetables 5%	Sauces, stuffing, etc. 7%
	Baked beans 6%		Prepared potatoes 7%
			Frozen fish and products 7%
			Yoghurts/chilled pots 7%
			Sausages 7%
			Cream/substitutes 6%
			Pork 6%
			Lamb/mutton 6%
			Fresh seed vegetables 6%
			Hot meat pies/puddings 6%
			Cold meats 6%

Source: adapted from Taylor Nelson (1990)

Appendix 11.3   Foods most reliant on meal occasions (% of consumption across meal occasions)

Breakfast	Lunch	Tea-time	Evening meal
Cereals 96%	Slimming and low calorie 87%	Cakes, tarts, pastries 35%	Rice 65%
Marmalade 95%	Canned soups 69%	Canned fish 29%	Packet pasta 63%
Dried fruit 67%	Packet soups 62%	Fish and meat pastes 28%	Fresh continental fruit/vegetables 62%
Sweet and savoury spreads 59%	Game 55%	Fresh meat pies 28%	Sauces, stuffing, savoury additives 52%
Bacon 53%	Fish and meat pastes 49%	Prepared potatoes 28%	Soya products 52%
Jam 53%	Milk (rice) pudding 46%	Canned meat loaf 28%	Packet mixes 52%
Eggs 44%	Hot instant snacks 45%	Cold meats 26%	Prepared potatoes 51%
Yellow fats 44%	Canned pasta and risotto 45%	Canned pasta 24%	Ice-cream and mousse 51%
Bread and rolls 42%	Fresh seed vegetables 44%	Canned pie fillings 24%	Prepared (frozen) vegetables 51%
Diet bread 34%	Cold meats 44%	Frozen fish and products 23%	Prepared sweet dishes 51%
Canned tomatoes 23%	Prepared pâté/continental sausages 43%	Baked beans 23%	Packet desserts/toppings 50%
Fresh medium fruit 20%	Tinned fruit pie fillings 43%	Jelly 23%	Pork 50%
Sausages 20%	Fresh leaf vegetables 42%	Salad vegetables 23%	Beef 50%
Tinned fruit 16%	Steamed/baked puddings 41%	Convenience dishes 23%	Ham/bacon joints 49%
	Custard 41%	Jam 23%	Gravy products 49%
	Gravy products 41%		Poultry 49%
	Cheese 40%		Fresh potatoes 48%
	Soya products 40%		Frozen/fish products 48%
			Fresh root vegetables 48%
			Fresh/smoked fish 47%
			Carry home foods 47%

Source: adapted from Taylor Nelson (1990)

12 The role of eating environments in determining food choice

Rick Bell and Herbert Meiselman

A consumer's choice in the real world is affected most significantly by the context of the situation in which the choice is being made, including the person, the product and the physical environment.

Two decades ago Russell Belk (1975) stated that '...consumer behavior is a function of the interaction between the individual and the situation'. Belk presented a model of situational effects that incorporated the factors of time, task, place, people and the individual's antecedent mood at the time of the behavior. Over the past six years, a number of papers have appeared in food acceptability and food choice literature that have sought to model food acceptance and choice within a contextual framework. In a conference on Food Acceptability, Meiselman *et al.* (1988) noted the importance of context in understanding acceptance of food in real eating environments and as predictors of actual food consumption. More recently, authors have begun to organize the various contextual variables. Rozin and Tuorila (1993) organized context into those factors affecting attitude and choice at the time of eating (simultaneous factors) and both prior and subsequent (temporaneous factors) to the eating experience. Meiselman (1994) organized context into the variables surrounding the food, the individual and the consumption situation.

This chapter will provide an overview of the impact of situation or context on food choice. Choice in eating situations is determined by the combination of the state of the eating/provisioning environment prior to and during the choice experience, and the state of the individual prior to and during the choice experience. The chapter will provide an introduction to the area and offer a conceptual framework to examine the impact of eating environments on food choice. It will then go on to look at what individuals bring to eating environments and the way in which choice can be influenced situationally. The concluding section looks at areas for future development.

Where we eat, the characteristics of the surroundings in which we eat and with whom we eat may be equally if not more important in determining which foods and how much of them we choose to eat than are the actual foods themselves. Our eating environments have changed and are continuing to change. Today, people are less likely to eat meals at home than they would have been a generation ago, and there is growing opportunity to eat away from home. Eating at the place of work, transportation dining (e.g. trains, planes and automobiles), fast food and various types of ethnic restaurants have all increased (Farrell, 1989). As households become smaller, and with convenience foods, shelf-stable foods and individual portions more available, cooking may become an option in households rather than a requirement; and eating will be more situationally varied as a result. As families spend less time preparing food, less time eating food and less time entertaining guests, it follows that they will spend less time in the kitchen.

Most of the early work in understanding factors affecting food selection focused on the food items themselves – the taste, texture, other sensory components and the overall acceptability of the food (Kare and Maller, 1967, 1977). More recently, papers in the food choice arena have suggested that factors other than the food – including social factors (with whom a consumer is eating or purchasing (deCastro and deCastro, 1989)), temporal factors (at what time or the time available for a consumer to eat or purchase (Birch *et al.*, 1984)) and environmental factors (the physical surroundings) – are instrumental in guiding food choices and should be the focus of more research (Meiselman *et al.*, 1988; Meiselman, 1992; Rozin and Tuorila, 1993).

In order to communicate the focus of this research area, it is best to lay down a conceptual framework. When a consumer enters an eating environment in order to select food, both the state of the individual and the state of the environment can affect behavior. These two main influences are shown in the schematic in Figure 12.1, which combines elements of the Fishbein and Ajzen model (1975), as well as other marketing and consumer behavior models (Randall and Sanjur, 1981). The individual and the environment will interact differently depending upon whether the environment is the home, the office, an airplane, a restaurant, etc., and depending upon the degree to which the individual may be susceptible to changes in each of these environments. Individuals will have different expectations for different eating environments, their prior experience will differ for each eating environment, their habitual behaviors may be present in some but not all of the environments, their level of attention paid to their surroundings may differ, and even their mood at the time will differ. All of these factors can affect their attitudes toward the environment, and their subsequent food selections.

12.1
Introduction to the area and conceptual framework

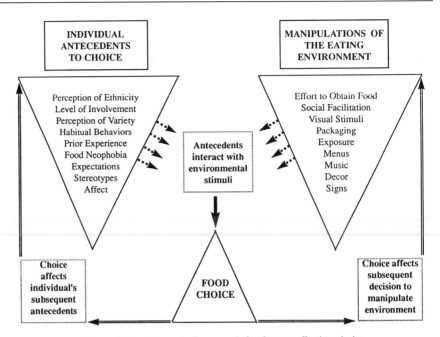

Figure 12.1 Schematic framework for factors affecting choice

**12.2
Antecedents to choice:
what individuals bring
to the eating
environment**

12.2.1   Expectations and stereotypes

One's expectation of a food has a profound effect on its acceptability. Cardello (1994) found that when the taste experience with a product fails to match the expectation for the product, the result is a disconfirmation of a subject's expectation. The direction of the disconfirmation can be either in contrast to the expectation (contrast effect) or in the direction of the expectation (assimilation effect). Most studies of expectations have supported an assimilation model. In this model, if one expects a food to be good, then the rating of the food is better than it might be were there no expectation. This expectation can function to effect food choice as well, and can interact with the eating environment. Before selecting a food, a consumer evaluates the merits of the food, and the source from which the food is being selected. For example, American consumers expect to like the steak they would get on board an airplane less than they would a steak from a restaurant (Cardello *et al.*, 1995). This lower expectation may be accompanied by a lower probability of selecting steak in certain environments owing to the expectation of the quality of that food available in that environment.

In another expectation study, Tuorila *et al.* (1994a) found that fat-free foods were expected to be less liked than their regular-fat counterparts and that this expectation was best predicted by prior experience with

these products. In a related study, Hellemann *et al.* (1994) presented subjects with low-fat meals; half of them were told the meals were low-fat,
while the remaining subjects received no information. Those subjects
whose expectations were disconfirmed were less likely to want to purchase
these items in the future, suggesting further that expectation can mediate
food selection behavior.

There are times when consumers assess food products before they
experience them; the result is a stereotype. These stereotypes are an uncri-
tical or unfounded judgment of a food that can influence ratings for
foods in particular institutions. Cardello *et al.* (1995) cite military food as
an example of this phenomenon. Though few civilians have had experience with military food, there is a strong negative stereotype for military
foods, and the ratings for the perceived quality of these foods is correspondingly lower.

These data suggest that both expectations and stereotypes for foods are
situation-dependent. As consumers enter an eating situation, they bring
with them certain expectations and stereotypes not only for the food
itself, but also for the environment in which the food is being offered, and
this in turn can affect selection (see Raats *et al.*, chapter 10, for a further
discussion on expectation).

12.2.2   *Prior experience, exposure and food neophobia*

Previous experience with food, avoidance of foods, and exposure to unfamiliar or novel foods can all function to mediate choice in eating environments. Studies suggest that familiarity can drive food selection (Pliner,
1982), and that consumers tend to avoid unfamiliar or novel foods, a
phenomenon labeled food neophobia.

Pliner (1982) systematically examined the role of exposure in encouraging trial of novel foods. She suggested that a food not tried is a food
not liked; but trying a food just once – a mere exposure – is likely to
increase the acceptance and retrial of the item. Recently, Pliner and
Hobden (1992) developed a scale to measure the trait of food neophobia,
and Pliner *et al.* (1993) applied this scale and showed that exposure to
novel foods can reduce a consumer's neophobia. Tuorila *et al.* (1994b,c)
further showed that by offering information about a novel product, its
ingredients, and its use context, expected liking and actual liking
increased. In other words, information could function like exposure. This
suggests that information might be the common ingredient in both exposure and neophobia studies. More research in cognitive determinants of
food choice is clearly needed.

Food markets have applied the concept of mere exposure to enhance
trial through the practice of providing shoppers in the store with free
samples. In one study (The Teen Market, 1982) the behavior of con-

sumers who tried free samples of an item was compared to a control group of consumers who had not tried the sample. Thirty-five per cent of those customers exposed to the free sample bought the item, whereas none of the control group consumers bought it. In restaurant settings, more facilities are putting out samples of meals for visual exposure to help increase appetite and encourage selection (*Consumer Reports*, 1992). The effects of this have not yet been systematically studied.

12.2.3   Habitual behaviors

Consumers are creatures of habit, and we tend to do things that are familiar and comfortable. Food habits are no exception. We shop in stores with which we are familiar or with which we have previously been satisfied; for many products we buy the same brand names or we habitually seek the lowest price among brands (Miller and Ginter, 1979). From a research perspective, four issues related to habit arise: (1) How do such habits develop? (2) Can habits be changed? (3) Are habits situation-dependent? (4) Do habits play a role in affecting food provisioning in various eating entertainments? In the recent psychological literature, Ronis *et al.* (1989) suggest that although attitudes may be important in mediating behavior change, habits may serve as a barrier to that change. Tuorila and Pangborn (1988) found a strong relationship between habit (measured by questions about behaviors performed habitually) and the frequency of consumption of sweet, salty and fatty foods, as well as coffee.

Bell and Mattes (unpublished) examined the habitual use of sugar located at the table of customers in a dining facility. They varied the amount of sugar (20% less or 40% more than usual volume) in the packets of sugar available at the table and tracked the use of sugar in coffee by regular diners for twelve weeks (which included the baseline and the two experimental conditions for each subject). Results suggest that regardless of the volume of sugar in the packet, packet use did not change. Subjects habitually used the same number of packets even after four weeks of a volume change.

12.2.4   Involvement

How important an item is to a consumer, i.e. the person's level of involvement, can affect behavior. Low-involvement products can be described as those for which very little time or thought is taken in making a decision about its choice. An example might be toothpaste, for which consumers either always buy a particular brand or make decisions based on lowest price. But the fear of making a bad choice in this scenario is minimal, because a mistake does not mean a great loss. Higher involvement products are those for which we consider several characteristics and weigh alternatives carefully before making a decision. For example, there is great fear of making a bad choice of an automobile or a bad choice of whom one votes for in an election for a high-ranking government office, because the cost of making a mistake is high and possibly long-lasting. The advertising and marketing literature has explored the involvement construct (Engel and Blackwell, 1982; Zaichkowsky, 1985), but most of the work focuses on differences among product categories, and food is grouped together into one category.

A consumer's level of involvement with particular food items or food categories could also affect choice and could mediate how easily the eating environment affects an individual's choice of foods. Unfortunately, there has been little published research on involvement and particular food categories. In a pilot laboratory study, Bell (unpublished) created an involvement scale and divided subjects into high and low involvement groups for particular products. The data suggest that subjects with higher involvement with a particular product category can distinguish sensory differences among products in that category better than those subjects who have lower involvement with that product category. Therefore, consumers might be more able to perceive environmental differences when they enter a dining facility if it contains foods for which they have a high involvement.

12.2.5 Perception, ethnicity and variety

If a variable in the eating environment were manipulated in an attempt to motivate a behavior change, that variable would have to be perceived by the consumer before it could have any effect on the consumer's behavior. The issue just described previously about involvement affecting perception suggests a co-mediating effect of these two antecedent variables. There are other examples of how perception can affect behavior. For example, a sign indicating a special in a cafeteria line needs to be seen in order to have an effect. What makes certain consumers attend to the sign and others not? It is likely a combination of several of the factors discussed previously: expectation, stereotypes, prior experience, habitual behavior and level of involvement.

Another variable that can interact with the environment to affect food choices is culture or ethnicity. There are several studies related to cultural differences in food preferences (Axelson, 1986; Pangborn, 1975; Rozin, 1976) that frequently relate to familiarity discussed earlier. There is also a substantial literature related to situational ethnicity (Cohen, 1978; Stayman and Deshpande, 1989). The definition of ethnicity for consumer behaviorists has been an evolving area. At first, researchers approached ethnicity from the ethnic perspective, wherein a subject's last name or the language spoken in the home was the determinant of the ethnicity of a subject. Cohen (1978) changed the approach by first having the subject provide a self-designated ethnicity followed by a valuation of the strength of identification with that group, or a felt ethnicity.

The term 'situational ethnicity' was first mentioned in the literature by anthropologist Joan Paden who noted in 1967 that 'particular contexts may determine which of a person's communal identities or loyalties are appropriate at a point in time' (cited in Okamura, 1981: 452). Paden claimed that ethnic identity depends upon the contextual or social situation and the individual's perception of the situation. This suggests that people who were born in India but who moved to London and have lived there for most of their lives may still consider themselves to be Indian; but their actions are more situationally dependent, and their consumer behavior more closely resembles that of other Londoners than that of consumers still in India.

Bell and Paniesin (1992) and Meiselman and Bell (1992) applied this concept to the question 'What makes a food ethnic?' in studies conducted in sensory laboratories. Bell and Paniesin (1992) varied the sauce, spice blend and name of ethnic items and found that the sauce was the major determinant of perceived ethnicity. Meiselman and Bell (1992) then varied larger components of recipes (sauce, meat and cheese) to focus on the perception of British and Italian foods. They found that adding recipe components to pasta could increase the perception of the food as being British, but adding Italian names increased perceived Italian ethnicity.

Another consumer perception that can function to interact with the environment to affect food choice is the perceived amount of variety available. There is little doubt that consumers seek variety in their food choices (Rozin and Markwith, 1991). The effect of variety on consumption has been documented (Rolls et al., 1982), and a scale to measure variety-seeking behavior has been constructed (Van Trijp and Steenkamp, 1989), but little is known about the factors affecting perceived variety. Bell et al. (1994a) considered perceived variety in the context of meals and diets, and, as will be presented later, also in the context of the eating environment (Bell, 1994; Bell and Finer, 1994).

12.2.6   Affect

Degree of liking for a food has been one of the most studied factors in motivating food choice. Hundreds of studies have used the nine-point hedonic scale (Peryam and Pilgrim, 1957) to measure a consumer's degree of liking for an item. Researchers have interpreted the mean hedonic rating for the item as a measure of the product's potential for market success. However, it is still not clear what hedonic ratings mean and what predictive value they have. Several recent reviews of hedonic ratings (Bell, 1992, 1993; Meiselman, 1992; Meiselman *et al.*, 1988) suggest that these ratings are poor predictors of consumers' behavior toward the product when available in the eating environment.

Having provided a description of the antecedent variables a consumer brings to the provisioning environment, we now turn our discussion to actual studies of environmental effects on choice.

12.3
Altering choice via the environment: situationally manipulating choice

In their model of food choice, Randall and Sanjur (1981) define the environment as comprising such factors as the season, employment of the individual, size of household and mobility of the consumer. Belk (1974, 1975) and Meiselman (1994) define the environment as the physical and social surroundings of the actual eating situation; and Rozin and Tuorila (1993) also include these elements in defining the context of the eating experience. In this section, eating environments will incorporate this latter definition as well. An examination of food choices begins by looking at how choice can be altered via the effects of cognitive information about specific foods (menus, packages and signs). After this, other aspects of the eating situation will be explored: appropriateness of food for a particular situation, the perception of the physical setting in which the food is offered, the location of food, the social factors involved in provisioning, the effect of exposure, visual stimuli and music and, finally, the perceived variety of foods available.

12.3.1 Cognitive information: menus, packages, signs

Before consumers make food choices, especially in a retail food market or catering facility, they are customarily exposed to cognitive variables meant to motivate selection. In a market there are signs about sales and particular items at reduced price, and there is information available on the food packages themselves; in restaurants there are signs about particular foods or specials, and there are the menus from which to choose items. Even within the home, food packages have labels, and cookbooks and newspapers and magazines have food advertisements and recipes.

Several studies have examined the effect of cognitive information on food-related behavior. Bell and Taub (unpublished) constructed a menu of novel and familiar foods from different cultures and experimentally varied the absence or presence of product names, product descriptions and country of origin information. They asked subjects to rate their likelihood of choosing each of the items and found that the country of origin information was the key variable driving behavioral intent. The explanation for this effect could be that subjects who responded to items for which there was an absence of information may have been more uncertain as to the item's identity and taste characteristics. Tuorila *et al.* (1994b,c) suggest that providing certain information about novel foods, especially to subjects who are more food neophobic, may function to reduce a consumer's uncertainty about a product's identity and increase the probability of its selection.

Pelchat and colleagues (Bell and Pelchat, 1994; Pelchat and Pliner, 1995) utilized product information in cafeteria settings to encourage purchase of novel foods over available familiar alternatives. They found that providing nutrition information alone or information about the positive taste characteristics of the item along did not encourage trial. However, Bell and Pelchat (1994) found that the combination of both types of information did increase trial of novel foods. They hypothesize that it may not be the type of information provided, but the amount of information provided that produced the effect. In fact, Edwards, as cited in Ogilvy (1963), stated that '... the more facts you tell, the more you sell ... an advertisement's chance for success invariably increases as the number of pertinent merchandise facts included in the advertisement increase.' This may be true for point-of-purchase display material as well, with the underlying reason being that more information not only reduces the uncertainty about a product, but also comprises more visual space that may attract a consumer's attention.

These studies pertained primarily to encouraging selection of novel foods. Other studies have also used messages about nutrition in order to motivate choices of more healthful foods both in-store (Jeffery *et al.*, 1982; Muller, 1984; Pennington *et al.*, 1988) and in cafeteria settings (Davis-Chervin *et al.*, 1985; Zifferblatt *et al.*, 1980).

For in-store studies, the data suggest that although knowledge about nutrition may increase because of point of purchase nutrition information, purchase patterns do not necessarily change. One study (Muller, 1984) showed that signage about the nutritive composition of certain foods placed next to the items themselves resulted in short-term shifts toward the selection of some of the more nutritive products. However, this effect was not consistent across products.

In dining facility studies, Davis-Chervin *et al.* (1985) showed positive effects on food choice of educating consumers on the tangible benefits of 'healthful' food choices, but the intervention was done at college dormitories, not in the dining facility itself. Zifferblatt *et al.* (1980) used a card game in the cafeteria line at a US National Institute of Health employee cafeteria to encourage lower-calorie food choices. During eight weeks of the trial, skim milk purchases increased, while dessert and bread choices, as well as the average number of calories purchased per day, declined. There was also evidence that this effect carried over during a ten-week follow-up period.

In an unpublished study, Bell and Kramer (unpublished) placed direct health messages (signs about low-fat foods and fruits and vegetables) and indirect health messages (presence of exercise equipment) at the start of a cafeteria line for one week in an attempt to encourage healthful food choices. Both approaches increased choice of fruits, vegetables and salads, and decreased the choice of meat items. They hypothesize that the presence of the exercise equipment may have produced a sense of guilt in consumers who do not regularly exercise, and their food choices changed accordingly.

12.3.2   Appropriateness of the eating environment

A basic tenet of situational effects on food choice is that foods are chosen in part because they are appropriate for a particular situation. Schutz (1988), who proposed the concept of appropriateness, was initially looking for a measure of consumer attitude that would better predict behavior than the often-used nine-point hedonic scale. Schutz' appropriateness measure asks consumers to rate how appropriate a product is for a given situation. These data are less variable than hedonic data, and appropriateness ratings appear to have a better relationship to predicting behavior. This makes sense intuitively, since although not everyone likes mashed potatoes equally, we all are more likely to agree that it is not an appropriate food to eat while driving in a car. And if a subject does not like mashed potatoes at all, he/she is unlikely to find them an appropriate food for any situation.

The perception of what is appropriate for a given situation could be one of the key factors underlying the effects eating environments can have on food choice. In a restaurant that has created a mood of elegance and style, it is probably less appropriate to order a hamburger than it would be in a restaurant that created the mood of a pub. It follows, then, that by changing the perception of the eating environment, we change what is deemed to be appropriate for that situation, and we ultimately change the behavior of the consumer.

No appropriateness studies have yet been reported from examining real food choices in actual eating establishments using a within-subjects design, but Marshall and colleagues (Marshall, 1993; Marshall and Bell, 1995) have looked at other facets of appropriateness via surveys. They looked at the effects of meal occasions, eating environments, and the characteristics of those meal occasions and environments on food choices, and compared these variables to hedonic ratings for the foods. Marshall and Bell had subjects rate fifty-one foods on a nine-point hedonic scale and then rate the characteristics of eleven different eating situations using Belk's (1974, 1975) situational elements, such as the time available for eating, the reasons for the eating occasion, their liking for the environment, the number of people with whom it is appropriate to eat in each environment and the appropriateness of the environment for particular meal occasions. The results suggest that the characteristics of situations are perceived to be distinctly different from one another and that meal occasion has a strong effect on food choice. Although less dramatic, the data also indicate that food choice is more situation-dependent than it is hedonically dependent – in other words, acceptance had little to do with choice. This supports Schutz' view that appropriateness ratings may be more predictive of behavior than are hedonic ratings.

12.3.3   Perception of ethnicity

Earlier we defined the term 'situational ethnicity' as being a consumer's behavior as a function of the environment they live in rather than the culture from which they originated. Meiselman and Bell, having previously altered the perceived ethnicity of foods in a laboratory (1992), hypothesized that this perceived ethnicity could be manipulated in the eating environment as well. They hypothesized that attributes of the environment could affect ethnic identification and food choice, as well as attributes of the food. This was demonstrated in a British restaurant (Bell *et al.*, 1994b), actually a training restaurant of a university catering department. Identical foods (both from British and Italian recipes) were offered on four days. On the first two days the restaurant was decorated as usual; on the next two days it was decorated with an Italian theme: red and white checkered table cloths, Chianti bottles on the table, Italian flags hung from the ceilings, and menus printed in Italian with English explanations. Results showed that the Italian theme increased pasta selection and decreased selection of fish and veal. In addition, more consumers ordered dessert on Italian theme days. Customer satisfaction surveys were used at the end of the meal to assess customers' perceptions of the ethnic origin of their meal and their liking for the food. In the Italian theme condition, more than twice as many customers perceived their meals as being Italian than did customers on non-Italian theme days.

Interestingly, hedonic ratings for all foods were the same – all fairly elevated. From a psychological perspective, this study shows an interesting effect of the environment on choice and perceived ethnicity of food; from an economic perspective it allows a restaurant to design more profitable menus with situational changes rather than food changes exclusively.

12.3.4   Location of food

Meiselman *et al.* (1994) took a different approach in modifying the environment. They did two experiments wherein they varied the effort needed to obtain either candy or potato chips ('crisps' in the UK) by moving them more than one hundred feet away from the cash register, where they had previously been placed. They found that with greater effort required to obtain these items, less were chosen. In the candy study, subjects substituted items from the dessert, fruit and accessory groups. In the crisps study, subjects substituted starch foods. The acceptability of the substituted items did not vary in any systematic way.

In another study of effort, Engell *et al.* (1994) varied the effort required to obtain water by placing water either at the table where subjects were eating or in another room. They found that less water was taken when the effort was greater. This study suggests that the provisioning of water and other meal items usually provided on the table (salt, sugar, etc.) could be greatly affected by their location in an eating environment.

Food retail stores have used location to vary food choice in their facilities through the use of end-of-aisle displays, providing certain products with more shelf space, and shelf position. The goal of the retailer is to encourage unplanned purchases – purchases not intended by consumers when they entered the market. Studies show that unplanned purchases occur more frequently with products found on the shelf at eye level, or those located in high-traffic areas such as the dairy and meat sections or the checkout counter (Assael, 1984). The data also suggest that these in-store manipulations are perhaps the most influential factor in an unplanned purchase decision, even more important when you consider that in a supermarket over fifty per cent of the purchases are considered by consumers to be unplanned (Kollat and Willet, 1967; *Consumer Reports*, 1993).

12.3.5 Social facilitation

Most of the literature on social facilitation has examined effects on intake (Bellisle, 1979; deCastro, 1987; Clayton, 1978), but several of these studies have direct implications for food choices. deCastro and deCastro (1989) had subjects keep dietary records of food choices, intake and a number of other variables, including the number of people present when eating. The number of people present was positively correlated with meal size even when meals eaten alone were excluded from the data analysis. They suggest that social factors are independently associated with an increase in meal size. Though this effect has only been found through dietary records, and as yet has not been observed experimentally, the phenomenon implies that the number of people with whom you eat has significant effects on how much you consume, and, hence, may have an effect on what you choose.

Pliner and Hobden (1994) have shown the effect of social modeling on novel food choice behavior. Subjects watched a video in which other consumers selected either a novel item or a familiar one when presented with a choice of the pair. Subjects were then presented with the same pairs from which to choose. Those exposed to the model were more likely to select the food that the model had selected, whether it was the novel or the familiar food.

In a similar design, Engell *et al.* (1994) instructed a person, called a 'confederate', to sit at the table with a subject and 'model' drinking behavior. The confederate would either restrict fluid intake or drink an excessive amount of water. Results suggest that the actions of the model clearly influenced the subjects' fluid intake. Even though this was a study of intake and not one of choice, it suggests that social facilitation can act as a signal to induce consumers to behave with another's actions in mind.

12.3.6   Visual stimuli and music

Other aspects of the physical surroundings of an eating environment can play a role in mediating food choice behavior, including the decor and lighting, and auditory stimuli, primarily music.

There is limited literature available on the effects of the decor and lighting that comprise visual stimulation in the surroundings of the retail eating environment (Singson, 1975; Stephenson, 1969). McDonald's and other fast-food chains tend to contain bright, bold primary colors in attention-grabbing combination with the intent to attract the attention of customers and draw them into the excitement of the environment. Once the consumer is there, the combination of the colors (perhaps in conjunction with other noise and music) acts to fatigue the senses, to increase anxiety and to encourage the customer to select his/her foods quickly, to eat quickly and then to leave promptly. This is done in order to keep the 'fast' in the fast-food operation. On the other hand, higher-priced restaurants want their customers to have a relaxing dining experience. The colors in these facilities tend to be subdued greens and subtle reds and browns, and the light is usually dimmer (Love, 1986).

The literature available on the use of music for changing behavior has focused primarily on music's effect on mood (for a review of this area, see Bruner, 1990). Marketers have embraced this relationship and, as a result, the advertising community has used music extensively to effect product image. But relatively little data is available on the effect of music on food choices. Milliman (1982, 1986) conducted studies wherein music tempo was manipulated in both a supermarket and in a restaurant. Slower music was accompanied by a slower flow of traffic in the supermarket and a higher sales volume (Milliman, 1982); and slower music was accompanied by a slower rate of eating and higher bar bills for customers in the restaurant (Milliman, 1986). Both of these studies suggest that the presence of slower tempo music leads consumers to spend more time in the provisioning environment, and the increase in time is accompanied by an increase in the number of items purchased. It follows that fast-food restaurants, wanting to attract customers and then move them out quickly, would likely play music with a faster tempo and probably at a louder volume; music at higher-priced restaurants, if at all present, is usually at a slower tempo, and the noise level is softened, often with the help of carpeting or sound-absorbing building materials. This area of visual stimulation and music has not received much attention in the food choice literature, and there is cause to believe that it deserves more. The type of music, volume of music and repetition of music in a provisioning environment have not been examined in relation to choice and could have significant effects.

12.3.7 Changing the perception of variety

Restaurant facilities and food markets want to provide consumers with ample variety. The usual manner in which to accomplish this objective is by ordering different items for the shelves, or by offering different items on a menu. Bell, having previously altered perception of ethnicity in eating environments without changing the food, and having previously examined what comprises consumers' definition of variety in their diets, hypothesized that the perception of variety could be manipulated in eating environments without changing the food.

First, Bell and Finer (1994) examined the amount of variety perceived to be available in different ethnic menus. They presented small, medium and large size menus from eight ethnic cuisines and matched the amount and type of dishes offered on each. They asked subjects to rate the variety available in each menu. Italian and Chinese menus (the two types of cuisine with which subjects were most familiar) were perceived as being lower in variety than the other cuisines. This suggests that the more familiar a consumer is with a cuisine, the more variety is needed in order to effect a change in perceived variety than would be needed with a lesser known cuisine. It is possible that a consumer entering one of these familiar ethnic restaurants expects there to be a certain number of choices available, and if this expectation is not met, his/her attitude could be affected and this, in turn, could affect behavior.

Bell (1994) then altered the eating environment to effect changes in perceived variety without changing the foods themselves. He introduced into a typically American lunch restaurant the same Italian theme variables described earlier (Bell *et al.*, 1994b), though the food items were different from those used in the earlier study. The Italian theme was introduced once a week for five weeks, and perceived variety and satisfaction ratings were collected at the end of meals. Results indicated that perceived variety available at the restaurant was higher than the baseline condition for all five weeks of the study, and it did not return to baseline levels for the next eight weeks following the experimental manipulation. Six months later, perceived variety and satisfaction did return to baseline levels. In the study, perceived variety was highly correlated with satisfaction, suggesting that consumer perceptions of both variety and satisfaction can be increased without altering available foods, but merely by manipulating the eating environment.

12.4
Applications, future directions and conclusions: where we can go from here

It is clear that changes in eating environments can have profound effects on behavior. This concluding section summarizes ways in which these findings can be applied and suggests future research in this area.

These findings can be useful for those involved in changing food choices in a more 'healthful' direction. Without getting involved in the ongoing discussion of what is healthy, it should be clear from this material that the eating environment is as important, or even more important, than food products or the individual eaters. Most attention in the past concerning healthful eating has been directed at altering food products (e.g. reduced fat and salt) and altering human attitudes and behavior (e.g. dieting) (see Anderson *et al.*, chapter 5). Altering the environment by manipulating physical and social dimensions such as effort, and the presence of companions, are inexpensive and simple alternatives which need to be better understood for their healthy potential.

Further application of eating environment research lies in the product development area. This research strongly suggests that full understanding of a product's potential success is not possible without an understanding of its position in the eating environment or occasion eating as Marshall suggests in chapter 11. Factors such as consumers' expectations, impact on perceived variety, requirements for increased or decreased effort, impact on social occasions and many others all play into a product's success. Yet current product development techniques focus more on the product and on the consumer than on the environment. We need to modify market research paradigms to include eating environment concerns.

Perhaps the easiest application of eating environment research comes in two different food environments: the retail outlet and the food service or catering outlet. In both of these one can see more closely the interaction of the food, the consumer and the environment. While there has been research in retail outlets, there has been relatively little in food service outlets. This remains an area of potential growth. Direct research on how eating environments affecting eating is indeed possible, as has been argued elsewhere (Meiselman, 1992). The results of such research can present clear guidelines for operation of food service outlets and retail outlets with more predictable consumer outcomes.

Future studies on the effects of the eating environment will require a better balance of laboratory and non-laboratory research. Laboratory studies will continue to provide controlled environments for detailed analyses of one to several variables at a time. Field studies provide natural, uncontrolled environments to observe and document the interaction of many complex variables operating in the food service situation. A discussion of field and laboratory approaches can be found in *Appetite*, **19** (1), August 1992.

The literature reviewed in this chapter on the studies of eating environment effects on behavior come from several different areas of research interest, including marketing and consumer behavior, psychology, sociology, nutrition, public health, retail and even the field of music. This shows the interdisciplinary nature of the research and the wide-ranging application of this research. Given the broad potential application of knowledge about eating environments' effects, it is surprising that we do not know more about these effects. The area is rich with unexamined possibilities, and by conducting research in both laboratory and real-world eating environments, we will uncover new factors underlying the consumers' choices of food in eating environments. The eating environment has emerged as a topic of research of equal interest to those of the food and the consumer.

References

Assael, H. (1984) *Consumer Behavior and Marketing Action*, 2nd edition, Kent, Boston.
Axelson, M. L. (1986) The Impact of Culture on Food-Related Behaviour. *Annual Review of Nutrition*, **6**, 345–363.
Belk, R. W. (1974) An exploratory assessment of situational effects in buyer behavior. *Journal of Marketing Research*, **11**, 156–163.
Belk, R. W. (1975) The objective situation as a determinant of consumer behavior. *Advances in Consumer Research*, **2**, 427–437.
Bell, R. (1992) The relationship of hedonics to behavior – the limitations and future of acceptance testing. *Activities Report of the R & D Associates*, **44** (2), 74–83.
Bell, R. (1993) Some unresolved issues of control in consumer tests: the effects of expected monetary reward and hunger. *Journal of Sensory Studies*, **8** (4), 329–340.
Bell, R. (1994) Short-term and longitudinal effects of ethnic theme manipulations on perceived variety in a restaurant setting. Submitted.
Bell, R. Effects of consumer involvement on the ability to perceive sensory differences between products in a single category. Unpublished data.
Bell, R. and Finer, G. (1994) The role of familiarity, gender and experience in affecting perceived variety for ethnic restaurant menus of different sizes. Submitted.
Bell, R. and Kramer, F. M. The effects of direct and indirect point-of-purchase health messages on healthful food choices in a cafeteria setting. Unpublished data.
Bell, R. and Mattes, R. M. The roles of habit and sensory adjustment in sugar usage. Unpublished data.
Bell, R. and Taub, S. Effects of name, food description and country of origin on cross-cultural food consumption intent. Unpublished data.
Bell, R. and Paniesin, R. (1992) The influence of sauce, spice, and name on the perceived ethnic origin of selected culture-specific foods. In L. S. Wu and A. D. Gelinas (eds), *Product Testing with Consumers for Research Guidance: Special Consumer Groups, 2nd Vol., ASTM STP 1155*, American Society For Testing and Materials, Philadelphia. pp. 22–36.
Bell, R. and Pelchat, M. L. (1994) Strategies for motivating the selection of novel foods: immediate and carryover effects of information in a dining facility. In I. Taub and R. Bell (eds), *Proceedings of the Food Preservation 2000 Conference*, Science and Technology Corporation, Alexandria, VA. In press.
Bell, R., Meiselman, A. and Bertolami, J. (1994a) Relationship between cultural and demographic variables and neophobia, variety seeking and perceived dietary variety: U.S. vs. Ireland subjects. Submitted.
Bell, R., Meiselman, H. L., Pierson, B. J. and Reeve, W. (1994b) The effect of adding an

Italian theme to a restaurant on the perceived ethnicity, acceptability, and selection of foods. *Appetite*, **22** (1), 11–24.

Bellisle, F. (1979) Human feeding behavior. *Neuroscience Biobehavioral Review*, **3**, 163–169.

Birch, L. L., Billman, J. and Richards, S. S. (1984) Time of day influences food acceptability. *Appetite*, **5**, 109–116.

Bruner, G. C. (1990) Music, mood, and marketing. *Journal of Marketing*, October, 94–104.

Cardello, A. V. (1994) Consumer expectations and their role in food acceptance. In H. J. H. MacFie and D. M. H. Thomson (eds), *Measurement of Food Preferences*, Blackie Academic, Glasgow. pp. 253–295.

Cardello, A. V., Bell, R. and Kramer, F. M. (1995) Attitudes of consumers toward institutional food. *Food Quality and Preference*. In press.

deCastro, J. M. (1987) Macronutrient relationships with meal patterns and mood in the spontaneous feeding patterns of humans. *Physiology and Behavior*, **39**, 561–569.

deCastro, J. M. and deCastro, E. S. (1989) Spontaneous meal patterns of humans: influence of the presence of other people. *American Journal of Clinical Nutrition*, **50**, 237–247.

Clayton, D. A. (1978) Socially facilitated behavior. *Quarterly Review of Biology*, **53**, 373–392.

Cohen, R. (1978) Ethnicity: problems and focus in anthropology. In B. J. Siegal *et al.*, *Annual Review of Anthropology*, Volume 7, Annual Reviews, Palo Alto CA. 379–403.

Consumer Reports (1992) June, 356–362.

Consumer Reports (1993) September, 560–569.

Davis-Chervin, D., Rogers, T. and Clark, M. (1985) Influencing food selection with point-of-purchase nutrition information. *Journal of Nutrition Education*, **17** (1), 18–22.

Engel, J. F. and Blackwell, R. D. (1982) *Consumer Behavior*, Dryden Press, New York.

Engell, D. E., Kramer, F. M., Malafi, T., Lesher, L. and Solomon, M. (1994) Economic and social effects on drinking in humans. *Appetite*, in press.

Farrell, F. (1989) Ethnic food sales show rapid growth. *Retail News*, October, 34–35.

Fishbein, M. and Ajzen, I. (1975) *Belief, Attitude, Intention and Behavior. An Introduction to Theory and Research*, Addison-Wesley, Reading MA.

Hellemann, V., Aaron, J. I., Evans, R. E. and Mela, D. (1994) Effect of expectations on the acceptance of a low-fat meal. In I. Taub and R. Bell (eds), *Proceedings of the Food Preservation 2000 Conference*, Science and Technology Corporation, Alexandria VA, in press.

Jeffery, R. W., Pirie, P. L., Rosenthat, B. S., Gerber, W. M. and Murray, D. M. (1982) Nutrition education in supermarkets: an unsuccessful attempt to influence knowledge and product sales. *Journal of Behavioral Medicine*, **5**, 189–200.

Kare, M. R. and Maller, O. (eds) (1967) *The Chemical Senses and Nutrition*, Johns Hopkins Press, Baltimore.

Kare, M. R. and Maller, O. (eds) (1977) *Chemical Senses and Nutrition*, Academic Press, New York.

Kollat, D. T. and Willet, R. P. (1967) Customer impulse purchasing behavior. *Journal of Marketing Research*, **4**, 21–31.

Love, J. F. (1986) *McDonald's: Behind the Arches*, Bantam, New York.

Marshall, D. W. (1993) Appropriate meal occasions: understanding conventions and exploring situational influences on food choice. *The International Review of Retail, Distribution and Consumer Research*, **3** (3), 279–301.

Marshall, D. W. and Bell, R. (1995) The relative influence of meal occasion and situations on food acceptability and choice. In T. Worsley (ed), *Food Choice*, (forthcoming) Adelaide: Food Choice Conference, Adelaide.

Meiselman, H. L. (1992) Methodology and theory in human research. *Appetite*, **19**, 49–55.

Meiselman, H. L. (1994) Contextual influences on food acceptance: the role of the food, the situation and the individual. In Flair-Sens (ed.), *Proceedings of Food Quality – Consumer Relevance Meeting*, Norway, in press.

Meiselman, H. L. and Bell, R. (1992) The effects of name and recipe on the perceived ethnicity and acceptability of selected Italian foods by British subjects. *Food Quality and Preference*, **3**, 209–214.

Meiselman, H. L., Hirsch, E. S. and Popper, R. D. (1988) Sensory, hedonic and situational factors in food acceptance and consumption. In D. M. H. Thomson (ed.), *Food Acceptability*, Elsevier, London. pp. 77–87.

Meiselman, H. L., Staddon, S. L., Hedderley, D., Pierson, B. J. and Symonds, C. R. (1994)

Effect of effort on meal selection and meal acceptability in a student cafeteria. *Appetite*, **23**, 43–55.

Miller, K. E. and Ginter, J. L. (1979) An investigation of situation variation in brand choice behavior and attitude. *Journal of Marketing Research*, **16**, 111–123.

Milliman, R. E. (1982) Using background music to affect the behavior of supermarket shoppers. *Journal of Marketing*, **46**, 86–91.

Milliman, R. E. (1986) The influence of background music on the behavior of restaurant patrons. *Journal of Consumer Research*, **13**, 286–289.

Muller, T. E. (1984) The use of nutritive composition data at the point of purchase. *Journal of Nutrition Education*, **16** (3), 137–141.

Ogilvy, D. (1963) *Confessions of An Advertising Man*, Atheneum, New York.

Okamura, J. Y. (1981) Situational ethnicity. *Ethnic and Racial Studies*, **4**, 452–465.

Pangborn, R. M. (1975) Cross-Cultural Aspects of Flavor Preference. *Food Technology*, **29** (6), 34–36.

Pelchat, M. L. and Pliner, P. (1995) 'Try it. You'll like it.' Effects of information on willingness to try novel foods. Submitted. *Appetite*, **24**, 153–166.

Pennington, J. A. T., Wisniowski, L. A. and Logan, G. B. (1988) In-store nutrition information programs. *Journal of Nutrition Education*, **20** (1), 5–10.

Peryam, D. R. and Pilgrim, F. J. (1957) Hedonic scale method of measuring food preferences. *Food Technology*, **11**, 9–14.

Pliner, P. (1982) The effects of mere exposure on liking for edible substances. *Appetite*, **3**, 283–290.

Pliner, P. and Hobden, K. (1992) Development of a scale to measure the trait of food neophobia in humans. *Appetite*, **19**, 105–120.

Pliner, P. and Hobden, K. (1994) Effect of exposure to models on food neophobia in humans. In I. Taub and R. Bell (eds), *Proceedings of the Food Preservation 2000 Conference*, Science and Technology Corporation, Alexandria VA, in press.

Pliner, P., Pelchat, M. L. and Grabski, M. (1993) Reduction of neophobia in humans by exposure to novel foods. *Appetite*, **20**, 111–123.

Randall, E. and Sanjur, D. (1981) Food preferences – their conceptualization and relationship to consumption. *Ecology of Food and Nutrition*, **11**, 151–161.

Rolls, B. J., Rolls, E. T. and Rowe, E. A. (1982) The influence of variety on human food selection and intake. In L. M. Barker (ed.), *The Psychobiology of Human Food Selection*, AVI, Westport. pp. 101–122.

Ronis, D. L., Yates, J. F. and Kirscht, J. P. (1989) Attitudes, decision, and habits as determinants of repeated behavior. In A. Pratkanis, S. Breckler and A. Greenwald (eds), *Attitude Structure and Function*, Erlbaum, Hillsdale NJ. 213–239.

Rozin, P. (1976) Psychobiological and cultural determinants of food choice. In T. Silverstone (ed.), *Dahlem Workshop on Appetite and Food Intake*, Dahlem Konferenzen, Berlin.

Rozin, P. (1982) Social determinants in human food selection. In L. M. Barker (ed.), *The Psychology of Human Food Selection*, AVI, Westport. pp. 225–245.

Rozin, P. and Markwith, M. (1991) Cross-domain variety seeking in human food choice. *Appetite*, **16**, 57–59.

Rozin, P. and Tuorila, H. (1993) Simultaneous and temporal contextual influences on food acceptance. *Food Quality and Preference*, **4**, 11–20.

Schutz, H. G. (1988) Beyond preference: appropriateness as a measure of contextual acceptance of food. In D. M. H. Thomson (ed.), *Food Acceptability*, Elsevier, London. pp. 115–134.

Singson, R. L. (1975) Multidimensional scaling analysis of store image and shopping behavior. *Journal of Retailing*, **51**, 38–52.

Stayman, D. M. and Deshpande, R. (1989) Situational ethnicity and consumer behavior. *Journal of Consumer Research*, **16**, 361–371.

Stephenson, R. (1969) Identifying determinants of retail patronage. *Journal of Marketing*, **33**, 57–61.

The Teen Market (1982) *Product Marketing*, 1–26.

Tuorila, H. and Pangborn, R. M. (1988) Behavioural models in the prediction of consumption of selected sweet, salty and fatty foods. In D. M. H. Thomson (ed.), *Food Acceptability*, Elsevier Publishers, London. pp. 267–279.

Tuorila, H., Cardello, A. V. and Lesher, L. L. (1994a) Antecedents and consequences of expectations related to fat-free and regular-fat foods. *Appetite*, **23**, 247–263.

Tuorila, H., Meiselman, H., Bell, R., Cardello, A. and Johnson, W. (1994b) The role of sensory and cognitive information in the enhancement of certainty and liking for novel and familiar foods. *Appetite*, **23** (3), 231–246.

Tuorila, H., Meiselman, H., Bell, R., Cardello, A. and Johnson, W. (1994c) Effect of label and information on perceived uncertainty and hedonic ratings of novel foods by subjects with varying neophobia. In I. Taub and R. Bell (eds), *Proceedings of the Food Preservation 2000 Conference*, Science and Technology Corporation, Alexandria VA, in press.

Van Trijp, H. C. M. and Steenkamp, J-B. E. M. (1989) Consumers' variety seeking tendency with respect to foods: measurement and managerial implications. *European Review of Agricultural Economics*, **19** (2), 181–195.

Zaichkowsky, J. L. (1985) Measuring the involvement construct. *Journal of Consumer Research*, **12**, 341–352.

Zifferblatt, S. M., Wilbut, C. S. and Pinsky, J. L. (1980) Changing cafeteria eating habits. *Journal of the American Dietetics Association*, **76**, 15–20.

Disposal

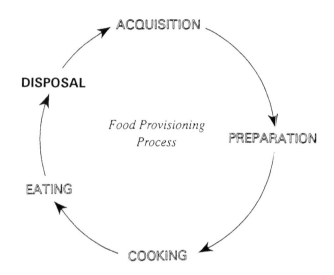

ACQUISITION

DISPOSAL

*Food Provisioning
Process*

PREPARATION

EATING

COOKING

The disposal of the meal 13
Rolland Munro

Remember those turkey sandwiches last Christmas or Thanksgiving? Day after day, turkey chiplets, turkey pie and turkey curry. How you never wanted to see another turkey again? ... Or think of opening the fridge to see what could be for supper. And hurriedly closing the door when you smell yesterday's fish.

Or think of thin passageways to a blocked sink, filled with old crusted milk bottles. The kitchen piled high with the debris of old fish 'n' chip papers, glistening cellophane wrappers from cut spam slices, broken styrofoam packets commemorating baked potatoes, red-stained foil mementoes of Indian take-aways. Beer bottles of all colours, shapes and sizes. Galleon sized plastic supermarket bottles, emptied of their coke and lemonade. The thought of cooking here could never occur spontaneously. The only possibilities for food are spam slices from the local store, baked potatoes, fish 'n' chips, and, for that more special occasion, an Indian or Chinese take-away. Which means more crushed up newspapers, more greasy cellophane wrappers, more mayonnaised styrofoam, more betel stained foil. Even making a cup of tea has begun to seem impossible here: the thought of anything but coke, lemonade, or beer would soon be defeated by the sheer weight of material stacked against it.

Think too of the heavyweight food champions of the world, eating to keep their thirty stone alive. Five hamburgers, three plates of chips, seven eggs and ten sausages. Just enough to keep one going between breakfast and lunch ... And think of the young girl, measuring her thigh against a wrist in the mirror of her mind. And her devices to deal with the dangers of food. Explaining to her parents that she is to eat at her friend's house only to arrive there to explain that she has already eaten. Then bingeing on chocolate in the small hours, when she can vomit in peace.

The theme I wish to explore in this chapter is the relation between the provision of food and its disposal. If my analysis is right, an adequate theorising of disposal may upset extant thinking about food choice. Far from 'choosing' what we eat, as is the presumption in both the 'production' view and the 'consumption' view, eating is governed in part, by an availability of 'conduits' for disposal.

In this chapter I wish to do more than simply draw attention to disposal as a topic. A move towards more processed and packaged 'convenience' food has already alerted many commentators to thinking about disposal. But this emphasis has tended to confine it to an environmental debate that, as yet, has failed to theorise disposal as anything more than an addendum to production and consumption. Environmental matters aside, we are left to assume that disposal can be viewed unproblematically, either as the inevitable 'waste' from a production of food, or as a hedonistic 'surplus' from its consumption.

Somewhere between the 'production' view, with its biophysical and spatial metaphors, and a 'consumption' view, with an ethereal emphasis on symbols, the importance of disposal has dropped out and the topic has been overlooked, or forgotten. Neither the 'production' view of the provision of food, nor the alternative 'consumption' view, *theorise* disposal and take cognisance of its effects. It is important to make clear, however, that a lacuna over disposal in the academy cannot eliminate an everyday attention to the matter. Rather a lack of theorising helps to *translate* this attention as integral to either production or consumption. For example, rather than being read as illustrative of attempts by food producers and consumers alike to wrestle with some acute problems of disposal, the explosion of promotion and advertising (Wernick, 1992) is seen instead as exemplifying a 'shift' in society towards the symbolic.

An absence of theorising on disposal is thus acute. There is surely more to the development of 'augmented product features' than an adding of 'value' to consumption. For example, one does not feel so 'cheap' buying one's groceries cut price in Sainsbury's as one might feel conspicuous doing so in the local shop. The prestige of Sainsbury's not only masks from view one's need to economise; it mutates it in ways that can make local shopping seem sloppy and lazy. It is important to add quickly here that I do not wish to dismiss either the production or the consumption view. As the present example makes clear, the 'conduits' I have in mind for disposal are drawn as much from what is termed the *symbolic* as from examples implicating a *physical* dispersal of food, such as garbage bags, or ingestion. Where I depart from either a production or a consumption view is in refusing to press a distinction between the symbolic and the physical. To see each of these as involving different types of disposal would merely propagate the production/consumption divide.

In raising the theme of disposal, therefore, I intend to raise questions over the propriety of this very division between the physical and the symbolic. The disposal of a meal, as I shall illustrate, does not divide up well between the elongated physical events of the stomach and a symbolic mastication in the memory or mind. I want, therefore, to go rather further than analysing disposal as a post-prandial exercise, whereby cigars, coffee or brandy are enjoyed as useful, if largely 'symbolic', *supplements* to the 'physical' disposal of a large, rich meal. To simply distribute examples of disposal between the physical focus of production and the symbolic emphasis of consumption would only polarise a theme of disposal and perpetuate its obviation.

Mary Douglas's lasting contribution to understanding about the provision of goods has been to make much more sophisticated, and much more social, the notion of consumption. Against economics, with its primary focus on production, Douglas proposes an 'anthropology of consumption'. Where economics invents a world of *goods*, Douglas and Isherwood (1980) emphasise a *world* of goods.

For Douglas (1966, 1975), belonging is pivotal to 'consumption'. According to Douglas and Isherwood (1980: p. 88), the world is organized in a 'recursive system of metaphors dealing with admission to bed, board and cult'. As they frame this world:

> The meanings conveyed along the goods channel are part and parcel of the meanings in the kinship and mythology channels.

In this way Douglas theorises 'the social' as being active in the sharing of goods. The social is not merely something that is added to an economic order; it is constitutive of that order. Every exchange, each purchase, is also an act of *affiliation*. Consumption is, in Fernandez's (1986) fine phrase, a 'returning to the whole'.

In contrast, the 'production' view atomises consumption as being no more than the result of choice among many goods. There is no sense of 'the social', except as something epiphenomenal to production. And the notion of choice is already displaced through an atomising of persons. Culture enters only by the back door. It is no more than a recognition that, in responding to market forces, 'individuals' reveal their preferences.

At first sight therefore the consumption view could hardly be more different. Any choice between goods, in the view of Douglas and Isherwood (1980: p. 76) is 'the result of, and contributes to, culture'. As Douglas and Isherwood (1980: p. 88) insist:

> No one likes to recognise that the capacity to share all three [bed, board and cult] is socially endowed, a result of current decisions, and not an ineluctable fact of nature.

By arguing that goods are consumed because they are 'good to think', Douglas rescues choice, and hence 'culture', from its displacement in western economies.

Culture, in the consumption view, is much more therefore than the economist's sum of revealed preferences. A key link between culture and consumption is identity-making. Friedman (1990: 327) develops the relations to identity thus:

> Following a line of argument that began with the recognition that goods are building blocks of life-worlds, we have suggested, as have others, that they can be understood as constituents of selfhood, of social identity.

Against their marginal status in the 'production' frame of economics, where culture has no place in analysis (beyond a status as another good), meanings are made central in understanding the consumption of goods.

13.3
A world of food

It is, at this point, that I wish to express some misgivings over the consumption view. Yes, it explains, in ways that the production view cannot, why we sit down to eat turkey at Christmas or Thanksgiving. Yes, it can explain, through arguments about a 'need' for belonging, why poor US immigrant families spend so much money at McDonald's, or other fast food places 'selling' America. It explains all this and much, much more.

The difficulty for the consumption view is that it can explain too much. Too much, too easily. There is no eating pattern it can't explain. Coq au vin cooked in red burgundy, or bacon sandwich take-aways; fresh flown strawberries in winter or baked beans on toast. Fancy silver candles at the table, or egg sandwiches reading page three of the Sun in the lorry cab. Wallpaper opera to sip by, or Meatloaf heavy metal to chew on. Place settings to select, or fighting over a cardboard box; a prelude to copulation, or a vain attempt to loosen the bowel. The variety of material for re-presenting identity is all grist to a symbolic mill. There is no *end* to the consumption theorist's ability to theorise.

So what is going wrong? It is, I suggest, over the entry of 'meanings', that the symbolic frame of consumption can founder. Douglas's 'world of goods', unless carefully handled, seems in danger of dropping quickly into a world of food. Food for thought. In an unkind interpretation, to which we return later, the consumption of goods turns into a production of meanings. Simply that. At a level of parody, the 'world of goods' begins to represent no more than the symbols that the social feeds off. Without some attention to disposal, symbols become just fodder for another version of theory building.

13.4
Dirt and dinner

To appreciate the importance of disposal in relation to the themes of identity and consumption, it is important to note the otherness of belonging, the dark side of affiliation. To return to Douglas and Isherwood (1980: p. 88):

> Sharing goods and being made welcome to the hospitable table and to the marriage bed are the first, closest fields of inclusion, where exclusion operates spontaneously long before political boundaries are at stake.

Exclusion is also intimately tied to the consumption thesis.

This theme of exclusion is a startling insight into how meanings could appear to be 'exchanged'. It is mere presence or absence that invites meaning. We need not think then of meanings as literally moving between minds. Rather, depending on how it is made present or absent, something is either 'in' or 'out'. The rest, what is 'added' or 'interpreted', is a complex of possibilities around this theme.

As I read it, the key phrase is one we have already met: 'a recursive system of metaphors dealing with *admission* to bed, board and cult' (emphasis added). Exclusion operates to 'define a category of outsiders'. So one aspect of disposal, deletion, is central to the consumption thesis. But more is entailed. As Douglas and Isherwood (1980: p. 88) go on to point out, the 'naturalness' of exclusion has to be asserted:

> ... whenever exclusion is operated to define a category of outsiders, the segregated category tends to be accredited with a different nature.

What is kept outside is always 'dirt', the excluded is always monstrous. Equally, to be kept outside, the excluded must be perceived of *as* dirt, *as* monstrous.

These preambles might seem to take us far from a disposal of the meal, but the point can now be made more quickly. What is not dinner is dirt. What is never dinner is not to be eaten. Similarly, what is seen as dirt, what has been excluded, can only, with some transformation, become dinner. Think of cannibalism and how rare the phenomenon is, with earlier reports now being disputed. What is good for the dinner may not be so good for the diner, especially if it became your turn next. But as dirt, as *unthinkable* for consumption, you need have no fear of your fellow diners turning on you to dispose of their remaining hunger.

Of course, inversions to the rule of dirt and dinner are possible. Industries are built around 'food disorders' and these survive on food becoming dirt. But in contrast to the slimmer's anxious refusal of food, the bulimic can stow away a dozen chocolate bars, confident of her ability to throw up. In extending this theme of disposal, I want now to suggest how difficulties and devices in disposal affect both where things are eaten and when eating may take place.

13.5
Passages for disposal

There are other conduits than ingestion, other sensibilities. We begin with smell. It is not all bad news with smell. I have lost count of the times, for example, that I have been told to sell my house by making fresh coffee. Now that is disposal! But for the smaller fry we can start with the case of the malingering meal.

The problem of fish we have already mentioned. The problem was one noted by Marshall (1990) as part of a team of researchers in Newcastle investigating the low sales of fish in that part of the country. The problem is one of leftovers. Those interviewed did not buy fish, they said, because they did not know what to do with the leftovers.

The problem was not only a 'culture' that lacks the know-how that goes all the way down to what to do with the herring bones. The problem, in part, turned out to be one of smell. Fish, it seems, pollute through smell. Presumably those with refrigerators have already discovered this, long before they open the fridge door. The smell of fish seeps through into the other food. By keeping the fish, other food becomes wasted, unusable. It loses its difference. The fridge is a device to preserve difference and defer disposal, but over fish it fails.

So into the bin with it. But here too there is a problem. Two problems. First, the fish is still fresh. So it is still a much more visible *waste*. This has its own effects. The disposal of a physical material can produce other, more symbolic, material for disposal. The scraps of fish, for example, make present the starving millions. As I scrape the fresh fish into the bin, I recall that feckless child who would not eat his fish pie while Biafra starved.

Second, as we know, the fish will still smell. It will smell till bucket day. So if bucket day is not till Friday, better not to have fish till Thursday evening. Or if bucket day is Monday, then with the fishmonger only open on Saturday, not Sunday, better not to buy fish at all?

So in respect of fish we have a story of the failure of two familiar devices that work well for other foods. Fridges and garbage buckets. For fish, the only device that works well is freezing. Hence the success of frozen fish, especially that which pre-empts disposal by arriving in meal-sized sizes.

Other examples of smell affecting consumption, and hence food provision, abound. A favourite of mine is the secretary at the office. She acts as gatekeeper to the consumption of garlic. In particular, she monitors breath; and those who repeatedly offend become literally shunned. As has happened to several members of staff and several postgraduate students, they lose membership. They become outcasts of the office, unable to make their way, not only socially through lack of contact and gossip, but academically, through being cut off from such essential benefits as photocopier facilities.

Turning now to the visual, we can see the advantages of particular forms of food. Meat, especially a good ham on a bone, can look well in the fridge. Unless one is Jewish, it is a bone of comfort, rather than a bone of contention; one that is as handy for making sandwiches as it is for making a light supper or lunch. It may even do for breakfast.

Vegetables too can work visually. They are often attractive to have around the kitchen well ahead of a meal, brightening up the worktops or the fridge. Onions especially can dangle winter long on a string. And that fruit bowl! Who could ever touch that? Now there's a case of food passing from bush to compost heap, without touching the lips of anyone on the way.

Sight works for disposal in other ways. In Edinburgh, sight clearly aids digestion. The food in restaurants is notoriously so awful that the only explanation given for the high prices is that, in Edinburgh, people 'eat the decor'. Equally in New York restaurants, I have been appalled at the grotesque proportions served up as meals. We queued for two hours to get into one pancake restaurant, only to eat about a third of the topmost pancake. Not because the food was poor, it was good enough. But because we were all too full to be able to digest any more. How comforting for the greedy to see that even they can't make inroads on a meal!

I like to think of those small family restaurants on Greek islands where inspecting the kitchen was part of ordering the meal. How helpful for disposal to know that the kitchen was kept clean. And, in reverse, how helpful for disposal that, in order to allow everyone in, it was kept clean enough. Who, apart from the Romans and bulimics, ever wants to 'revisit' a meal in the night?

In the auditory realm, consider how the after-effects from a meal affect consumption. Burp, and we are back to the malingering meal. And how many more people could enjoy bean dishes, were it not for a problem of flatulence in the office? The fear of breaking wind surely affects the sales of pulses for all who have to work indoors.

Talking about meals, conversations about recipes, memories of favourite meals, recommending favourite restaurants, these are all very pleasant ways to dispose of a meal. You can enjoy the experience of talking, ahead of eating and long afterwards. And the exercise of this conduit can be so enjoyable that you do not have to have actually enjoyed the meal. For some, it seems, it's worth paying fifty pounds a head just to slate a fancy restaurant. For others, the cost of joining in on the office outing will have to be paid for with sacrifices long afterwards.

A much underrated conduit of disposal is touch. Touch, not taste. Here I am not thinking only about something being too spicy or too salty. Other feelings abound. And have their effects. For example, what about the sales of chilli in the west? How many more people would order their curry 'vindaloo' were it not for its effect on the stomach and bowels?

So what about the sales of 'positive food', like 'nutrition', the sales of food that leaves you feeling good: salad, yoghurt. Some of these may not taste so good, especially when eating them every day, but presumably their after-effects are either good or neutral. But feelings can play tricks. I remember day after day eating junk food as an undergraduate and feeling bloated, but only wanting more junk food. The heaven when I began cooking and first tasted freshly cooked green vegetables! That disposed of my hunger for junk food, but only then did I realise what I had been hungry for.

In this respect, feeling may be a much more important conduit of disposal than even sight in affecting food provision. In considering the popularity of the office sandwich, for example, I have in mind a particular contrast. How soothing that slothful seeping of blood seems after lunch on highdays and holidays and how debilitating it appears when facing the afternoon work from Monday to Friday!

These brief examples suggest that theorising food choice in terms of a production and consumption of meanings fails to take materiality seriously. As we have seen, problems over disposal, together with the availability of devices to aid disposal, have profound effects over food choice. If not yet convinced, think of the most famous after-effect of all: fat. How the failure to dispose of calories affects the provision of food: as starch, pasta is 'out' one decade; and, as fibre, virtually a fad the next. Or the way that a discourse on calories stimulates the provision of a food range: from all those slimmers' biscuits, posing as whole meals, to extra large cream cakes, when extra naughty equates to extra nice.

13.6
A disposal turn to consumption

Before any more adequate notion of disposal can be developed, a crucial distance has to be placed between possible interpretations of Mary Douglas's work. As has been hinted, Douglas's work lends itself too easily to a dualist interpretation of material *and* symbols; of physical goods *and* meanings. A parallel world of words *and* things. To avoid this interpretation, we have to be careful not to take too literal a view of meanings as being 'conveyed along the goods channel'. Meanings do not literally travel anywhere. Only inscriptions travel.

If inscriptions are material, then, like other material, they also require disposal. It is important to see that any exchange of meanings cannot take place *supra* the world of action. There is not one world of goods and another world of expressions. Rather, expressions *inscribe* a world of goods and goods incite expressions. Both goods and expressions are inscriptions of material. The cry 'not turkey again!', as I view it, does not exist *elsewhere*, in say a plane parallel to that of ingestion. Expressions exist *alongside* other material effects.

For the present analysis, all that need be avoided is a tendency to reserve the term material for that physical matter which is privileged *as* more prominent or *as* more solid. It may, for example, take a parent-cook longer to digest the unwonted response of 'not turkey again!', than for the child to finish off the unwanted plate. If the obtrusively physical were the only matter for disposal, we could simply take an implicit plumbing metaphor in the notion of conduit more seriously. For example, the Romans gorged themselves in long banquets, but in order to do so they used a feather to tickle their throats. By adding vomiting to their repertoire, they doubled the physically available number of conduit for the disposal of food.

Emphasising a materiality to disposal, then, amounts to more than a current emphasis in sociology for bringing back 'the body'. In insisting that inscriptions are material, and not (only) referential, we have to think again about language. The critical theme I want to emphasise is a lack of homogeneity in language. Language is always heterogeneous. This is why it cannot live elsewhere, forever divided between material and meaning.

The child crosses her fingers as she says she will be good and I smile as I promise to try and deliver this chapter by the deadline.

With this emphasis on materiality, we also have to overcome a similar tendency to make homogenous a remedy and an effect. There is no reason to suppose, for example, that a quarrel between friends can only be resolved through verbal expression. A quiet meal together may be as equally healing. The principal point to see is that only some conduits are 'internal' to the body. Or rather, only *part* of the conduit may be internal to the body.

It follows that, in thinking about disposal, we also have to be on guard not to insist on a homogeneity in *conduit*. Not all conduits for disposal are the most obvious, even when some, like exercise, have in other ways become everyday. My favourite example of disposal is from a Canadian study: the woman who took up exercise because she 'couldn't stand the feeling of jello walking behind her' (Ray, 1992). Presumably, to come to this view, she had already divided off from her 'self' that unwanted part of body as 'jello'. But this is not merely to suggest horses for courses: physical ingestion for the base food and symbolic mastication for 'augmented product features'. I have already suggested that in theorising disposal we should move beyond distributing disposal down the parallel worlds of the body's arithmetic and a mental defecation. To recognise the importance of discourse for disposal is to suggest that conduits for disposal may be *extended* across different and various materials.

13.7
Food and forgetting

This matter of variety in extension brings me to a final theme. We sometimes use the phrase 'I forgot about food.' What can such a phrase mean if, as I am about to suggest, we also use food as a means of 'forgetting'? A way of making absent; a *means* of disposal.

Food may not only require conduits for its disposal, the very consumption of food may well act as a major conduit for 'forgetting'. Think how *relaxing* eating can be. Agreed, not always, and not for everyone. But often, and, even within a society that has reduced its communal meals to one a day, perhaps surprisingly often.

Is eating then a form of extension? Is eating 'a return to the whole'? One form of extension that breaks us free from the desperate knots of symbolic orders? Sometimes taking us 'back' to community and, occasionally, in 'forgetting', bringing us back to the body. As itself a conduit, does not the disposal of a meal at least defer, if not dispose of, the existential angst of being, the endless anxiety of existing as an inchoate, interpretative being?

For food understood as commodity surely is a failing solution. This is not only because the more the one half of the human race eats, the more the other half grows hungry. It is to recognise the limits of food provision. Food eaten alone may be filling, but it is never *fulfilling*. Is this then what eating disorders presage: that we now inhabit little more than a world of foods? The more we rely on the 'physical' nature of food for 'forgetting', the more food appears to fill the world.

Understanding the matter of choice, therefore, is not one of food being less or more important today: of being a 'foodie'; or finding food 'boring'. Epithets are always reductive, regardless of whether they magnify us as important, or diminish us as waste. The difficulty is to recognise how it is that efforts by manufacturers and distributors to talk up food accomplishes more than an 'augmentation' of product features. Food promotion has *discourse effects* that bring in to being artefacts that alter the availability and distribution of resources for identity making. According to Strathern (1991), cutting this figure, or that figure, draws on the potential to move among (various) artefacts through an attachment to (different) artefacts. Is it Earl Grey? Jack Daniels on the rocks, please. Mayo *and* tomato, thanks. Does food, then, in its choice, engage a whole new range of artefacts in ways that *multiply* our extension? Configuring us as sovereign in the supermarket and, simultaneously, as 'individualistic' in what we eat.

The smell of food, the taste on our lips, the weight in our stomachs, all these require disposal. But these are also conduits that seem (however misleadingly) to bring us *out*. As personalities. I don't like strawberries. I'm not a tea person. No sugar, thanks. No, I never have lunch. Sorry, I'm allergic to onions. How *clever* to find this wine in Safeways. Mm, what a *wonderful* smell!

So why *food*? Why this attachment to, and through, foods? Surely there are other moments that we sit down together, kith and kin? Moments of communion, when we are free of food? Or, in the form of Baudrillard's (1983) simulacrum, has a parody of the 'consumption' view caught up with us?

13.8
Disposal of the debris

Are there *reasons* for what we eat? Is food choice a rational activity, one whose patterns are amenable to analysis and advertising? Or is eating the last refuge of the irrational? An island of emotion, where desire and greed, insatiability and fear, snobbery and impulse reign supreme?

In re-examining the 'consumption' view, I have explored the disposal of the meal as a matter of belonging. Without belonging, the much vaunted 'world of choice' reduces to a heap of goods. This, I take it, is the key insight of Mary Douglas. In stressing belonging, however, I also have wanted to convey a feeling of belonging that is not rooted in the idea of habit. For Douglas, there is a dynamic to food choice that sits ill with a 'puppet' consumer, or the dull lifelessness of a 'collective consciousness'. Food choice then seems integral with *feelings* of communion, a communion in the spirit of a Bund (Hetherington, 1994).

But here there are severe difficulties with theorising who (or better what) constitutes communion or community today. Yes, the type of food we eat can advertise identity: 'By all means buy the wine from Safeways but *don't* serve their mints – economy is the wrong note to end on.' But what if eating excites longing as much as belonging? As a conduit of disposal for *that* longing, perhaps only the companion at the table matters. Perhaps this is why we end up not dining on our companions; the conduits cannot work both ways – we cannot have our company and eat it. Perhaps, but we may mistake communion if we think too humanistically here. For today that companion may be a mirror image for the anorexic; and for that sandwich in the lorry cab, 'she' may well be page three.

These are complex matters. Nevertheless, they help to show that we cannot upset extant ideas of the economic order just by a symbolic inversion, replacing an overemphasis on materials with one of meanings. In challenging the production view, therefore, I have tried to do more than suggest the presence of symbolic inversions in the food process, whereby a meal may be eaten in anticipation, even justification, of the after-lunch cigarette. To conflate the phenomenology of belonging with the matter of (symbolic) identity is to make the further mistake of turning a 'world of goods' into a 'world of foods'. Belonging does not reduce to the theme of 'identity', a reduction that feeds (sic) on symbols of one kind or other. A feeling of belonging depends more on the processes of extension and exclusion and these, I have been arguing, depend not only on affiliation but also on disposal. Deletion alone involves not only a making of absence, but also a presencing of absences. For this reason, disposal always involves (perverse as it may sound) attachment, a making of presences.

For this making of presences and absences, there must be a materiality to the processes of attachment and detachment that make up extension and exclusion. Food offers itself, as such, as a medium; it acts as a prosthetic device for the manufacture of identity. This alone would not explain why food appears so pervasive. Instead, although there is too little space to pursue the suggestion here, one is left to wonder if there may be something about the very 'disposability' of food that lends itself to a making of identity. Food is so temporary, so replaceable. In the *moment* of food choice, I enjoy a feeling of being able to move my identity about. We are so seldom stuck with our choice in ways that we can be stuck with a house, a car, or an ornament. With the purchase and display of food, identity seems so immanently *reversible*.

Contrast this theme of belonging with a conventional analysis of food provision. There are several points to make. First, conventional analysis, as some of the chapters in this book illustrate, situates eating as a consequence, in part, of 'decisions' over production, where price and availability dominate, and, in part, 'decisions' over consumption, where taste and habit are regarded as critical. What we eat, ingest, in the production view, is no more than an effect of the rigours of demand and supply.

Eating is a mere incidental, and temporal, moment between the 'real' places of action, the so-called decision centres of the factory and the shop. Indeed, on this analysis, far from being central, the meal is sometimes no longer even necessary to the processes making up the food chain, since much that is purchased in the west is for display (fruit bowls, freezer compartments) not ingestion.

Second, in an attempt to accommodate something of both the 'production' view and the 'consumption' view, I hope to have illustrated how the supposed last stage in the food cycle, a disposal of the debris – in the washing of plates and glasses, wiping the table and mats, and filling up the garbage bags – has implications for the acquisition, cooking and ingestion stages in the process. To illustrate this theme, I have considered how problems with disposal 'feed' back into the choice process. In the case of fish, for example, the disposal problems are manifest at one extreme in the absence in some parts of the UK of fish from the table. Or, alternatively, in the 'use' of fish only in frozen or canned versions that contain the polluting quality of smell. Problems of disposal 'feed' forward in ways that bring the food cycle to a full circle. The point of my examples, however, is not just to give emphasis to a circularity of consumption.

My third and main point has been to illustrate the sheer wealth of different conduits involved in consumption. However, rather than treating the matter of living simply as an action replay of extrusion from one-off consumer evaluations, I draw on examples to stress the continuous and extended nature, not only of consumption, but also of the self. A theorising of disposal has to go beyond an emphasis that we consume now, and pay through extrusion later. We neither live to consume, nor consume to live. We live nowhere else but in a consumption of artefacts. And here, in a world that demands rapid movement in identity, food has a comparative advantage over other goods. In food the artefacts of our consumption are, more likely than not, momentary. As such, the identities that they carry seem so readily disposable.

References   Baudrillard, J. (1983) *Simulations*, trans. P. Foss *et al.*, Macmillan, New York.

Bourdieu, P. (1984) *Distinction: a social critique of the judgement of taste*, trans. R. Nice, Routledge, London.

Douglas, M. (1966) *Purity and Danger: an analysis of the concepts of pollution and taboo*, Routledge, London.

Douglas, M. (1975) *Implicit Meanings: essays in anthropology*, Routledge, London.

Douglas, Mary and Baron Isherwood (1980) *The World of Goods: towards an anthropology of consumption*, Penguin, Harmondsworth.

Fernandez, J. (1986) *Persuasions and Performances: the play of tropes in culture*, Indiana University Press, Bloomington.

Friedman, J. (1990) Being in the World: globalization and localization. *Theory, Culture & Society*, **8**, 4, 311–328.

Hetherington, K. (1994) The Contemporary Significance of Schmalenbach's Concept of the Bund. *The Sociological Review*, **42**, 1–25.

Marshall, D. (1990) A Study of the Behavioural Variables Influencing Consumer Accept-
ability of Fish and Fish Products. *Ph.D Thesis*, University of Newcastle-upon-Tyne.
Ray, M. (1992) I Don't Like Feeling There's Jello Walking Behind Me: women's exercisers,
control and discipline. Paper presented at the *Theory Culture & Society 10th Anniversary
Conference*, Champion, Pennsylvania.
Strathern, M. (1991) *Partial Connections*, Rowman & Little, Maryland.
Wernick, A. (1992) *Promotional Culture: advertising, ideology and symbolism*, Sage, London.

Index